the chemistry and technology of furfural and its many by-products

SUGAR SERIES

Vol. 1. Standard Fabrication Practices for Cane Sugar Mills (Delden)
Vol. 2. Manufacture and Refining of Raw Cane Sugar (Baikow)
Vol. 3. By-Products of the Cane Sugar Industry, 2nd edition (Paturau)
Vol. 4. Unit Operations in Cane Sugar Production (Payne)
Vol. 5. Noël Deerr: Classic Papers of a Sugar Cane Technologist (Payne, Compiler)
Vol. 6. The Energy Cane Alternative (Alexander)
Vol. 7. Handbook of Cane Sugar Engineering (Hugot, 3rd edition)
Vol. 8. Management Accounting for the Sugar Cane Industry (Fok Kam)
Vol. 9. Chemistry and Processing of Sugarbeet and Sugarcane (Clarke and Godshall, Editors)
Vol. 10. Modern Energy Economy in Beet Sugar Factories (Urbaniec)
Vol. 11. By-Products of the Cane Sugar Industry, 3rd edition (Paturau)
Vol. 12. Cogeneration in the Cane Sugar Industry (Payne)
Vol. 13. The Chemistry and Technology of Furfural and its many By-Products (Zeitsch)

sugar series, 13

the chemistry and technology of furfural and its many by-products

KARL J. ZEITSCH
Formerly consultant to Illovo Sugar Ltd, Durban, South Africa
Dürener Str. 393, D-50935 Köln, Germany

Elsevier
Amsterdam - Lausanne - New York - Oxford - Shannon - Singapore - Tokyo 2000

ELSEVIER SCIENCE B.V.
Sara Burgerhartstraat 25
P.O. Box 211, 1000 AE Amsterdam, The Netherlands

First edition 2000

Library of Congress Cataloging-in-Publication Data

Zeitsch, Karl J.
The chemistry and technology of furfural and its many by-products / Karl J. Zeitsch.
p. cm. -- (Sugar series ; 13)
Includes bibliographical references and index.
ISBN 0-444-50351-X
1. Furfural. 2. Furfural--Derivatives. I. Title. II. Series.

QD405 .Z45 2000
547'.592--dc21

99-058936

ISBN: 0 444 50351 X

⊗The paper used in this publication meets the requirements of ANSI/NISO Z39.48-1992 (Permanence of Paper).
Printed in The Netherlands.

This book is dedicated to Mr. Jon L. Buzzard, formerly General Manager of the ILLOVO furfural plant in Sezela/South Africa, whose indefatigable search for innovation was the driving force for many of the processes described here for the first time.

Table of Contents

Page

Page

Page

Page

1. Introduction

In 1921, the QUAKER OATS COMPANY of Cedar Rapids, Iowa, had some redundant pressure cookers no longer needed for an abandoned cereal process. To make use of this equipment, and of massive quantities of oat hulls from their manufacture of rolled oats, they started an industrial production of furfural.

The circumstances of this historic initiative are splendidly described in an article by Harold J. Brownlee and Carl S. Miner [1] which contains the following passages:

> "This company never had a pilot plant for furfural.Work moved directly from the laboratory experiments to what was essentially a full scale unit. This was the result of the fact that the QUAKER OATS COMPANY had available in the plant at Cedar Rapids, where the first furfural processing plant was to be operated, a number of iron pressure cookers about 8 x 12 feet, which had been used in the manufacture of a cereal product which did not prove profitable. Since these cookers were available and since the process was to consist of the treatment of oat hulls with acid under pressure, it seemed advisable to try to use these digesters at least for the first attempts at large scale operation.
>
> When the decision was reached to line the iron digester at Cedar Rapids with an acid-resistant material, a study was made of a number of metals. A consultation with the various purveyors indicated the use of 20-gage Monel, the edges being turned up and welded. The digester lining was finished, and the first digestion made in November 1921.
>
> Prior to the actual experiment, a blank test was run on the Monel metal lining by applying steam pressure of 60-70 pounds blowing off the steam, and putting on a vacuum immediately following, but when the cooker was opened for inspection after the test, the lining had caved in badly. Apparently there were a few pinholes which permitted air and steam to seep slowly behind the lining. This experience was repeated at least once before this method of installing a tight lining was abandoned. Finally, one hole was left in each end of the cooker in the Monel shell to act as a breather and equalize the pressure on both sides of the liner. This

> prevented further cave-ins but had the disadvantage that the corrosive vapors from the cooked hulls penetrated behind the Monel metal and eventually corroded the iron badly.The lining served, however, for much of the preliminary work on furfural but never proved wholly satisfactory. Later liners were made of copper, but they were never quite satisfactory either, and ultimately linings were used of carbon brick with acidproof cement, which is the type still in use in the furfural plants today."

Thus, it should be noted that the reactor design was dictated by the old equipment available, and was not the result of a study of optimization. This is understandable as the kinetics of furfural formation were not established until many years later. Hence, it is all the more remarkable that this early process of 1921 is still in use. Nevertheless, the following treatise will show that since its inception described above, the furfural process, and the processes regarding its by-products, have undergone sweeping developments, and based on chemical insight they have now reached a stunning degree of sophistication.

Reference

[1] H. J. Brownlee and C. S. Miner, Ind. Eng. Chem. 40 (1948) 201-204.

2. The Reactions Leading to Furfural

Furfural is universally made from agricultural raw materials rich in pentosan. By aqueous acid catalysis, the pentosan is hydrolyzed to pentose, and this pentose is dehydrated to furfural in a unified process.

2.1. Stoichiometry

The stoichiometry of the two reactions reads as follows:

(1) Hydrolysis of pentosan:

PENTOSAN	+	n x WATER	→→	n x PENTOSE	
$(C_5H_8O_4)_n$	+	n H_2O	→→	n $C_5H_{10}O_5$	
n x 132.114	+	n x 18.016	→→	n x 150.130	g/mole

(2) Dehydration of pentose:

PENTOSE	-	3 x WATER	→→	FURFURAL	
$C_5H_{10}O_5$	-	3 H_2O	→→	$C_5H_4O_2$	
150.130	-	54.048	→→	96.082	g/mole

Thus, the overall reaction can be said to be

PENTOSAN	-	2 x WATER	→→	FURFURAL	
132,114	-	36.032	→→	96.082	g/mole

so that the theoretical yield of furfural from pentosan is

$$Y_{th} = 96.082/132.114 = 0.72727$$

This is the sacred figure against which all furfural plants are measured.

2.2. Mechanisms

Pentosan (polypentose) consists predominantly of rings linked by oxygen bridges (ether bridges) as shown below:

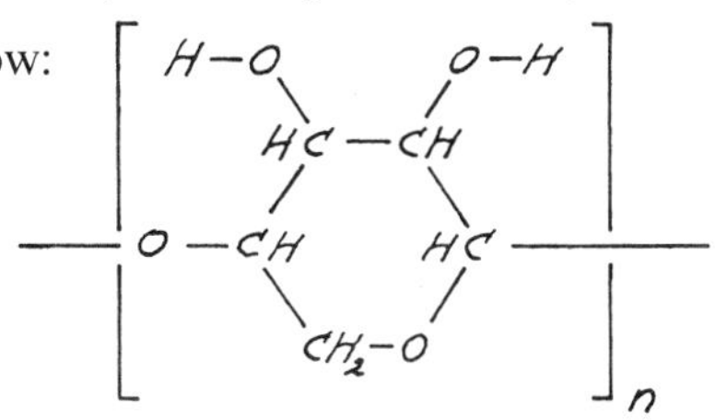

Actually, the rings are not plane but have a chair conformation [2]:

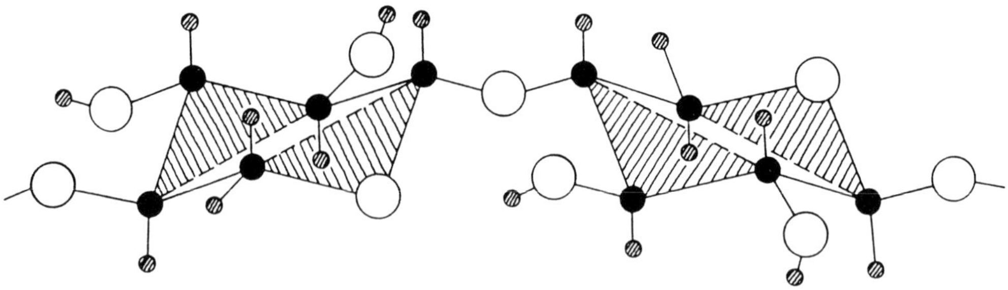

The large white circles symbolize oxygen, the intermediately sized black points symbolize carbon, and the small hatched circles symbolize hydrogen.

Inasmuch as the pentosan structure represents a polyacetal, the acid hydrolysis of pentosan corresponds to the hydrolysis of acetals [3].

The mechanism of the acid hydrolysis of pentosan is shown schematically in Figure 1. It comprises the following steps:

(1) Protonation of an oxygen link, thus leading to trivalent oxygen (first and second lines of the diagram).

(2) Cleavage of a carbon/oxygen bond leading to a carbocation on one side of the oxygen bridge, and to a hydroxyl group on the other side of the oxygen bridge (third line of the diagram).

(3) The carbocation takes up water (forth line of the diagram).

(4) The resulting H_2O^+ group liberates a hydrogen ion, thus leaving a hydroxyl group behind (fifth line of the diagram).

This sequence of processes is repeated until all oxygen bridges have disappeared so that the rings have become individual pentose molecules.

The subsequent formation of furfural from pentose involves the liberation of three molecules of water per molecule of pentose. Any such major transformation of a molecule does not take place concertedly but in steps. A plausible mechanism is illustrated in Figure 2. The initial pentose is shown in its prevalent ring form representing an intramolecular hemiacetal. The open-chain aldehyde form in equilibrium with the ring form can be disregarded as it amounts to less than one percent of the total pentose present. The transformation steps shown consist of two 1,2-eliminations and one 1,4-elimination of water. The 1,2-eliminations must

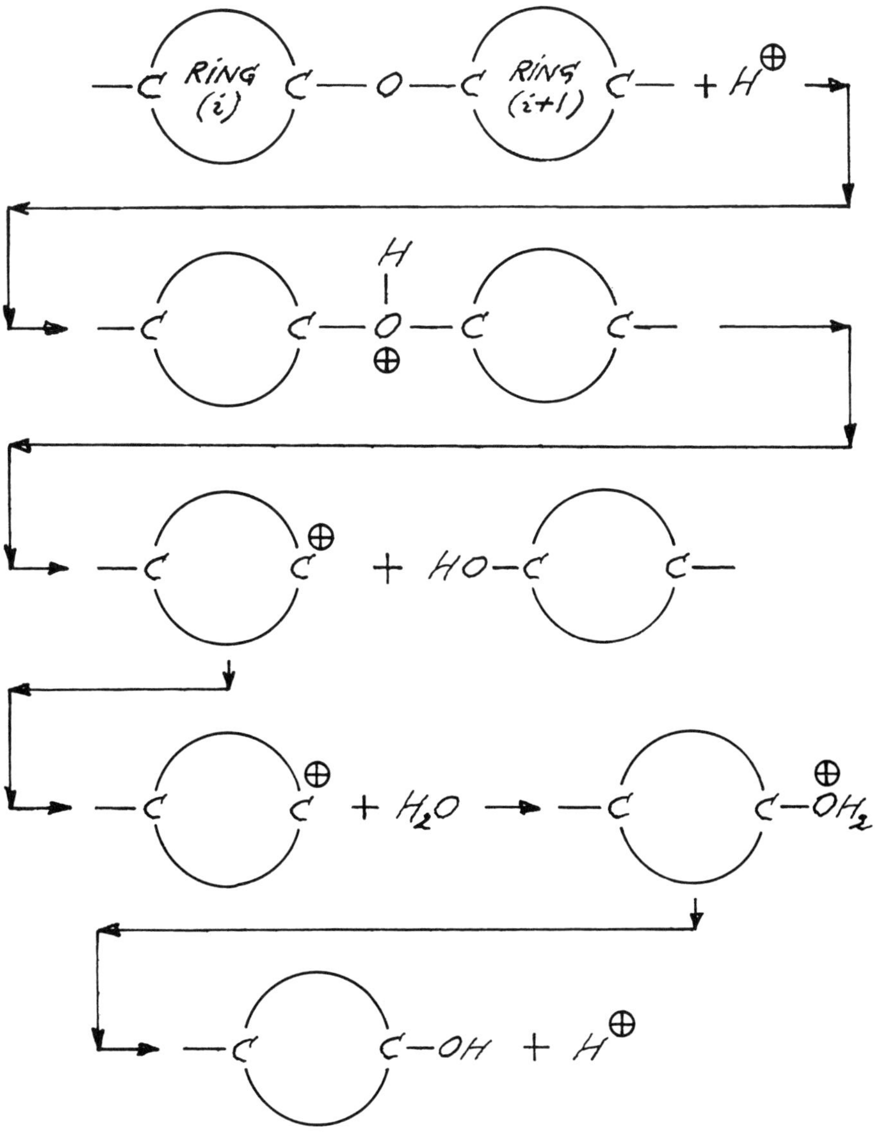

Figure 1. The Mechanism for the Hydrolysis of Pentosan.

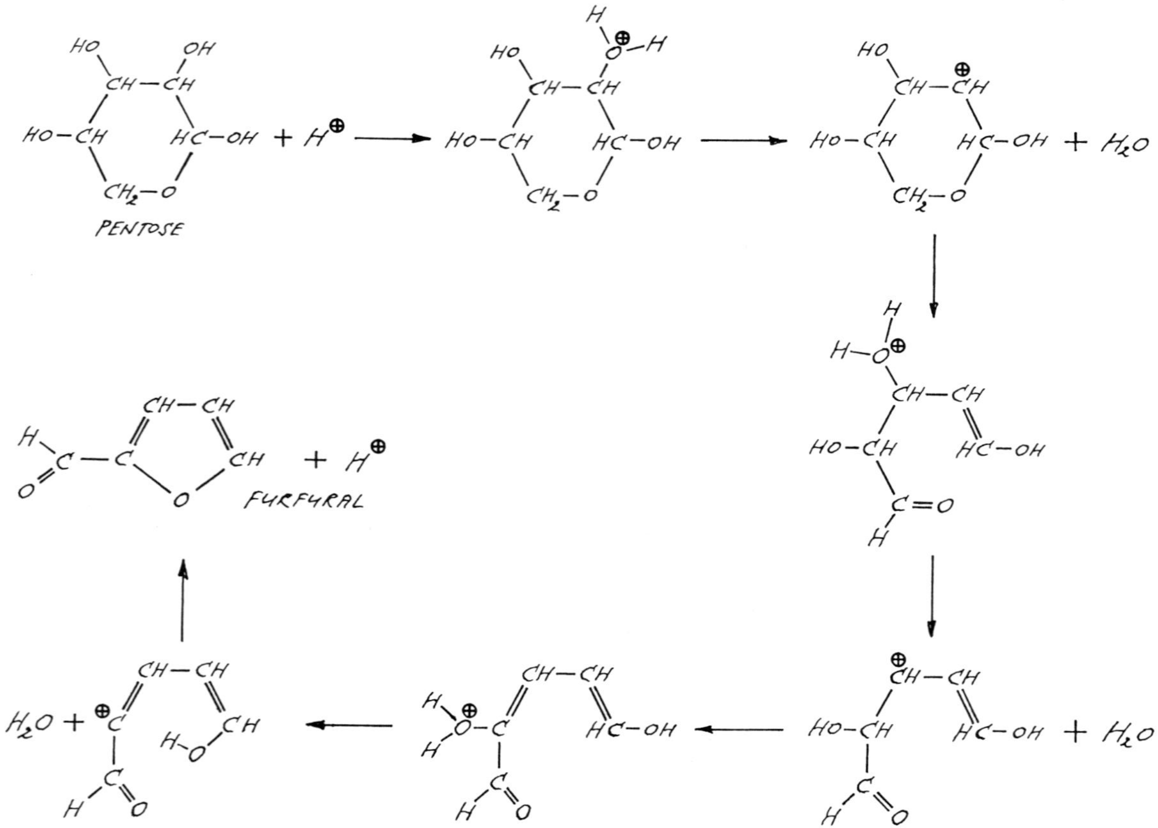

Figure 2. The Mechanism for the Dehydration of Pentose to Furfural.

imply the involvement of two neighboring carbon atoms and the formation of a double bond between them, while the 1,4-elimination involves two carbon atoms separated by two other carbon atoms and the formation of a ring.

In detail, when a hydrogen ion attacks a lone (nonbonding) electron pair of a hydroxyl oxygen bonded to a carbon atom, the result is a transition state with a trivalent positively charged oxygen atom. As oxygen is more electronegative than carbon, the positive charge (electron deficiency) immediately shifts to the neighboring carbon atom before a fission of the C-O bond leads to a positively charged fragment and to the liberation of a neutral water molecule. In the fragment, the positively charged carbon atom is trivalent. On account of this unusual situation, the two electrons from a neighboring C-X bond are sucked into the space between the two carbon atoms to form a double bond. This causes a fission of the C-X bond from which the electrons were taken, thus freeing a hydrogen atom for migration within the molecule. This hydrogen ion will seek out another lone (nonbonding) electron pair of a hydroxyl oxygen to trigger another liberation of water.

In the final 1,4-elimination, the trivalency of a carbon atom does not lead to a double bond formation but to a ring formation sterically facilitated by the fact that carbon atoms participating in double bonds form planar structures characterized by bond angles of 120° (plane trigonal orbitals). After the 1,4-elimination, ejection of a hydrogen ion completes the process.

In summary, the conversion of pentose to furfural is seen to be based on the fact that hydrogen ions transform hydroxyl groups of the pentose to H_2O^+ groups representing the prerequisite for the liberation of water.

A ring closure as the second step [4] instead of the third step is unlikely, for steric reasons.

References

[2] H. R. Christen and K. Freytag, Chemie organischer Naturstoffe, Verlag Sauerländer, Aarau, 1974.

[3] J. March, Advanced Organic Chemistry, John Wiley & Sons, New York, 1992.

[4] C. D. Hurd and L. L. Isenhour, J. Amer. Chem. Soc. 54 (1932) 317-330.

3. Acid Catalysis

As stated already in chapter 2, the hydrolysis of pentosan to pentose and the dehydration of pentose to furfural are both catalyzed by acids. It is, therefore, appropriate to give a brief summary of important features of acid catalysis.

3.1. The Temperature Dependence of Acidity

In the second half of the nineteenth century, Svante Arrhenius (1859-1927) found the rate of acid-catalyzed reactions to be proportional to the hydrogen ion concentration. Although it turned out later that this is only a special case of a more general proton transfer concept, the hydrogen ion concentration remains an important aspect in acid catalysis.

In general laboratory practice, acid catalysis is commonly carried out at only slightly elevated temperature. Under such conditions, strong mineral acids such as hydrochloric acid and sulfuric acid are usually considered as "completely dissociated". This in itself is erroneous, and, worse yet, at any given acid concentration the hydrogen ion concentration diminishes with increasing temperature, and the extent of this phenomenon differs from one acid to another. The reason for this lies in the fact that the dielectric constant of water, responsible for the dissociation of the acids, diminishes strongly with increasing temperature [5] as shown in Figure 3. For hydrochloric acid, sulfuric acid, and phosphoric acid, the resulting decrease in acidity [6] is illustrated in Figure 4. The same trend is seen in the temperature dependence of the dissociation constant of acetic acid [7] shown in Figure 5, and of the dissociation constant for the second dissociation step of sulfuric acid [8] shown in Figure 6.

For a kinetic appraisal of furfural processes, which are universally carried out at temperatures in excess of 150 °C, the decrease of acidity with increasing temperature brings about a major problem as in all kinetic studies made for furfural to date, for obvious reasons of convenience, it has been customary to formulate the reaction rate as being proportional to the <u>initial</u> hydrogen ion concentration measured before the reaction, at room temperature, although in reality, at the high reaction temperatures of interest, the acidities are quite different. In view of the temperature dependence of the acidity being different for different acids, this means that when a kinetic formulation with the initial hydrogen ion concentration is derived from reaction experiments with hydrochloric acid, this formulation cannot be

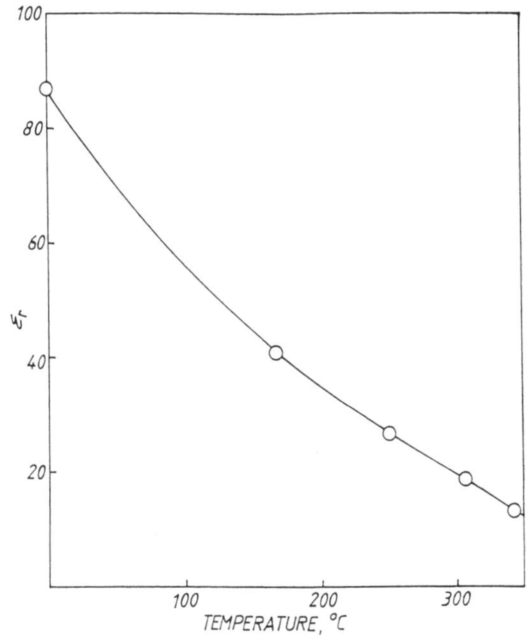

Figure 3. The Dielectric Constant of Water as a Function of Temperature.

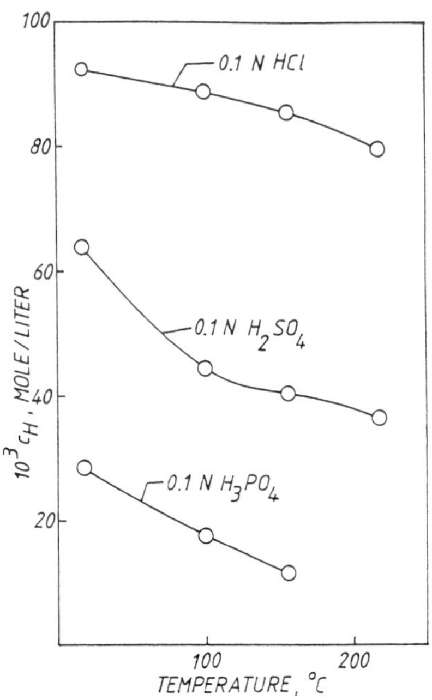

Figure 4. The Hydrogen Ion Concentration of Various Acids as a Function of Temperature.

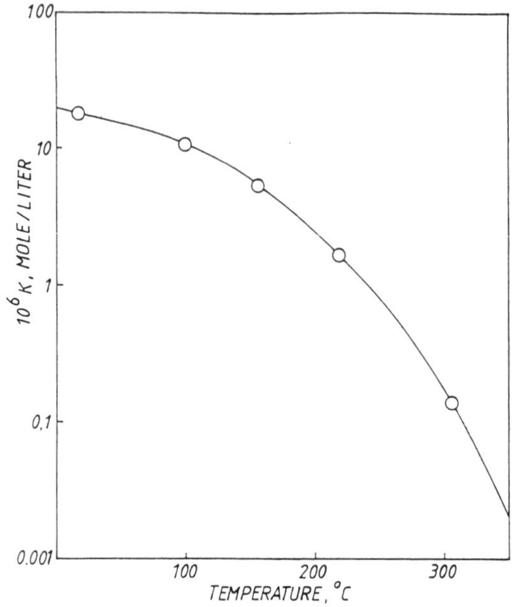

Figure 5. The Dissociation Constant of Acetic Acid as a Function of Temperature.

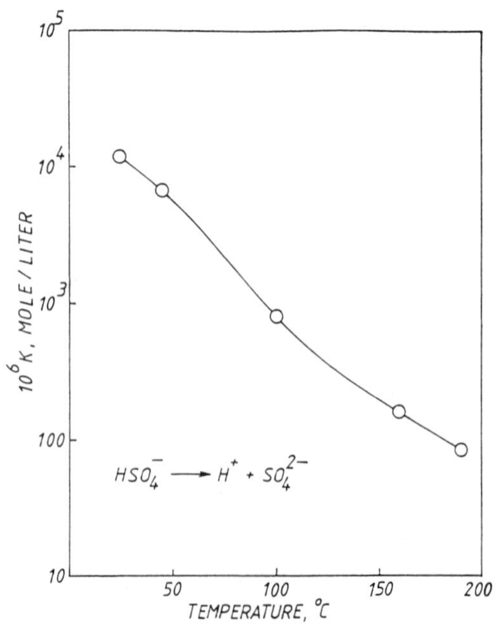

Figure 6. The Dissociation Constant for the Second Dissociation Step of Sulfuric Acid as a Function of Temperature.

applied to sulfuric acid, and vice versa, and it is fundamentally objectionable to pack the temperature dependence of the reaction rate into a single term (the "exponential factor" containing the activation energy) when in reality there are two different and opposing effects of increasing temperature, one being due to the given decrease of acidity, and the other being due to the growing energy of the molecules.

3.2. The Proton Transfer Concept

The claim of Arrhenius that the rate of acid-catalyzed reactions is proportional to the hydrogen ion concentration was soon found to require amendments as catalytic effects were discovered where the hydrogen ion concentration was negligible. In view of this predicament, T. M. Lowry [9] created a generalized proton transfer theory. For the most simple case of a mere rearrangement (isomerization) of a molecule, this theory can be outlined as follows:

Any acid-catalyzed reaction consists of three steps:

(a) An addition of a proton to the molecule to be converted.

(b) A rearrangement of the molecule activated (destabilized) by the added proton.

(c) A withdrawal of the added proton to yield a neutral product molecule.

The species adding the proton is called "proton donator", and the species withdrawing the proton is called "proton acceptor".

Against this background, in Lowry's words, the overall catalytic process can be seen as if a voltage is applied to the molecule to be converted (addition of a proton at one point of the molecule, and withdrawal of a proton at some other point of the molecule), so that an electric charge (an electron deficiency) is pulled through the molecule. Very instructively, Lowry speaks of "proton sources" and "proton sinks", thereby underlining the important fact that acid catalysis requires two agents ("terminals" in Lowry's words), whereas the concept of Arrhenius gave the erroneous impression that only one agent is involved (the hydrogen ion).

An illustration of Lowry's concept of acid catalysis is given in Figure 7, where S is the molecule to be converted (rearranged) to T. The proton donators cited as examples are the oxonium ion H_3O^+, an undissociated acid molecule HA, and water as a special case of HA, while the proton acceptors cited as examples are water (transformed to H_3O^+), and the acetate ion (transformed to acetic acid). Contrary to the concept of Arrhenius, Lowry's concept can explain why water as a proton donator and acetate ions as proton acceptors represent a power-

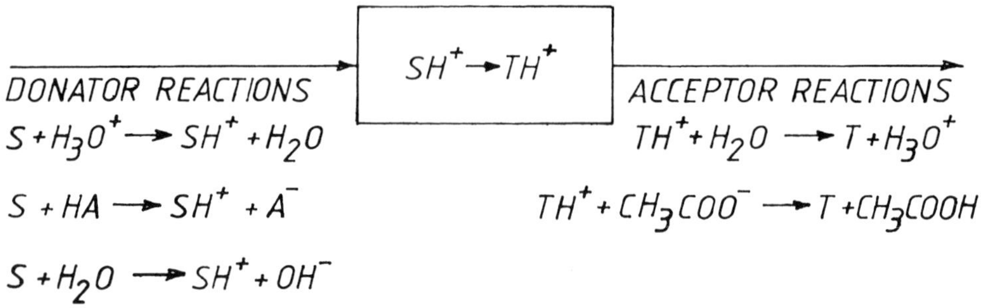

Figure 7. Lowry's "Voltage Model" of Acid Catalysis.

ful catalytic system even when the hydrogen ion concentration is insignificant. With acetate ions voraciously "sucking up" protons, this system has a high "catalytic voltage". Consequently, at high temperature (180 °C), furfural can be effectively produced with water as the "catalyst", even when the carboxylic acids liberated from the raw material are neutralized by an excess of calcium carbonate [10].

Lowry's ideas extended the notion of an acid to that of a substance capable of acting as a proton donator, so that, in his terminology, even pure water is an acid. On the other hand, the notion of a base was extended to that of a substance capable of acting as a proton acceptor, so that, in Lowry's terminology, water is a base as well. Hence, water turns out to be an amphoteric substance of central importance for catalytic processes.

When the catalysis is supported not only by hydrogen ions but also by other species X_i, the reaction rate is expressed as

$$r = k_0\,[H^+] + k_1\,[X_1]^a + k_2\,[X_2]^b + \ldots\ldots$$

which is commonly referred to as the equation of "<u>general acid catalysis</u>". For $k_1, k_2, \ldots = 0$, this relationship degenerates to

$$r = k_0\,[H^+]$$

known as the equation of "<u>specific acid catalysis</u>". Thus, "specific acid catalysis", representing the ancient finding of Arrhenius, is merely an approximation of "general acid catalysis", sometimes a fairly good approximation, but a totally unacceptable approximation in other cases.

In furfural technology, rate equations based on the assumption of "specific acid catalysis" are sufficiently accurate when use is made of high hydrogen ion concentrations produced by strong mineral acids such as H_2SO_4, but such formulations fail completely when the only "catalyst" used is water.

References

[5] E. U. Franck, Z. physik. Chemie, Neue Folge, 8 (1956) 107-126.

[6] A. A. Noyes, A. C. Melcher, H. C. Cooper, and G. W. Eastman, Z. physik. Chemie 70 (1909) 335-377.

[7] A. A. Noyes, Y. Kato, and R. B. Sosman, Z. physik. Chemie 73 (1910) 1-24.

[8] E. U. Franck, D. Hartmann, and F. Hensel, Discuss. Faraday Soc. 39 (1965) 200-206.

[9] T. M. Lowry, J. Chem. Soc. 1927, 2554-2567.

[10] S. I. Aronovsky and R. A. Gortner, Ind. Eng. Chem. 22 (1930) 264-274.

4. The Kinetics of Pentose Formation from Pentosan

The rate of pentose formation from pentosan is proportional to the pentosan concentration, but is diminished by the sequence reaction of pentose to furfural, so that

$$d[PENTOSE]/dt = k_0[PENTOSAN] - k_1[PENTOSE]$$

where t is the time coordinate. From measurements by Dunning and Lathrop [11], referring to a digestion of corncobs with aqueous sulfuric acid at two acid concentrations (1.9 and 4.4 % by weight) and two temperatures (100 and 121 °C), the first proportionality factor is found to be

$$k_0 = 7.832 \times 10^4 \text{ liter/(mole min)} \times c_H \times \exp(-5163\ ^\circ K/T)$$

where c_H is the initial hydrogen ion concentration (mole/liter) at room temperature (attention is drawn to the decrease of acidity with increasing temperature as discussed in the preceding chapter), and where T is the absolute temperature (°K). Data reported by Bryner, Christensen, and Fulmer [12], referring to a digestion of oat hulls with aqueous hydrochloric acid at various concentrations (0.025 to 0.275 N) and various temperatures (100 to 164.4 °C) are similar but less consistent.

Inasmuch as this hydrolysis involves a solid (pentosan), it is obvious that the degree of comminution of this solid plays a role and may change k_0 to some extent. It is equally probable that the nature of the raw material will have some effect on k_0 in that pentosan firmly embedded in cellulose or resin will react more slowly than pentosan readily accessible. However, all of these effects are only of minor significance as at equal hydrogen ion concentration the hydrolysis of pentosan to pentose is very much faster than the subsequent dehydration of pentose to furfural. Consequently, in any appraisals of furfural reactors the period of time required for the pentosan hydrolysis to pentose is usually disregarded.

References

[11] J. W. Dunning and E. C. Lathrop, Ind. Eng. Chem. 37 (1945) 24-29.

[12] L. C. Bryner, L. M. Christensen, and E. I. Fulmer, Ind. Eng. Chem. 28 (1936) 206-208.

5. The Kinetics of Xylose Disappearance

The pentosan of the raw materials commonly used for the production of furfural consists mostly of xylan and of a small portion of arabinan. In sugarcane bagasse, the ratio of xylan to arabinan is 18.0 to 2.4 [13]. With this being so, essentially all studies of the kinetics of furfural formation were made by starting with pure xylose. There are indications that furfural formation from arabinose proceeds with similar ease but somewhat more slowly [14]. Thus, in view of the percentage of arabinan in the total pentosan being small, no great mistake is made by considering all of the pentosan as xylan, and to limit the further discussions to the furfural formation from xylose.

The most easily determined effect in the formation of furfural is the disappearance of xylose. Although not necessarily all of the xylose consumed is transformed to furfural, the process of xylose disappearance is of central importance for the design of furfural reactors as it permits determining the residence time of the raw material in the reactor. Obviously, it would not make sense to continue the process when essentially all of the pentose has been consumed.

Using the xylose disappearance as a measure for determining the residence time is a valid procedure as it was established beyond any doubt that the xylose disappearance rate is not affected by loss reactions to be discussed later. In other words, irregardless of whether loss reactions occur or not (which depends on the design of the reactor), the xylose disappearance rate is the same.

The first kinetic work in this regard was published by Dunlop [15] who studied the rate at which pure xylose in aqueous solution disappeared at 160 ^{o}C in the presence of two different concentrations of hydrochloric acid. He found the rate of xylose disappearance to be proportional to the hydrogen ion concentration.

Many years later, a more detailed study of the subject was published by Root, Saeman, Harris, and Neill [16]. Using sealed glass ampoules heated by immersing them in oil of various temperatures, these authors reacted aqueous xylose solutions acidified by sulfuric acid, and measured

(1) the xylose disappearance rates,

(2) the furfural appearance rates, and

(3) the furfural yields

over a temperature range from 160 to 280 °C. The initial xylose concentrations ranged from 3.125 to 200 g/liter (0.02082 to 1.33218 mole/liter), and the sulfuric acid concentrations ranged from 0.00625 to 0.8 N (0.03065 to 3.831 % by weight).

The ampoules were evacuated before flame-sealing. For samples reacted for less than half a minute, very small ampoules carrying approximately 0.02 ml of solution were used, whereas for all other samples the ampoules were larger, carrying 0.25 ml. The mass of the sample was determined by weighing the ampoule before and after it was filled.

The rate of xylose disappearance was found to be

$$-d[XY]/dt = 9.306 \times 10^{15} \text{ liter/(mole min)} \times c_H \times [XY] \times \exp(-16894\ ^{\circ}K/T) \qquad (1)$$

where [XY] is the xylose concentration (mole/liter),

t is the time coordinate (minutes),

c_H is the initial hydrogen ion concentration (mole/liter), and

T is the absolute temperature (°K).

A similar but less comprehensive study was published by Schoenemann and Hofmann [17]. There is agreement that the rate of xylose disappearance is proportional to the hydrogen ion concentration, and that there is an exponential temperature dependence in accordance with the law of Arrhenius, but there are minor differences in the numerical values of the proportionality factor and the exponent.

Equation (1) can be used to calculate the residence time required for xylose to disappear. It is customary to calculate the time τ required for the xylose to be reduced to 1 % of its original concentration. As equation (1) yields

$$d[XY]/[XY] = -\, k_1\, dt \qquad (2)$$

where

$$k_1 = 9.306 \times 10^{15} \text{ liter/(mole min)} \times c_H \times \exp(-16894\ ^{\circ}K/T) \qquad (3)$$

it follows that

$$[XY]/[XY]_0 = \exp(-k_1 t) \qquad (4)$$

where $[XY]_0$ is the initial xylose concentration. Thus, setting

$$[XY]/[XY]_0 = 0.01 \text{ for } t = \tau \qquad (5)$$

results in

$$\exp(k_1\, \tau) = 100 \qquad (6)$$

yielding

$$\tau = 4.60517/k_l \quad (7)$$

This is an important equation for designing simple batch-type furfural reactors featuring a constant and uniform hydrogen ion concentration enforced by sulfuric acid.

A graphical representation of equation (7), using k_l of (3) with the hydrogen ion concentrations of various aqueous sulfuric acid solutions, is shown in Figure 8, with various temperatures as parameter.

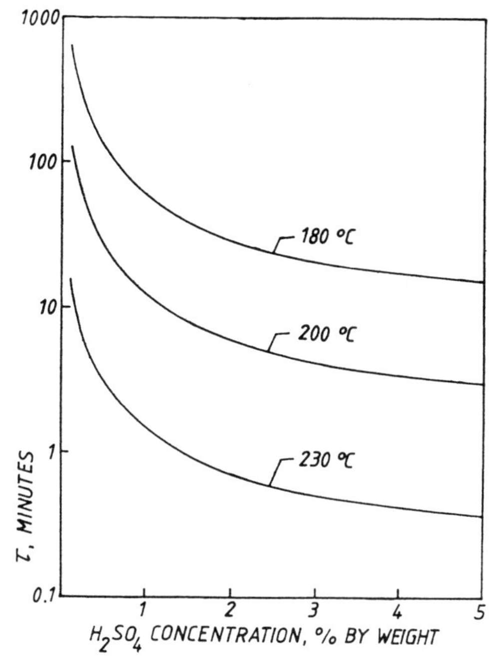

Figure 8. The Nominal Reaction Time τ as a Function of the Sulfuric Acid Concentration and the Temperature.

The ordinate is logarithmic to cover the huge range needed, the actual τ values for 0.1 % sulfuric acid being as follows:

230 °C:................15.7 minutes

200 °C:..............132.1 minutes = 2.201 hours

180 °C:..............638.7 minutes = 10.645 hours

Thus, only a few degrees C cause enormous differences in the reaction time, and inasmuch as the latter determines the throughput of a reactor, it is obvious that high temperatures are desirable. In practice, they are limited only by considerations of corrosion.

References

[13] M. Saska and E. Ozer, Ann. Progress Rpt. Sugarcane Res., Louisiana Agric. Expt. Sta., 1993, 241-253.

[14] C. D. Hurd and L. L. Isenhour, J. Amer. Chem. Soc. 54 (1932) 317-330.

[15] A. P. Dunlop, Ind. Eng. Chem. 40 (1948) 204-209.

[16] D. F. Root, J. F. Saeman, J. F. Harris, and W. K. Neill, Forest Products Journal 9 (1959) 158-165.

[17] K. Schoenemann and H. Hofmann, Chemie-Ing.-Techn. 29 (1957) 665-674.

6. Furfural Loss Reactions

As pointed out already in the preceding chapter, not all of the pentose consumed will necessarily end up as furfural, the reason being that in addition to the dehydration of the pentose two sequence reactions, both involving the furfural product, can take place:

(1) A reaction of furfural with itself, commonly called "furfural resinification".

(2) A reaction of furfural with an intermediate of the pentose-to-furfural conversion, this reaction being commonly called "furfural condensation".

The reactions (1) and (2) may or may not take place, depending on whether or not the furfural formed by the dehydration of pentose is permitted to stay dissolved in the liquid phase. Reactions (1) and (2) represent "loss reactions" in that they consume furfural and lead to products other than furfural. Hence, when the reactions (1) and (2) are avoided, by measures to be discussed later, then all of the disappearing pentose is converted to furfural. In this case, furfural is obtained at theoretical yield. By contrast, when the reactions (1) and (2) do take place, then the quantity actually produced will be smaller than the theoretical yield, and the extent of the losses will depend on how long the furfural is permitted to stay and react in the liquid reaction medium.

The "loss reactions" are possible only in the liquid phase, whereas they cannot take place in the vapor phase as the latter is devoid of catalytically active species. Thus, if furfural is instantly vaporized as it is formed, no "loss reactions" occur, and the yield will be 100 percent.

This chapter will deal with the case when "loss reactions" are permitted to proceed.

6.1. Furfural Resinification

With their "ampoule process" described in the preceding chapter, Root, Saeman, Harris, and Neill [18] determined the resinification loss by starting with aqueous furfural solutions of various acidities. According to

$$-d[FU]/dt = k_2 [FU]$$

with [FU] as the molar furfural concentration, the rate constant was found to be

$$k_2 = 1.685 \times 10^9 \text{ liter/(mole min)} \times c_H \times \exp(-11108\ ^{\circ}\text{K}/T)$$

Combining this expression with equation (3) of the preceding chapter leads to

$$k_2/k_1 = 0.181 \times 10^{-6} \times \exp(+5786\ ^{\circ}\text{K}/T)$$

As can be seen, this ratio is only a function of temperature. A graphical representation is shown in Figure 9. With increasing temperature, k_2/k_1 is seen to diminish strongly, so that above 200 °C resinification plays only a minor role.

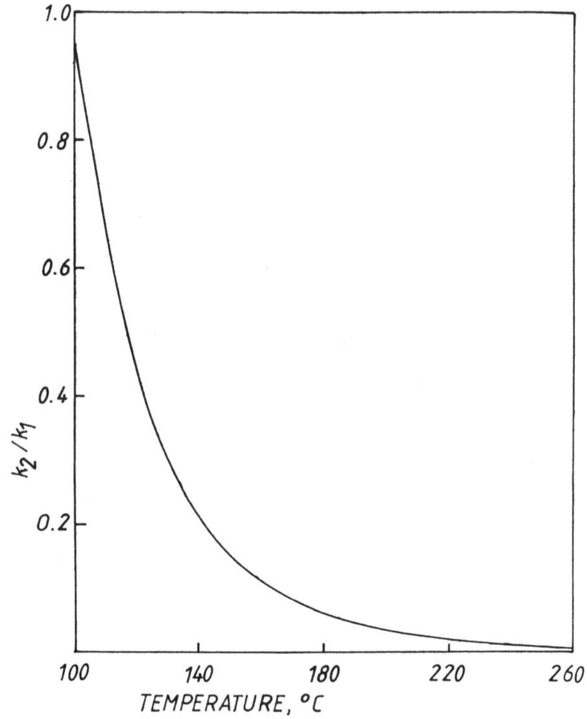

Figure 9. Dependence of the Ratio k_2/k_1 on the Temperature.

The reason for this phenomenon lies in the "entropy effect": Increasing temperature favors disintegration of molecules, so that inversely a combination of molecules, as in resinification, is being inhibited.

6.2. Furfural Condensation

When furfural is added to a xylose solution, and when this mixture is then submitted to a xylose-to-furfural catalysis, the furfural yield is smaller than without the added furfural. However, a reaction of furfural with xylose can be excluded since the addition of furfural to a xylose solution does not increase the rate of xylose disappearance. Consequently, the furfural must react with one of the intermediates of the xylose-to-furfural conversion shown in chapter 2 (Figure 2). One possibility is that one molecule of furfural reacts with the first intermediate to give furfural xylose:

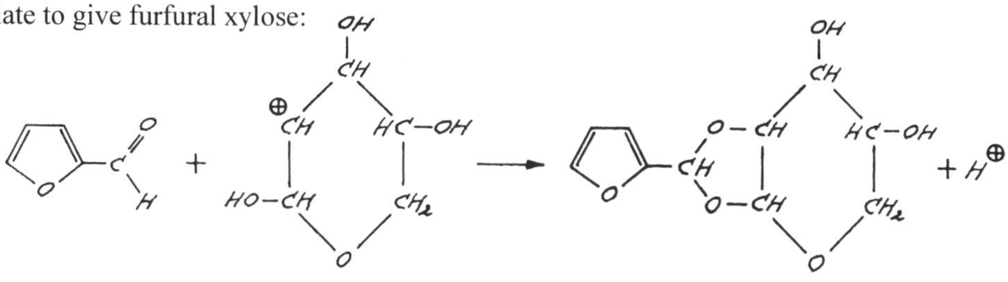

FURFURAL INTERMEDIATE FURFURAL PENTOSE

However, it is also possible that two furfural molecules react with an intermediate to give difurfural xylose:

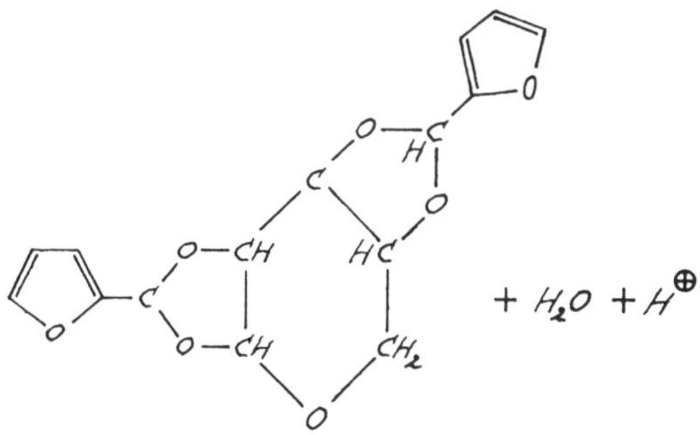

Thus, according to the laws of kinetics, the overall rate of these furfural condensation reactions should be

$$d[CP]/dt = k_a [FU] [INT] + k_b [FU]^2 [INT]$$

where [CP] stands for the concentration of the condensation products, and [INT] stands for the concentration of the intermediate.

Although [INT] can be taken to be proportional to the xylose concentration, there is no known experimental way to determine k_a and k_b explicitly. What is possible is to measure the actual yield as a function of time, xylose concentration, acidity, and temperature, for the experimental setup chosen, and to use these yield curves, together with the known pentose disappearance rate and the known furfural resinification rate, as a graphical interpolation basis for determining the losses by the condensation reactions. Such a procedure, reported by Root, Saeman, Harris, and Neill [18], is given in an appendix chapter, but it is usually considered too complicated and too unreliable to be used for yield prognoses.

6.3. General Loss Appraisal

Whenever furfural losses are permitted to occur, two findings are absolutely certain:

(1) The loss by condensation is massively greater than the loss by resinification.

(2) The loss by condensation and the loss by resinification both diminish strongly with increasing temperature, due to the "entropy effect": Increasing temperature favors disintegration of molecules, so that inversely a buildup of larger molecules is inhibited.

6.4. Additional Loss Reactions in Sulfite Liquor

All available publications on the kinetics of furfural formation are based on xylose in water. Thus, it is hardly surprising that these kinetics are found to be far from correct when they are applied to the pentose contained in sulfite liquor, the obvious reason being that this liquor contains substances known to react with furfural and with intermediates of the pentose-to-furfural conversion [19], with lignosulfonate being the main culprit, so that the quantity of furfural produced per unit mass of pentose is very much smaller than what kinetics in water predict. In other words, the kinetics of furfural formation in water must be supplemented by further loss terms. So far, none of the respective rate constants have been determined. Only an overall yield for special circumstances can be given in a later chapter.

References

[18] D. F. Root, J. F. Saeman, J. F. Harris, and W. K. Neill, Forest Products Journal 9 (1959) 158-165.

[19] E. Hägglund, Chemistry of Wood, Academic Press Inc., New York, 1951.

7. The "Paradox" of Furfural Yields

With the "sealed ampoule process" used for their kinetic studies, Root, Saeman, Harris, and Neill [20] achieved furfural yields well in excess of 70 % at temperatures above 220 °C, whereas industrial furfural processes, operating at lower temperatures and featuring a continuous removal of the furfural by steam stripping, have typical yields below 60 %. By contrast, in analytical chemistry, at a proven yield of 100 % [21], the formation of furfural from xylose or pentosan is routinely used for the quantitative determination of these substances. It is of great importance to elaborate the reasons for this "yield paradox".

In the "analytical furfural process" for the quantitative determination of xylose or pentosan, the substance to be analyzed is added to 12 % aqueous hydrochloric acid saturated with sodium chloride, the latter being used to raise the boiling point. This mixture is brought to boiling, and is maintained in the state of boiling, at atmospheric pressure, throughout the digestion period. Simultaneously, from a separate flask with a separate heat source, a small stream of water vapor is fed into the acid to serve as an entrainer for the vapor released from the reaction mixture. As the 12 % hydrochloric acid saturated with sodium chloride boils at approximately 110 °C, the water vapor undergoes superheating.

On the other hand, industrial furfural reactors are heated by steam injection, i.e. by condensation of water vapor, and inasmuch as at any pressure a condensing water vapor cannot boil a xylose solution, because of the boiling point elevation caused by the xylose, the reaction medium in industrial furfural reactors does not boil. Hence, the furfural generated and remaining in the liquid phase endowed with an acid catalyst can react, in solution, with itself and with intermediates of the xylose-to-furfural conversion, thus leading to high-boiling loss products, and the extent of the loss reactions depends on the period of time spent by the furfural in the liquid phase. By contrast, in the "analytical furfural process", where the reaction medium is in the state of boiling, the furfural generated cannot go into solution but is "rejected" into the vapor phase, where it is "safe", i.e. incapable of undergoing loss reactions, since the vapor phase does not have catalytically active species.

In more detail, this fundamental difference between the "analytical furfural process" and the "industrial furfural processes" is illustrated schematically in Figure 10 showing

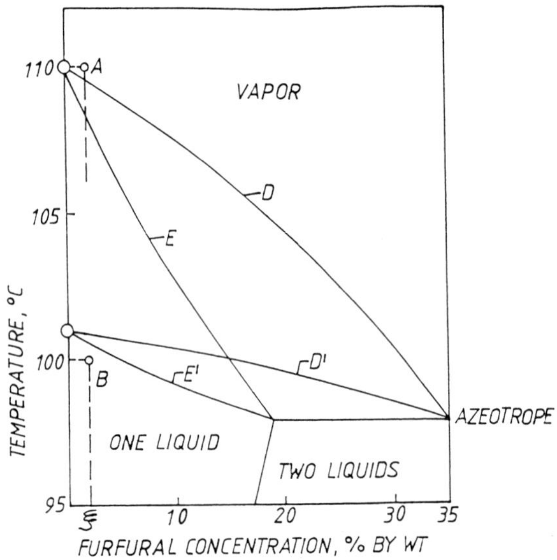

Figure 10. Phase Diagram for Explaining the Difference between Analytical and Industrial Furfural Processes

D and D': Dew Point Curves

E and E': Boiling Point Curves

the phase diagram for furfural in an aqueous solution having a boiling point of 110 °C (12 % aqueous hydrochloric acid saturated with sodium chloride), and in an aqueous solution having a boiling point of 101 °C (xylose solution). If a small furfural concentration ξ is generated in the first case (analytical furfural process), this results in point A lying in the vapor field. Thus, in this case, any furfural formed is indeed converted to vapor where it cannot react with anything. Hence, no "loss reactions" are possible, so that the analytical furfural process has a yield of 100 %.

The situation is quite different when a small furfural concentration ξ is generated in the second case, and when the heating is effected by condensing steam of 100 °C (atmospheric analogue of an industrial furfural process) as this results in point B lying in the liquid field where furfural can react with itself and with the first intermediate of the xylose-to-furfural conversion.

The crucial point is that loss reactions can be completely avoided in a boiling liquid but can not be avoided in a non-boiling liquid, and it is the inherent disadvantage of the conventional industrial furfural processes that at any pressure the condensing steam used for heating and stripping is thermodynamically incapable of boiling a xylose solution.

It is instructive to compare the formation of furfural in a boiling xylose solution (analytical furfural process) with an injection of some ether into boiling water. The ether/water phase diagram is shown schematically in Figure 11. When ether is injected into boiling

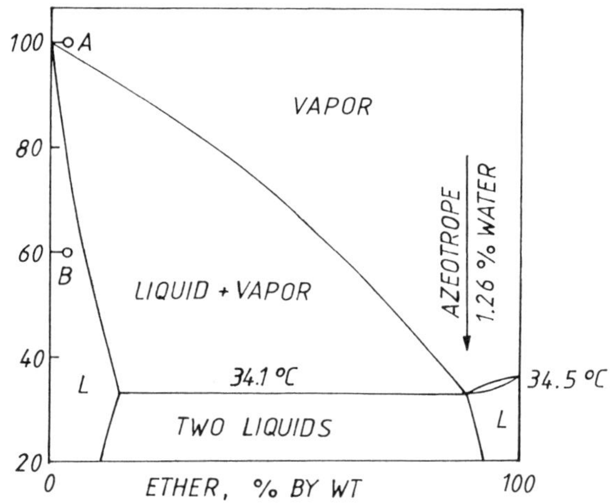

Figure 11. Schematic of the Ether/Water Phase Diagram. Not to Scale.

water, this leads to point A lying in the vapor field, which means that at 100 °C a liquid mixture of ether in water does not exist, so that ether injected into boiling water cannot dissolve but is "rejected" by being instantaneously and completely converted to vapor. In the same fashion, furfural generated in a boiling xylose solution cannot dissolve in the latter but is "rejected" instantaneously and completely as vapor ("analytical furfural process").

On the other hand, if some ether were injected into non-boiling water of say 60 °C, then point B of Figure 11 shows that the ether will dissolve in the liquid phase although 60 °C is far above the boiling point of ether or of the ether/water azeotrope. This case corresponds to the industrial furfural processes where due to heating by condensing steam the reaction medium does not boil, so that any furfural generated (injected) does dissolve in the liquid phase, thus undergoing loss reactions with itself and with the first intermediate of the xylose-to-furfural conversion.

In simple terms, it can be said that the yield tragedy of industrial furfural reactors is due to the fact that the steam used for heating and stripping is simply not hot enough to bring the reaction medium to boiling.

In the "sealed ampoule process" used by Root, Saeman, Harris, and Neill [20], almost all of the furfural generated remains in the liquid phase. After a rapid establishment of a chosen temperature by immersion of the ampoule in an oil bath, the xylose solution is in equilibrium with its vapor at this temperature, so that, after attainment of this equilibrium, no further boiling takes place, with almost all of the ampoule content being in the liquid phase, and only a tiny fraction of water being in the vapor phase. Then, when furfural starts forming, the boiling point of the mixture diminishes, because of the furfural/water azeotrope, so that, on account of the constant temperature of the oil bath, some more vapor is formed until, at a higher pressure, a new equilibrium is established, but irregardless of these phenomena, due to the ampoule leaving only little room for vapor, there is always a large percentage of liquid and only a tiny percentage of vapor. Hence, almost all of the furfural generated is in the liquid phase, where reactions with itself and with intermediates of the xylose-to-furfural conversion necessarily cause losses.

In view of this situation, it may seem surprising that in the "ampoule process", without any removal of furfural, the losses are hardly greater than in the industrial processes with their huge expense for steam stripping. The explanation lies in the simple facts that at any time the loss reactions are slower than the furfural formation, and that the principal loss, which is furfural condensation, diminishes as the xylose concentration diminishes, so that it comes to a halt when all of the xylose is consumed.

Against this background, the "ampoule process" is convincing proof for the contention that the continuous steam stripping used in the conventional industrial furfural processes is by no means essential. Of course, contrary to the "ampoule process", where the input is a liquid phase, in the industrial processes the furfural must be separated from the solid residue of the raw material, to get hold of it as a product, but there is no compelling reason for doing this stripping continuously. In a batchwise operation, it is perfectly possible to do the stripping at the end of the process, at only a minor yield disadvantage but at great savings in steam.

On an industrial scale, the "ampoule process" is duplicated by a continuous "plug flow" operation, where the raw material enters at one end and exits at the other end of the reactor. During the residence time in such a reactor, there is absolutely no removal of the furfural generated, as in the ampoules.

In conclusion, although all presently used industrial furfural reactors exhibit rather high losses, since their reaction medium does not boil, it must be kept in mind that a 100 % yield is possible, as in the "analytical furfural process", when the reaction medium is maintained in a state of boiling throughout the digestion, with the vapors continuously removed by a condenser. How this can be achieved in a simple fashion will be shown later.

References

[20] D. F. Root, J. F. Saeman, J. F. Harris, and W. K. Neill, Forest Products Journal 9 (1959) 158-165.

[21] E. E. Hughes and S. F. Acree, Journal of Research of the National Bureau of Standards 21 (1938) 327-336.

8. The Discoloration of Furfural

Freshly distilled furfural is colorless, but when it is exposed to the atmosphere for some time, it turns via yellow and brown to black. This color buildup is known to be triggered by oxygen as proven by the fact that when freshly distilled colorless furfural is stored under its own vapor pressure (after pulling a vacuum), it stays colorless indefinitely [22]. Three further facts shed light on the nature of the phenomenon:

(1) The discoloration is accompanied by a formation of water [22]. The more color builds up, the more water is formed.

(2) No discoloration takes place when the hydrogen on the 5-position of the furan ring is replaced by a less reactive (more firmly bonded) group.This is shown by the fact that 5-methyl furfural

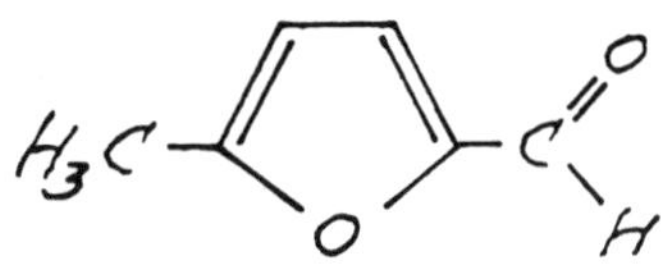

as well as furan dialdehyde

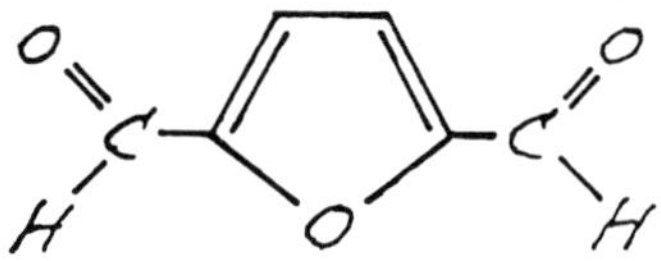

where the 5-positions are occupied by stable groups, can be stored indefinitely without any color buildup [23]. Thus, it can be concluded that the extremely reactive hydrogen in 5-position plays a pivotal role in the discoloration process.

(3) Color buildup is known to be due to an increasing number of conjugated double bonds. The longer the system of conjugated double bonds, the longer Schrödinger's box model of the molecule, thereby increasing the wavelength of light absorption [24]. This can be exemplified by the following comparison:

Furfural with three conjugated double bonds is colorless.

Difurfural (5,5'-diformyl-2,2'-difuran) with 6 conjugated double bonds is yellow.

Consequently, the discoloration of furfural must be due to the formation of increasingly longer systems of conjugated double bonds. This rules out polymers of the type shown in Figure 12 [25] as in these structures with increasing n (increasing degree of polymerization) the number of conjugated double bonds does not increase. Irregardless of n, all of these polymers are colorless.

Against this background, we can revert to the role of oxygen as the proven trigger of the discoloration process. As oxygen is known to be a diradical

• O – O •

capable of abstracting hydrogen atoms weakly bonded to carbon, oxygen abstracts the hydrogen atom at the 5-position of furfural, thus producing two radicals:

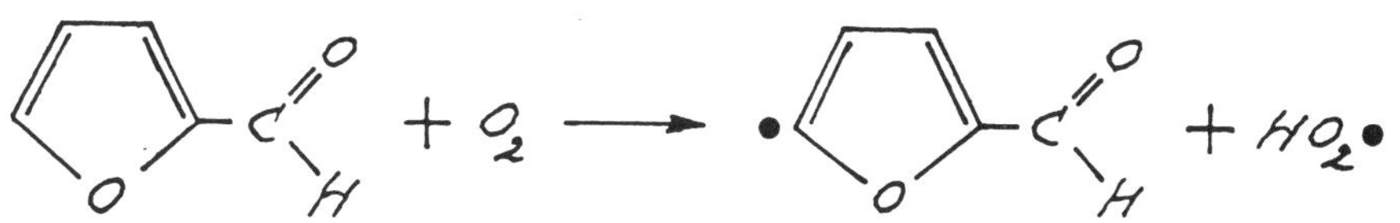

When the first radical on the right-hand side, henceforth designated as FU•, attacks the C=O double bond of a furfural molecule, the result is the sequence of reactions shown in Figure 13. The liberation is accompanied by an electronic rearrangement ("double bond migration") enforced by the "conjugation energy" released when a system of conjugated double bonds is extended [26]. As can be seen, this leads to the formation of an uninterrupted sequence of conjugated double bonds, which explains the discoloration by an increase in the wavelength of light absorption, according to Schrödinger's equation.

In the presence of further FU• radicals, the process of color buildup continues by a sequence of reactions further extending the system of conjugated double bonds as shown in Figure 14.

When Mother Nature has a choice between conjugated double bonds (alternation of double bonds and single bonds) and nonconjugated double bonds, she chooses the first alternative as the resulting system has a greater thermodynamic stability (lower energy content). This is revealed by the conjugated alternative having a lower heat of combustion [27]. On account of this, it is observed that nonconjugated double bonds migrate to become conjugated. This is what happens in the formation of the polymers causing the discoloration of furfural. The double bond migration in every other furan ring of the chain to achieve uninter-

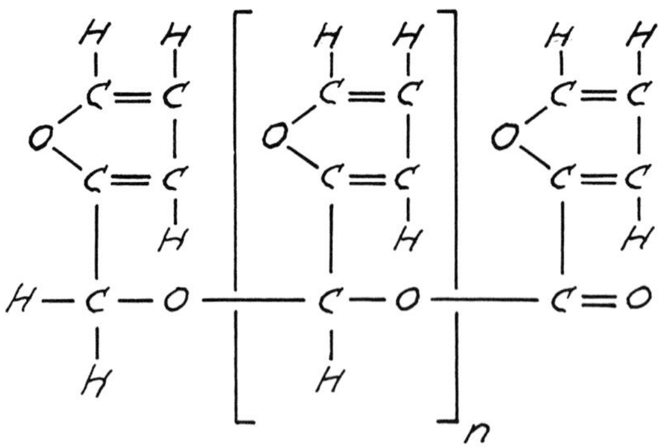

POLYFURFURAL (1)

BRIDGES BETWEEN THE ALDEHYDE OXYGEN AND THE ALDEHYDE CARBON

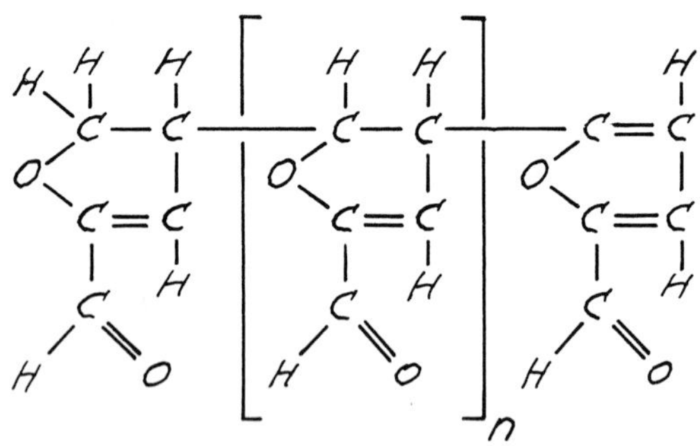

POLYFURFURAL (2)

BRIDGES BETWEEN CARBON ATOMS 2 AND 3

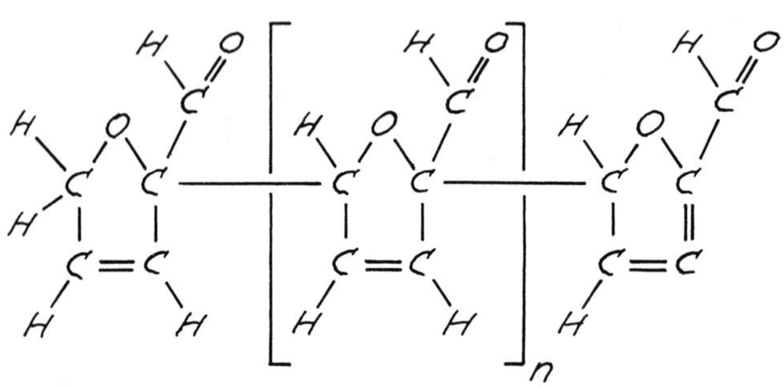

POLYFURFURAL (3)

BRIDGES BETWEEN CARBON ATOMS 2 AND 5

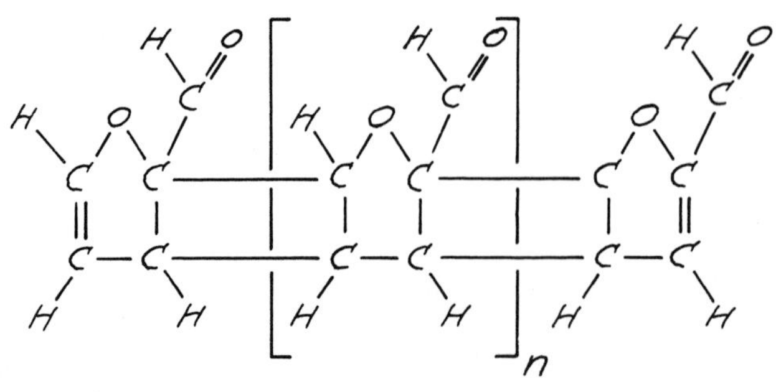

POLYFURFURAL (4)

BRIDGES BETWEEN CARBON ATOMS 2 AND 5 AS WELL AS 3 AND 4

Figure 12. Colorless Polymers of Furfural.

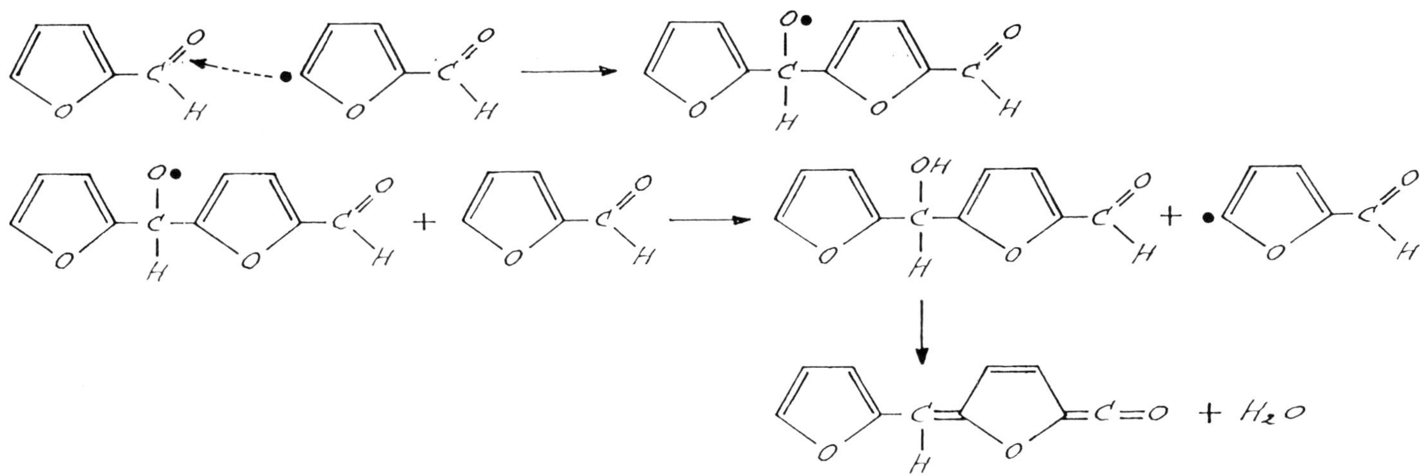

Figure 13. The First Step of the Color Buildup.

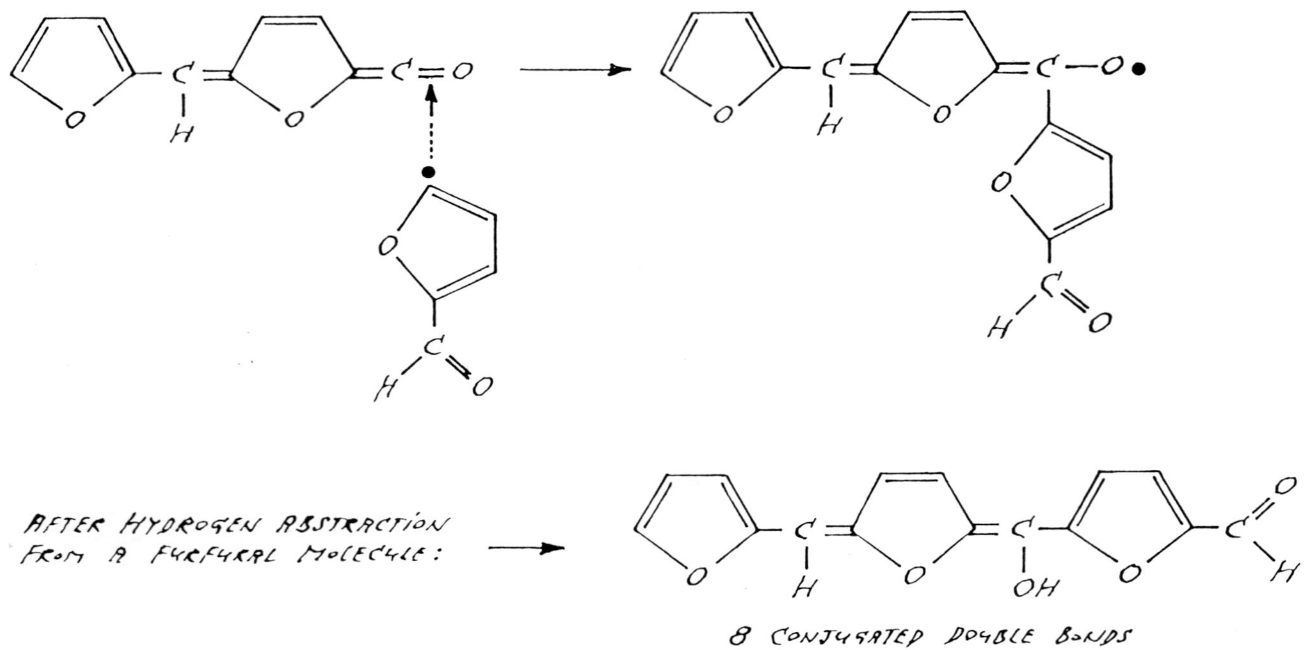

Figure 14. The Second Step of the Color Buildup.

rupted conjugation is necessarily imposed by the energy principle of thermodynamics. Without this electronic rearrangement, oxygen would still trigger a polymerization of furfural, but the effect would not be visible.

The relative quantity of color-causing substances in furfural is extremely small.The naked eye can detect colored substances in concentrations as low as 10^{-5} M. When a commercial 98 % pure furfural has turned totally black, and is then submitted to a vacuum distillation, it will still yield more than 97 % of pure furfural. In other words, in all but extreme cases, the color-causing substances amount to less than 1 %. Even if furfural is permitted to solidify to a gel after storage for many years, vacuum distillation of this gel will still yield 90 % of pure furfural [28].

References

[22] A. Gandini, Adv. Polym. Sci. 25 (1977) 47-96.

[23] Encyclopedia of Polymer Science and Engineering, Volume 7, 466, John Wiley & Sons, New York, 1987.

[24] G. Karagounis, Einführung in die Elektronentheorie organischer Verbindungen, Springer-Verlag, Berlin, 1959.

[25] W. S. Penn, Brit. Plastics 16 (1944) 286-292.

[26] C. K. Ingold, Structure and Mechanism in Organic Chemistry, G. Bell & Sons, London, 1953.

[27] P. Sykes, Mechanism in Organic Chemistry, Longman Scientific & Technical, Harlow, 1986.

[28] A. P. Dunlop and F. N. Peters, The Furans, Reinhold Publishing Corporation, New York, 1953.

9. Raw Materials

The production of furfural requires raw materials rich in pentosan. The pentosan content of some materials is given in Table 1. From these figures, it is readily understood why most furfural plants use corncobs. Bagasse, employed widely in hot climates, has not only less pentosan but also a very low bulk density, so that plants using this inferior raw material must accept the significant disadvantage of operating with less mass per unit of reactor volume.

Table 1. The Pentosan Content of Various Raw Materials in Percent of Dry Mass [29] [30] [31].

Corncobs:	30 to 32 %
Oat hulls:	29 to 32 %
Almond husks:	30 %
Cottonseed hull bran:	27 to 30 %
Birch wood:	27 %
Bagasse:	25 to 27 %
Sunflower husks:	25 %
Beech wood:	24 %
Flax shives:	23 %
Hazelnut shells:	23 %
Residues of olive extraction:	21 to 23 %
Eucalyptus wood:	20 %
Quebracho wood after tannin extraction:	19 %
Balsa wood:	18 %
Rice hulls:	16 to 18 %
Spruce wood:	11 %
Pine wood:	7 to 9 %
Douglas fir wood:	6 %

The pentosan content is measured by converting the pentosan to furfural, and by then determining the furfural, usually by precipitation with barbituric acid. The procedure is described in an appendix chapter. When the correct experimental conditions are employed, the furfural yield of this procedure was proven to be 100 percent [32].

References

[29] W. Jaeggle, Escher Wyss Mitteilungen 2 (1975) 38-51.

[30] J. W. Dunning and E. C. Lathrop, Ind. Eng. Chem. 37 (1945) 24-29.

[31] E. Hägglund, Chemistry of Wood, Academic Press Inc., New York, 1951.

[32] E. E. Hughes and S. F. Acree, Journal of Research of the National, Bureau of Standards 21 (1938) 327-336.

10. Furfural Processes

10.1. The Batch Process of QUAKER OATS

The batch process of QUAKER OATS is the oldest way of making furfural.As described in the introduction, this process was conceived in 1921, employing available cookers formerly used for the manufacture of an unprofitable cereal product.

The cookers were cylindrical vessels, 12 feet long by 8 feet in diameter, arranged horizontally, and rotating on a longitudinal axis, with steam entering through one trunnion and the product vapor leaving through the other trunnion. The overall process is illustrated in Figure 15.

The raw material mixed with sulfuric acid is introduced through a manhole, and after closing the latter, rotation of the cooker and passage of steam to give 153 °C are applied for 5 hours. The temperature of 153 °C was imposed by the pressure rating of the available cookers. After trying various materials, QUAKER OATS ended up lining the cookers with carbon bricks sealed by an acid-proof cement. This is the process as it is still used today.

Using air-dry oat hulls with a moisture content of 6 % as the starting material, and aqueous sulfuric acid as the catalyst, Brownlee [33] studied the effect of the initial water content on the furfural yield. The quantity of sulfuric acid per unit weight of dry substance was kept constant at 2.246 kg per 100 kg, but by adding increasing quantities of water the initial dilution of the acid was varied from 12.338 % at 13.500 % initial water content to 2.819 % at 43.092 % initial water content. The furfural obtained in the distillate varied from 40.271 % to 52.262 % of the theoretical yield, and the furfural lost in the residue varied from 1.25 % to 3.80 % of the theoretical yield. The final moisture content (water + furfural + volatile by-products) after 5 hours of treatment increased from 28.7 % at the lowest initial water content to 53.5 % at the highest initial water content. The results are plotted in Figure 16. As can be seen, the furfural obtained in the distillate, which is the only furfural quantity counting for the production, has a maximum at 25.4 % initial water content, the decrease below this value being due to a yield decrease with increasing xylose concentration, and the decrease

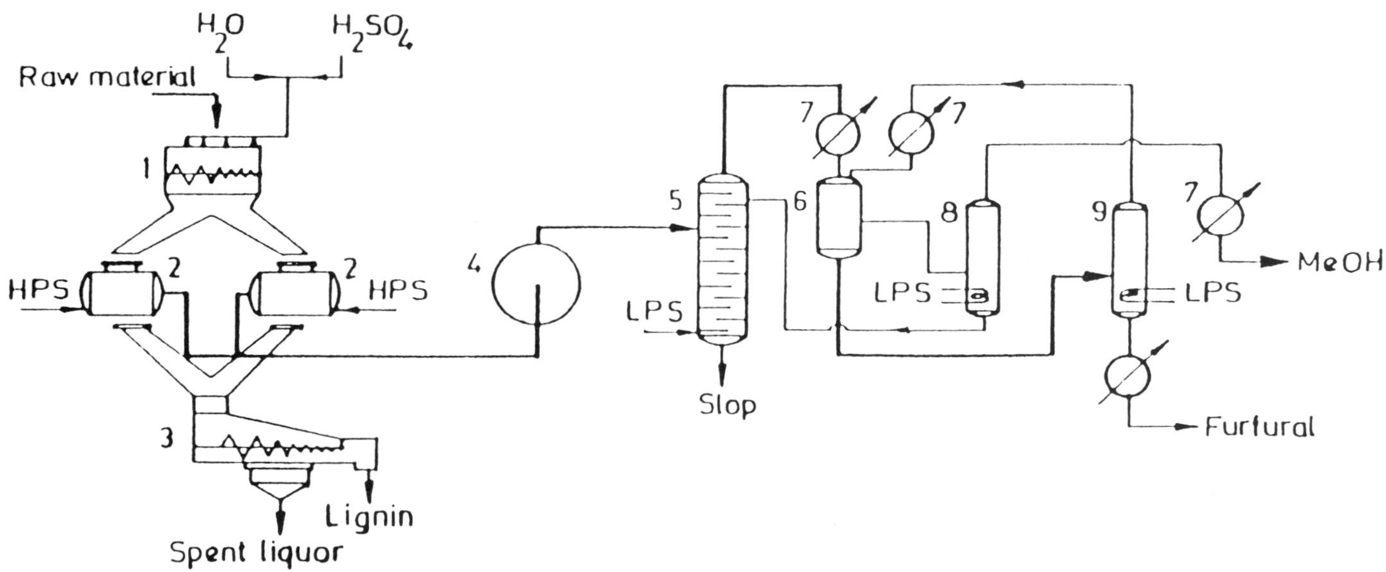

Figure 15.The Batch Process of QUAKER OATS.

(1) Mixer. (2) Reactors. (3) Screw Press. (4) Secondary Steam Generator. (5) Azeotropic Distillation Column. (6) Decanter. (7) Condensers. (8) Recovery Column for Low Boilers. (9) Furfural Dehydration Column.

HPS = High Pressure Steam. LPS = Low Pressure Steam.

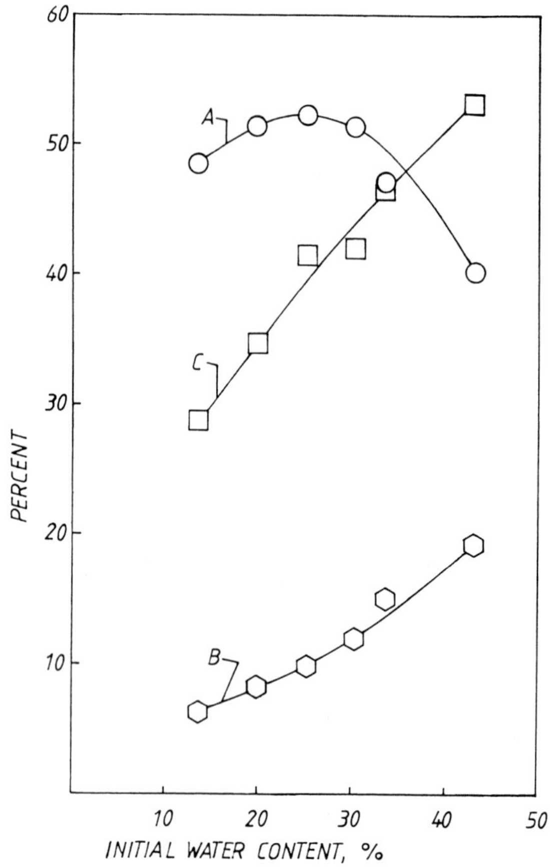

Figure 16. Operational Data of the QUAKER OATS Process
as Reported by Brownlee [33].
A - Furfural in the Distillate in % of the Theoretical Yield
B - Final Moisture Content of the Residue
C –Furfural in the Residue in % of the Theoretical Yield

above this value being largely due to the increasing furfural losses in the residue as more moisture retains more furfural.

For Brownlee's optimum conditions at 25.4 % initial water content, the initial sulfuric acid concentration was 6.05 % by weight, the furfural in the distillate was 52.3 % of the theoretical yield, and the furfural in the residue was 9.9 % of the theoretical yield. Thus, the overall chemical yield was (52.3 + 9.9) % = 62.2 %, and the furfural lost in the residue amounted to a whopping 15.9 percent of the total furfural formed. Of course, some more furfural could have been recovered from the residue if steam stripping would have been continued, but the respective benefit, diminishing exponentially, did not justify the steam cost thereby incurred.

In spite of its venerable history, the batch process of QUAKER OATS does have significant disadvantages which can be summarized as follows:

(a) A long residence time because of the low temperature.

(b) A high requirement for sulfuric acid to somewhat compensate the low temperature.

(c) Special measures against corrosion (carbon bricks).

(d) An extremely acid residue.

(e) Problems in processing fines which tend to be blown out by the steam as the tumbling of the charge invites such an effect.

(f) The rotation of the reactor requires a rather complicated design hardly warranted by its marginal benefit for stripping.

10.2. The Batch Process Used in China

China has adopted a very simple inexpensive batch process shown schematically in Figure 17.

All reactors have a diameter of 1.5 m and a height of 8 m. Rather oddly, they are made of mild steel and have an enormous wall thickness of 50 mm to sustain the corrosion. There is no lining, but the inside wall gets covered, and somewhat protected, by furfural polymers naturally formed in the process. The raw material usually consists of ground corncobs from which the fines were removed by sifting. The particle size is between 20 and 30 mm, and the initial moisture content of the raw material is in the order of 15 percent.

In the feed screw, made of stainless steel, 4 % aqueous sulfuric acid is sprayed over the incoming corn cobs to give 1.5 kg of acid per 100 kg of liquid phase. The acidified raw material is charged through a hatch in the top until about 75 % of the reactor is filled. After closing of the hatch, 1 to 1.5 tons/h of steam of 6 to 7 ATM is passed through the charge from the bottom to the top for a period of 4 to 5 hours. After this time, the residue is discharged by opening a flap valve at the bottom. In this rather brutal fashion, the reactor is emptied in 20 seconds.

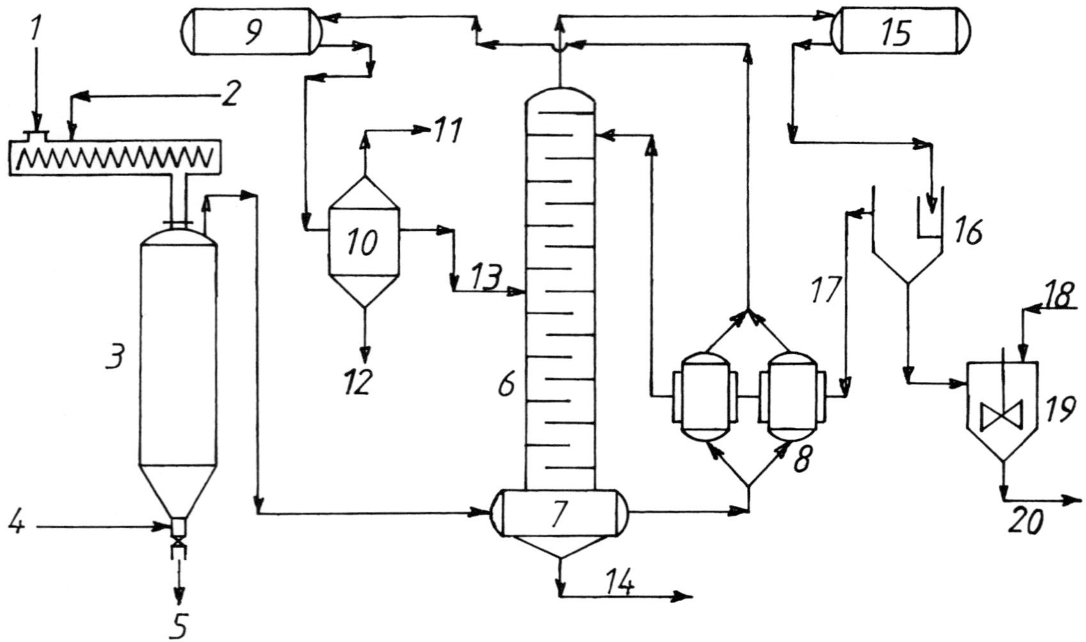

Figure 17. Schematic of the Chinese Furfural Process (Plant in Shanying, 2500 tons/a with 6 reactors).
(1) Ground corncobs. (2) Sulfuric acid. (3) Reactor. (4) Steam. (5) Residue. (6) Azeotropic distillation column. (7) Reboiler. (8) Heat exchangers. (9) Cooler. (10) Flash tank. (11) Noncondensible vapors. (12) Solids. (13) 5 % Furfural. (14) 2 % Acetic acid. (15) Condenser. (16) Decanter. (17) Aqueous phase. (18) 7 % Aqueous sodium carbonate. (19) Neutralizer. (20) Raw furfural to refining.

Instead of using a secondary steam generator, the Chinese process passes the reactor vapor directly through the reboiler of the azeotropic distillation column, and instead of using a recovery column for low boilers, most of the latter are flashed into the atmosphere before distillation starts.

As in the case of the batch process of QUAKER OATS, the furfural yield in the distillate is in the order of 50 percent.

10.3. The Batch Process of AGRIFURANE

The AGRIFURANE process, also known as PETROLE CHIMIE process, uses several batch reactors operating in series as shown in Figure 18.

The raw material is mixed with the filtrate of a belt filter press dewatering the residue. This leads to a slurry having a solid-to-liquid ratio of 1 : 6 by weight.

The first reactor gets a mixture of primary and secondary steam to attain 177 °C (9.35 bar). The vapor from this first reactor is fed into the second reactor, but in addition to this vapor, the second reactor also gets some primary steam to partially make up for the pressure loss in the first reactor. Analogously, the vapor from the second reactor is fed into the third reactor, and so forth. As there must be a pressure drop from reactor to reactor to ascertain flow, the last reactor has only 161 °C (6.34 bar), but by an intricate pipe and valve system the batch period is split up in such a way that each reactor is run at each of the different temperatures for an equal interval of time, so that each charge is treated equally.

The latest AGRIFURANE process uses sulfuric acid to give 1 % of this catalyst in the liquid portion of the charge, but the consumption of this acid is greatly reduced by the fact that in dewatering the residue by a belt filter press most of the acid is recovered in the filtrate, and this filtrate is recycled to be mixed with the incoming raw material to form the slurry mentioned above.

The idea behind using a slurry rather than loose raw material was to permit processing fines, and the reason for running the reactors in series was obviously the desire to reduce the steam consumption. Nevertheless, the AGRIFURANE process has serious disadvantages:

(a) A costly valve control system to permit "switching" reactors for the purpose of giving each charge the same treatment.

(b) An extremely costly belt filter press for dewatering the residue.

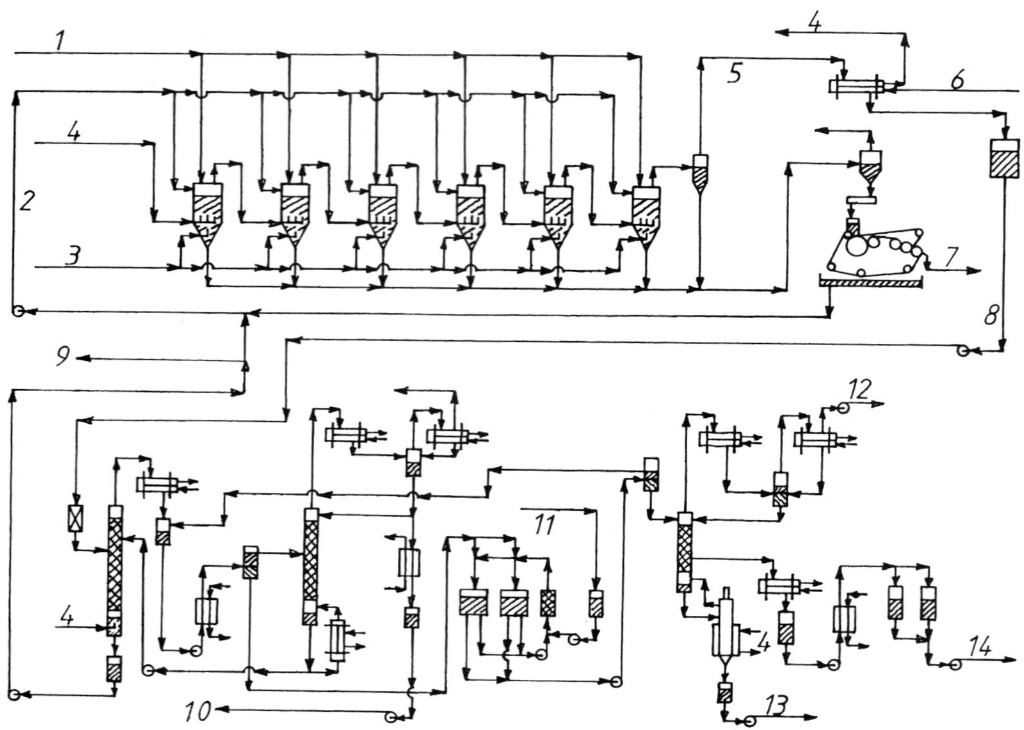

Figure 18. The AGRIFURANE Process.

(1) Raw material. (2) Acid water. (3) Primary steam. (4) Secondary steam. (5) Vapor of 5.5 % furfural. (6) Demineralized water. (7) Cake to drier. (8) Reactor condensate. (9) Waste water. (10) Low boilers. (11) Aqueous sodium bicarbonate. (12) Air. (13) Polymers. (14) Furfural.

(c) A drier to make the belt filter cake burnable.

Because of its high cost of investment, this process may be considered as obsolete.

10.4 The Continuous Process of QUAKER OATS

In the nineteen sixties, after having used their original batch process for some 40 years, QUAKER OATS built a revolutionary continuously operating furfural plant in Belle Glade, Florida. Although this plant ceased operation in 1997, it was a milestone in furfural technology.

The reactor system of this installation is shown schematically in Figure 19. The plant had three such trains, with two trains operating, and the third acting as a spare or undergoing maintenance. The raw material was bagasse from a neighboring sugar mill. The bagasse was first submitted to a pretreatment with low pressure steam to impart an increased moisture content claimed to be necessary as a "lubrication" in the subsequent auger press ("French press") used as both feeder and pressure lock for the reactors. The pretreatment was carried out in a "blender" 3 m in diameter by 5 m long, equipped with two horizontal paddles and a steam distributor in the bottom.

In the auger presses, each having a feeding capacity of 60 tons/h, some of the impregnation water was squeezed out again, so that the bagasse entered the reactor proper at 45 % moisture. Each reactor consisted of four horizontal sections in series, While the auger press was made of stainless steel, the reactor, 1.8 m in diameter and each section 16 m long, was made of mild steel lined with acid-resistant bricks. The transport paddles were made of stainless steel.

Through multiple nozzles, steam and sulfuric acid were added. The steam had a pressure of 160 psi = 10.888 ATM, but it was superheated to 650 °C. This had a drying effect, so that in spite of the water added with the acid, the residue left the reactor with a moisture content in the order of 40 percent.

The residue was ejected intermittently by a double lock discharge system illustrated in Figure 20. It featured two piston valves ("ram valves") and an intermediate chamber. With the first ram valve open and the second ram valve closed, the chamber was filled with both solids and product vapor at the operating pressure of the reactor. Then the first ram valve was closed and the second ram valve was opened, thus leading to an ejection of the chamber content into a cyclone, where the solids were separated from the vapor.

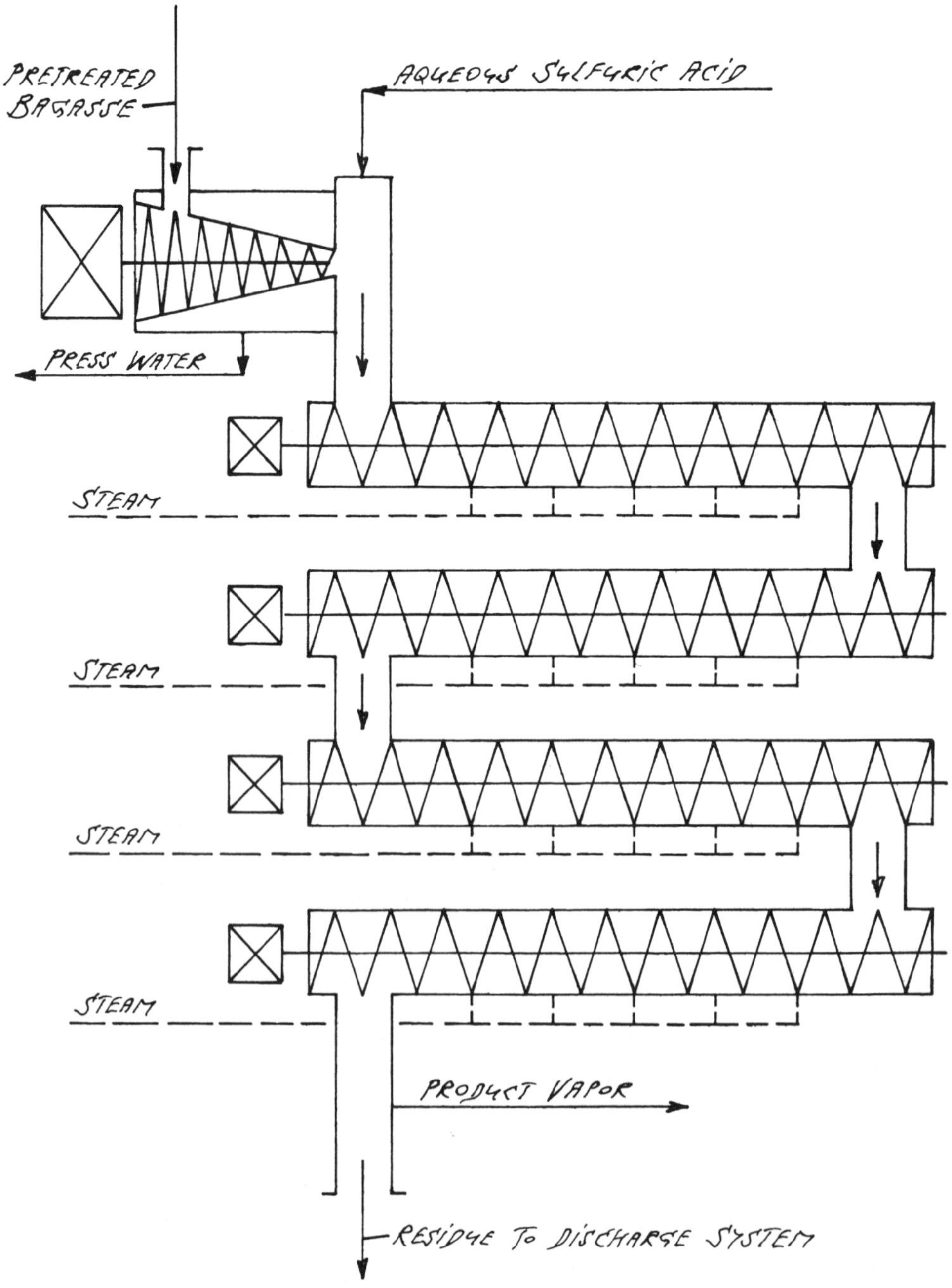

Figure 19. The Continuous Process of QUAKER OATS.

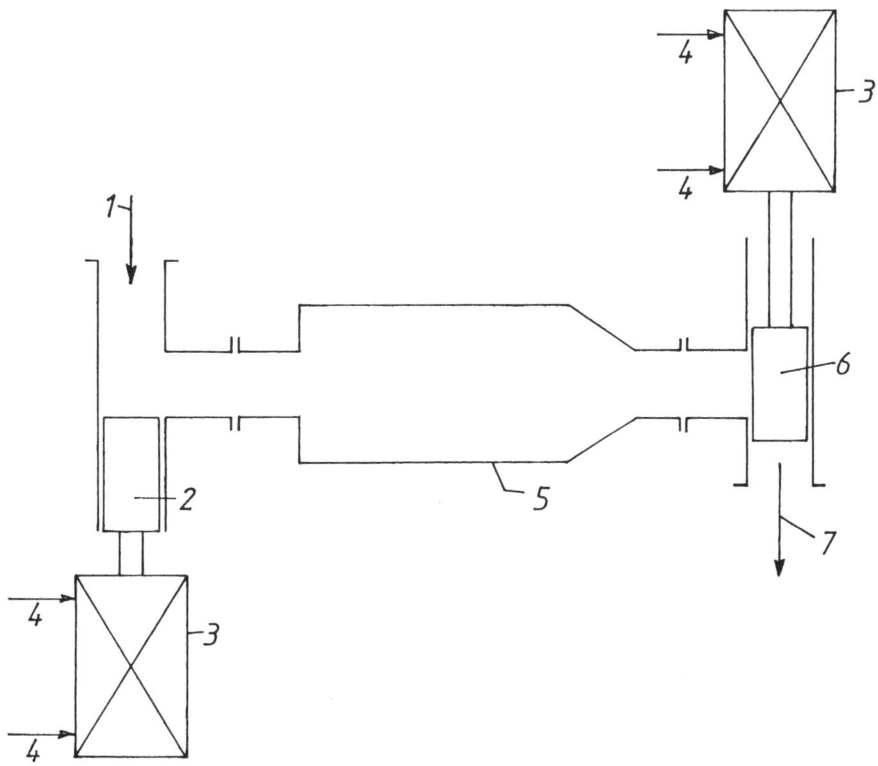

Figure 20. Schematic of the Discharge System.

(1) Residue and vapor from the reactor.

(2) First ram valve. (3) Pneumatic activator.

(4) Compressed air. (5) Chamber.

(6) Second ram valve. (7) Discharge to cyclone.

At a total residence time of one hour, the furfural yield of this process was 55 percent. Attempts to run the plant without sulfuric acid had to be abandoned as due to insufficient softening of the bagasse the reactors jammed.

There were only minor operational problems with this continuous process, and all of these problems would have been amenable to elimination by improved control equipment:

(a) If the bagasse feed was interrupted, the pressure seal would be lost, so that a "blow-back" would result in an emptying of the reactors.

(b) The bagasse had to be of uniform moisture content. If the feed to the auger press turned out to be too dry, this resulted in severe vibrations reducing the lifetime of the drive gears.

(c) Tramp iron entering the feeders would cause them to jam.

As to the reasons for shutting down the plant, they were manifold, and not exclusively of a technical nature:

(1) High cost of maintenance, primarily for the drives of the auger presses. An overhaul of these units was required every 1200 hours (50 days), which accounted for one third of the maintenance budget.

(2) Unfortunate relationships between QUAKER OATS and the sugar mill supplying the bagasse.

(3) Declining interest of the parent company (GREAT LAKES) in the furfural market.

10.5. The Continuous Process of ESCHER WYSS

The continuous process of ESCHER WYSS, now abandoned, used a fluid bed system illustrated in Figure 21. Having passed a rotary feeder, the raw material fell through a central pipe, where it was sprayed with aqueous sulfuric acid to give 3 % of this catalyst in the moisture portion of the feed. In the lower half of the reactor, steam emanating from a rotating distributor maintained the raw material in a state of suspension ("fluid bed") while carrying out the hydrolysis and the dehydration reaction desired. The level of the fluid bed, measured by gamma rays, was maintained by discharge valves controlled by the gamma ray signal. The temperature was 170 °C, and the mean residence time was 45 minutes.

The disadvantages of the ESCHER WYSS process were the following:

(a) A very low yield as in the random motion of a fluid bed there is no defined residence time, so that some particles leave the system immediately while others stay longer than necessary.

(b) An enormous sensitivity of the rotary feeder to abrasion by sand.

(c) Severe corrosion as even highly alloyed stainless steel (UDDEHOLM 904 L, 20 % Cr, 25 % Ni, 4.5 % Mo, 1.5 % Cu, 0.4 % Si, 1.7 % Mn) did not withstand the attack of the acid spray at the reaction temperature. As the spray must have a higher acidity than what is needed for the process, because of the

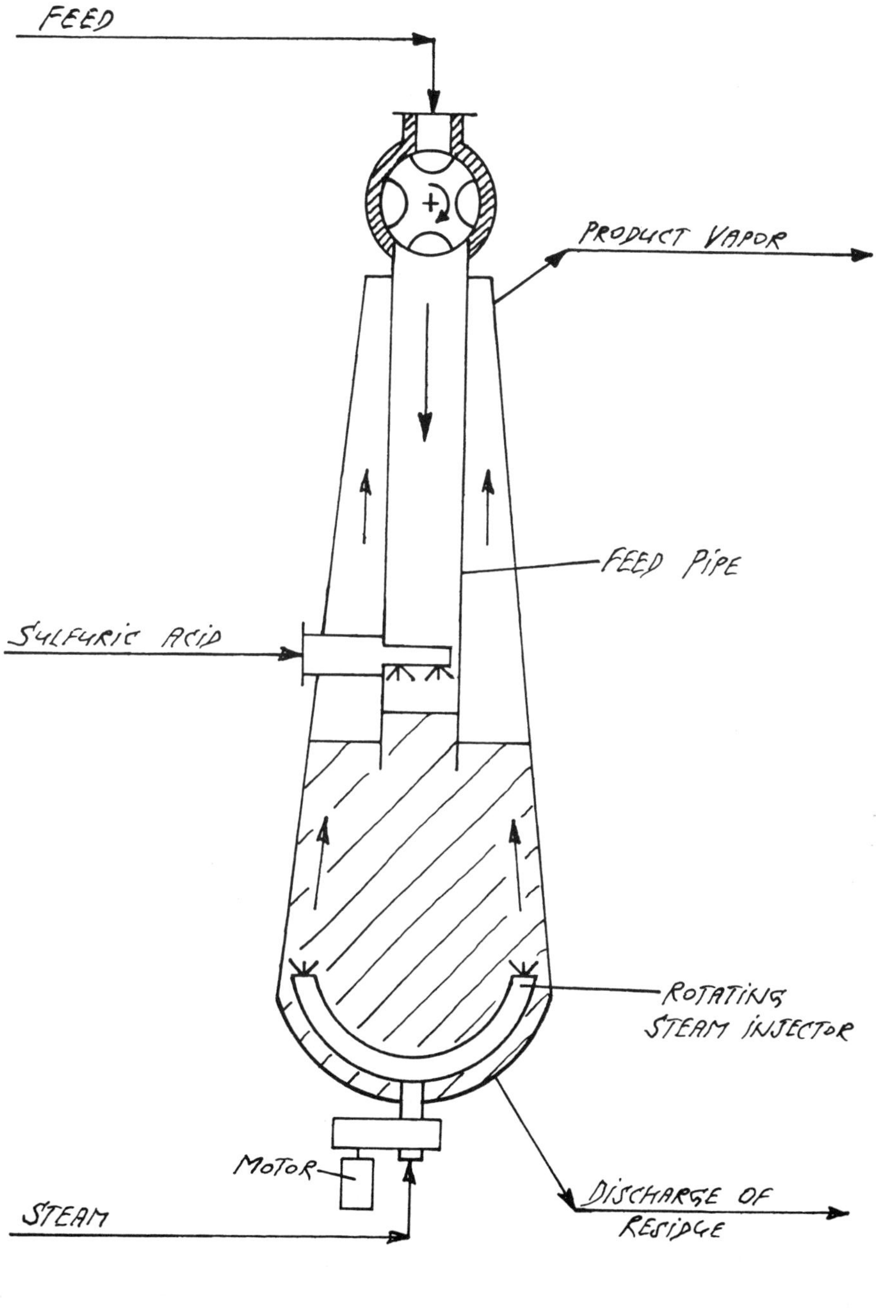

Figure 21. The ESCHER WYSS Process.

moisture content of the raw material, the choice of spraying into the reactor instead of preacidifying outside the reactor at ambient temperature was most unfortunate.

(d) In a fluid bed reactor, the steam input is not a free variable as when the flow is too low, the fluid bed collapses, and when the flow is too high, the particles are carried out of the system. It is, therefore, not surprising that the ESCHER WYSS reactor proved to be particularly sensitive with regard to losing fines, much to the detriment of subsequent equipment.

In favorable cases, when processing corncobs, the ESCHER WYSS reactor could be operated without sulfuric acid by relying for the catalysis on “innate” carboxylic acids (mostly acetic acid and formic acid) liberated from the raw material, but this required a lengthened residence time, thus reducing the throughput.

10.6. The Continuous Process of ROSENLEW

A typical ROSENLEW process operating on bagasse is illustrated schematically in Figure 22. The raw material is sifted in rotating screens, commonly called “trommels”, to remove fines. Normally 40 percent of the incoming bagasse must be rejected in these machines. The remaining coarse fraction enters the reactor at the top, by two intermittently opening hydraulic shutters, and the residue is ejected periodically from the bottom through several discharge valves, so that there is a slow continuous movement of gravitationally compacted raw material in downward direction, the residence time in the reactor being 120 minutes. Simultaneously, superheated steam of 10 bar is fed into the bottom, flows upwards, reacts with the raw material, picks up volatile reaction products, and leaves from the top. With the raw material moving downwards and the steam flowing upwards, this is a countercurrent mode of operation.

No foreign acid is added, the catalyst being a mixture of acetic acid, formic acid, and minor quantities of higher carboxylic acids formed from the raw material. This is called “innate catalysis” or “autocatalysis”. In addition to the rather modest hydrogen ion concentration produced by these poorly dissociated carboxylic acids, according to Lowry’s “general acid catalysis” discussed in chapter 3.2, undissociated acid molecules as well as water contribute to the catalysis in these reactors. Apart from this chemical complication, the mass transfer phenomena in the ROSENLEW reactor are intricate as well, for the following reasons:

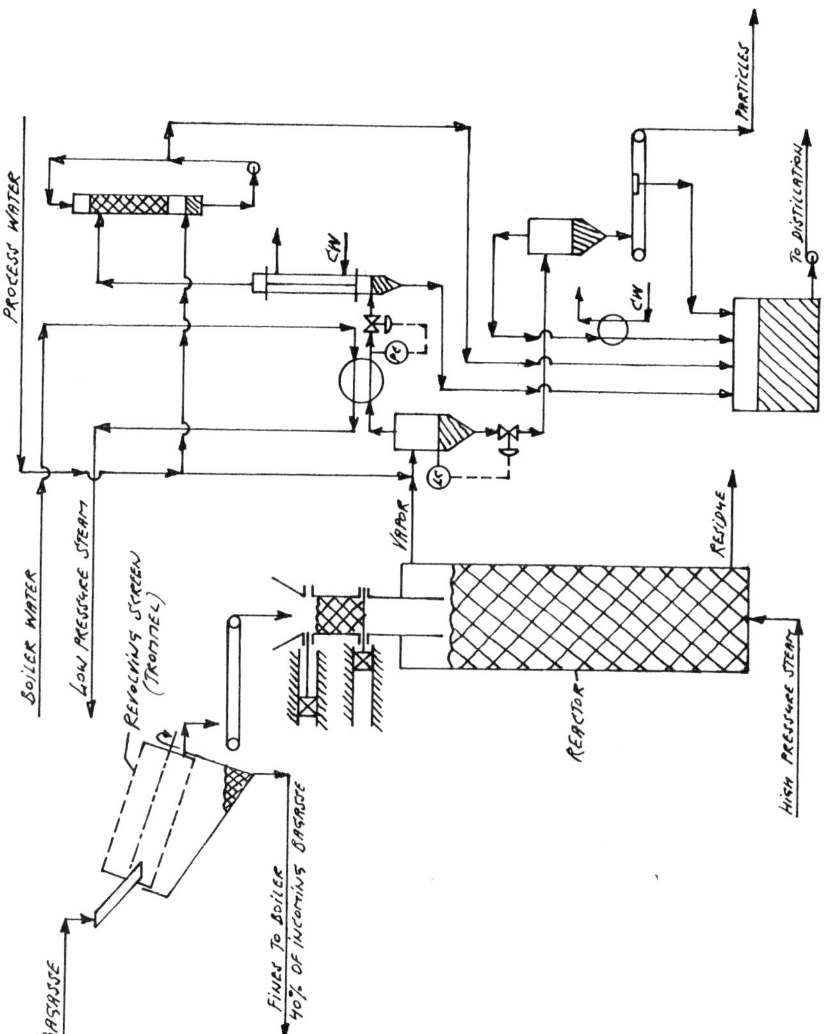

Figure 22. The ROSENLEW Process.

(1) The acid concentration is not uniform but features a "profile" in vertical direction. The incoming raw material has no acidity, but as it enters cold, acid vapors created further down condense on it and thus impart some acidity to the surface of the incoming particles, from where the acidity enters the particles by diffusion. This condensation process triggers the formation of more acid from the raw material, so that in the downward movement of the charge the acidity in any particle increases until it reaches a maximum at approximately one third down the height of the reactor. This maximum results from the fact that the steam introduced at the bottom produces a stripping effect in picking up volatile components from the liquid phase in the particles according to the respective vapor/liquid equilibria, so that the acid concentration at the bottom is close to zero and increases from the bottom in upward direction.

(2) The furfural concentration, depending on the acid concentration, also features a "profile" in vertical direction. The incoming raw material contains, of course, no furfural, but as it enters cold, furfural vapors created further down condense on it, and this imparts some furfural to the surface of the incoming particles, from where the furfural enters the interior of the particles by diffusion. Further downward, at higher acidities, where furfural is actually formed in the interior of the particles, and where the stripping effect of the steam removes the furfural from the surface of the particles, the diffusion of furfural is in opposite direction, from the interior of the particles to their surfaces.

(3) As formation of the "innate acids" requires acid catalysis, there is a problem in starting the reactor operation as initially the raw material does not contain any acid. Thus, unless the process is "triggered" by an initial addition of a "foreign acid", the startup is very slow as it relies on the small effect of water to catalyze the formation of tiny quantities of carboxylic acids, which in turn improve the catalysis, thus making more carboxylic acids, and so forth, until steady state is reached. Such a startup can take days, and it must be realized that any major interruption of the process, due to lacking raw material or other causes, also requires such a startup as it is not admissible to leave the charge in the reactor for any extended period of time since this is known to result in a

solidification of the residue, preventing its ejection.

(4) The furfural production depends on the steam input in a complicated fashion. At a low steam input, the furfural formed in the particles is not stripped sufficiently fast so that it can undergo loss reactions with itself and with intermediates of the pentose-to-furfural conversion. Hence, over a large range of steam input rates, the furfural output increases with increasing steam input, but this effect levels out as intensified stripping not only improves the desired rapid removal of furfural but also reduces the acidity throughout the column, thereby reducing the catalysis of furfural formation. Consequently, when a critical high steam input is exceeded, the furfural output drops off sharply as the catalyst is "blown out".

None of these complications materialize in a system with sulfuric acid as the latter is nonvolatile so that it cannot be stripped out by steam, and as it is so strong that the catalytic contribution by the "innate acids" is usually negligible, so that the acidity can be considered uniform throughout the charge of the reactor.

In steady state, the ROSENLEW reactor can be viewed as a stripping column energized by the steam injection in the bottom, with the charge of raw material representing a random packing, and with carboxylic acids injected in the upper range where these acids are formed in the reactor. By subdividing such a column into a large number of horizontal slices of finite height, and by applying distillation laws as well as some kinetic assumptions to each slice, an attempt has been made to simulate the operation of the reactor to the end of obtaining the concentration profiles and to show how the furfural output depends on the steam input [34], but the problem with such an approach is the sad fact that the kinetics of acid generation and of furfural formation by carboxylic acids, based on Lowry's "general acid catalysis", has never been studied so far, so that the most important input for such a calculation procedure is lacking completely.

10.7. Processes of the Future

As shown in chapter 5, increasing temperatures produce an exponential rise in the rate of xylose disappearance, thus reducing the reactor size dramatically. A process requiring a huge apparatus at 150 °C calls for no more than a pipe when carried out above 200 °C. In addition, in the present processes permitting furfural losses by employing a nonboiling reaction medium, the yield increases with increasing temperature, due to the "entropy effect" working against the loss reactions building larger molecules, so that the drive towards improved yields is a second argument in favor of ever higher thermal levels. Consequently, while so far the highest temperature ever used in an industrial furfural plant was 184 °C, referring to the continuous process of QUAKER OARS (chapter 10.4), all more recent process proposals focus on temperatures in excess of 200 °C, with particular preference for 230 °C.

Finally, on another front, an understanding of the reasons for the present huge losses in industrial furfural reactors has shown a way towards the 100 % yield routinely obtained in the "analytical furfural process" discussed in chapter 7.

All of these considerations give a colorful picture of potentially revolutionary new processes. Although so far none of these processes has gone beyond the pilot plant scale, they deserve particular attention as they hold the key to progress.

10.7.1. The SUPRATHERM Process

The SUPRATHERM process developed by KRUPP [35, 36, 37, 38] is a continuous hydrolysis where by using a high temperature between 200 and 240 °C the reactor is reduced to a simple pipe. An outline of the process in its original form is shown in Figure 23.

Chopped bagasse or "pith" (the fines fraction of bagasse) is fed into a tank 1 where it is mixed with a liquid recycle fraction and some highly diluted sulfuric acid to yield a flowable slurry. The latter is circulated through a high-speed rotor-stator comminution machine 2 converting the slurry to a smooth pulp. From this comminution circuit, a side-stream is withdrawn by an eccentric worm pump 3 and delivered to a flow reactor 4. In the region of the inlet end of this reactor, high pressure steam is injected to heat the pulp to 230 °C, thus rapidly effecting the desired hydrolysis. The reacted pulp is passed through a cooler 5 and a pressure control valve 6 before it enters a cyclone 7. In the cyclone, operated at reduced pressure, the reacted pulp is separated into a vapor fraction rich in furfural and an underflow representing the residual slurry. Due to the sudden decompression in the valve 6, furfural formed in the particles is released explosively by azeotropic cavitation. In this fashion, furfural is liberated from the particles much more completely than in the conventional stripping process based on transport by diffusion. In addition, the high reaction temperature greatly reduces losses by both resinification and condensation, thus leading to a substantially increased

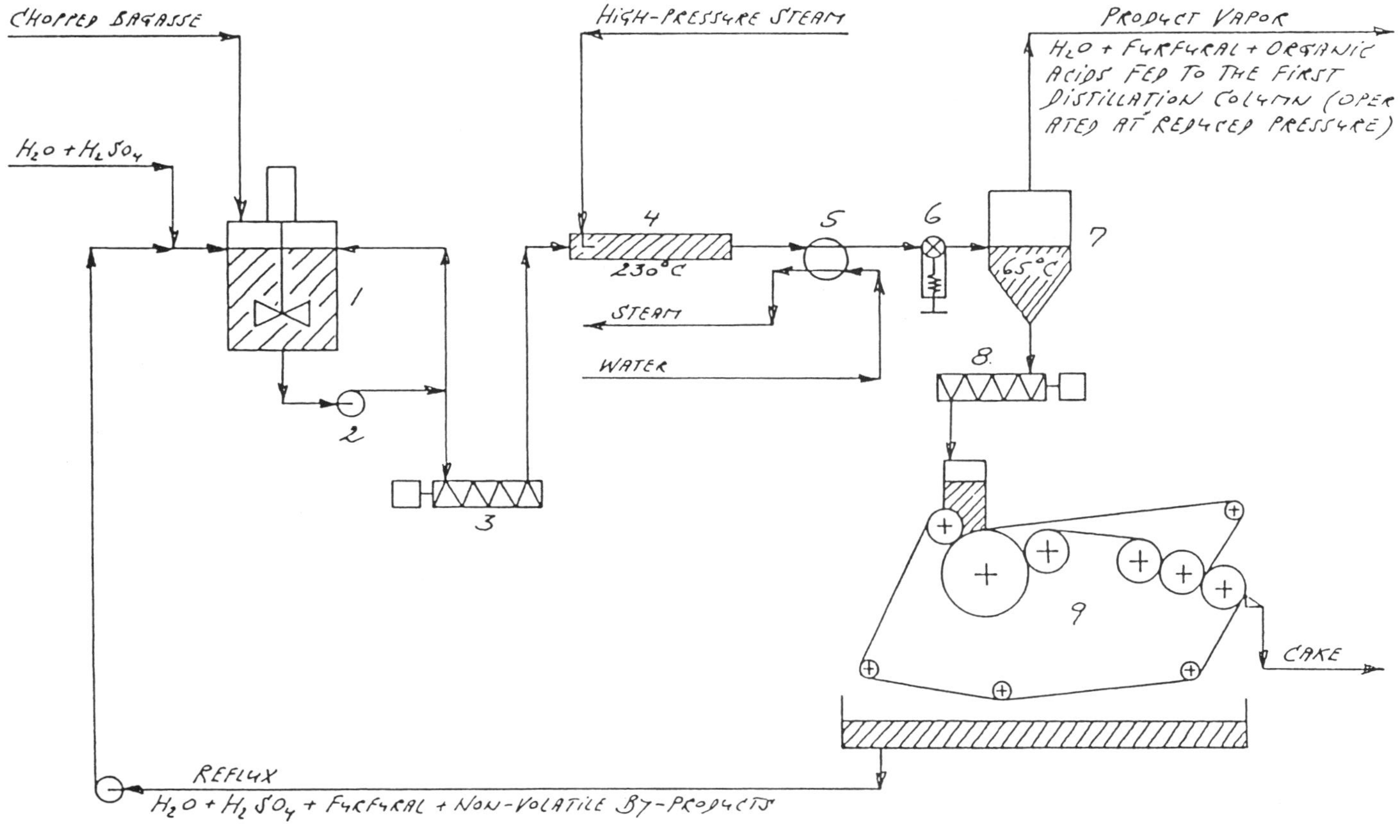

Figure 23. The SUPRATHERM Process.

furfural yield. It is noted that phenomenologically this process corresponds to the "closed ampoule process" discussed in the chapters 5, 6, and 7 as during the residence time in the reactor there is absolutely no removal of furfural.

The cooler 5 and the reduced pressure in the cyclone both serve to increase the furfural concentration in the vapor fraction. The latter enters the first distillation column without intermediate condensation, thus conserving its high enthalpy. Contrary to conventional processing, this vapor is entirely free of particles, so that encrustation problems, a well-known plague of orthodox furfural plants, are reliably avoided.

The underflow of the cyclone 7 is withdrawn by an eccentric worm pump 8 and delivered to a belt filter press 9 yielding a highly dewatered cake and a filtrate consisting essentially of water but loaded with small concentrations of sulfuric acid, furfural, and by-products. This filtrate is recycled to tank 1 for preparing the feed stock slurry. Due to this scheme, most of the sulfuric acid is recovered and reutilized, the only loss being the quantity contained in the cake. This loss is replaced in tank 1. Analogously, the water leaving the system with the cyclone vapor and the cake is also replenished in tank 1 so that the overall mass balance is satisfied.

Due to the recycle system, very little furfural is lost, and the furfural concentration in the cyclone vapor increases until it reaches a steady state limit. Undesirable nonvolatile by-products such as sugars cannot build up to a prohibitive concentration as a certain portion of them continuously leaves the plant with the cake. Thus, the cake discharge represents the "stabilizer stream" always required in recycle systems.

As compared to conventional furfural plants, the SUPRATHERM process is seen to feature the following advantages:

(1) A truly continuous mode of operation devoid of gate problems.

(2) A very short residence time permitting an extremely small reactor volume.

(3) A high reaction temperature reducing both resinification and condensation losses, thereby increasing the yield.

(4) Explosive release of the furfural by azeotropic cavitation within the particles of the raw material, with the effect of a further yield increase.

(5) A high concentration of furfural in the product vapor.

(6) A product vapor absolutely free of particles, thus avoiding encrustation problems.

(7) Extensive recovery and reutilization of the reactor acid.

(8) Outstanding simplicity. Startup within seconds.

The only disadvantage is the high cost for the investment and maintenance of the belt filter press, and for a drier to make the cake burnable.

When the SUPRATHERM process was conceived and tested in 1988, it was believed imperative to use sulfuric acid to render the reaction sufficiently fast. This implied that the belt filter press had to withstand this acid. Although this is no fundamental problem, as shown by the AGRIFURANE process described in chapter 10.3, it is a cost factor. It is, therefore, noteworthy that many years later, in view of findings regarding the high effectiveness of "general acid catalysis" as offered by the carboxylic acids generated from the raw material, it was calculated that the SUPRATHERM process could do without sulfuric acid, provided it is started with some foreign acid sufficient to initiate the "innate" acid generation by the raw material. If this concept is proven correct, it would greatly increase the attractiveness of the process as the corrosion load on the belt press would be much less severe, and the cake would be free of sulfuric acid so that any boiler could burn it.

Obviously, if the reactor size per unit of throughput is kept constant, omitting the sulfuric acid can be compensated by a slightly higher temperature, and inversely, if the temperature is kept constant, omitting the sulfuric acid can be compensated by a somewhat greater reactor volume.

The STAKE process, described in the following subchapter, has shown that at 230 ^{o}C working without any "foreign acid", only with the "innate acids" produced by the raw material, requires a residence time of 6.3 minutes, and that normal stainless steel is sufficient to withstand these "innate acids" at this temperature.

10.7.2. The STAKE Process

Since several decades, a Canadian company named STAKE TECHNOLOGY in Norval, Ontario, has been building plants capable of converting wood, bagasse, and other lignocellulosic raw materials into cellulose, lignin, and a xylose syrup. This process, called "Staketech Biomass Conversion" (SBC), involves three stages:

(1) A high temperature/high pressure hydrolysis of the raw material without the addition of any chemicals. This hydrolysis, typically carried out at 230 ^{o}C, is followed by flashing to atmospheric pressure ("steam explosion").

(2) A water extraction of the hydrolyzed raw material to produce a crude xylose-xylan solution.

(3) A dilute alkali extraction to remove the lignin, thus leaving cellulose behind.

Originally, the principal aim of this process was making cellulose, but it was realized that step (1) is eminently suitable to make furfural.

The core asset of STAKE TECHNOLOGY is a patented "feeder gun" as shown schematically in Figure 24.

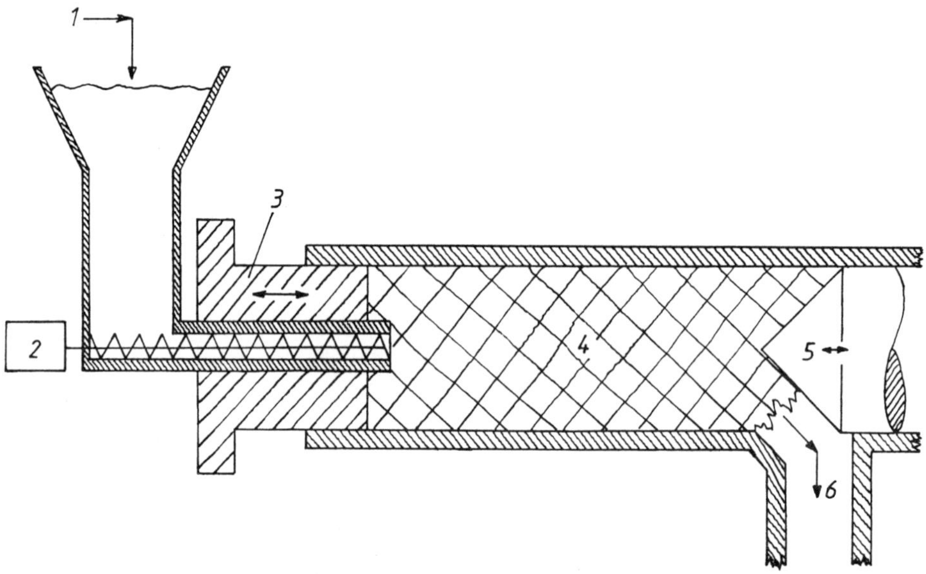

Figure 24. Schematic of the STAKE Feeder Gun.
(1) Input of the raw material. (2) Motor of the screw conveyor.
(3) Reciprocating piston. (4) Sealing plug of compressed raw material.
(5) Choke. (6) 28 bar to reactor.

A screw conveyor takes the comminuted raw material into a coaxial cylindrical chamber. On the left-hand side, this chamber features a reciprocating annular piston actuated hydraulically at a frequency of 120 strokes per minute to form a plug of compressed raw material serving as a pressure lock between the ambient atmosphere and the hydrolysis reactor. Due to its high frequency of operation, this feeder is likened to a machine gun. On the right-hand side, the chamber features a choke applying an adjustable pressure on the plug of raw material, and partially opening the chamber to permit the feed stock to be passed into the hydrolysis reactor.

The enormous advantage of the STAKE feeder lies in the fact that it can handle almost dry raw material. For the manufacture of furfural, water is required only to the extent as it is needed for hydrolyzing pentosan to pentose, and for dissolving this pentose. The subsequent conversion of pentose to furfural actually creates water. Thus, any excess input water is undesirable as it dilutes the acid catalyst and reduces the caloric combustion benefit of the residue. Against this background, a furfural process using the STAKE feeder has inherent advantages.

A measured characteristic of a STAKE process at 230 °C is illustrated in Figure 25. It shows the appearance and disappearance of pentose as a function of time. As can be

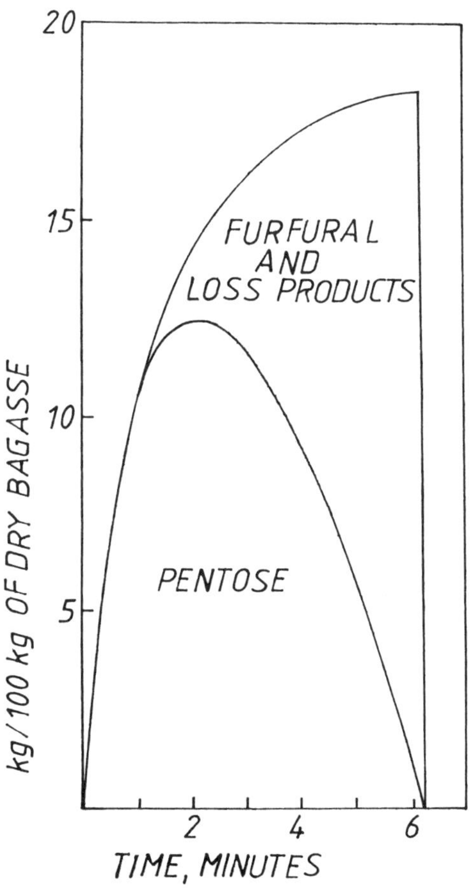

Figure 25. Measured Characteristic of a STAKE Process at 230 °C.

seen, the process was found to be completed after 6.3 minutes. As this result was obtained without addition of any foreign acid, it confirms the effectiveness of "general acid catalysis" produced by innate carboxylic acids. Thus, the same "general acid catalysis" can be expected to take place when the SUPRATHERM process described in the preceding subchapter is run without any foreign acid.

An overall concept of the STAKE process is depicted in Figure 26. The furfural yield of this scheme was calculated to be in the order of 66 percent.

In a screw conveyor 1, the incoming bagasse is wetted by a small quantity of water to effect "lubrication". Then the feeder gun 2 injects the bagasse into a digester 3 heated to 230 °C by means of steam. Corresponding to this temperature, the pressure in the digester is 27.7 ATM. If due to a lack of sufficient raw material the feeder gun fails to seal the system, there will be a "blow-back" of particles and vapor into a buffer bin 4 until the seal is restored.

In the digester 3, the raw material undergoes hydrolysis to furfural, and is then flashed through valve 5 into a blow bin 6 separating the residue from a product vapor rich in furfural. A screw conveyor 7 takes the residue into a rotary drier 8 energized by superheated steam. The exiting vapor is freed of entrained particles by cyclone 9 and enters a partial condenser 10 producing an aqueous effluent and a small vapor fraction containing some furfural. The latter fraction is added to the product vapor from the blow bin 6. The particles leaving the drier have a moisture content low enough to permit burning in the boiler.

10.7.3. The SUPRAYIELD Process

The SUPRATHERM process and the STAKE process both exploit the yield potential of high temperatures but do not do anything about the fundamental cause of the loss reactions, which is a nonboiling reaction medium. By contrast, a more recent concept pointedly called SUPRAYIELD process [39], marketed by BOSCH PROJECTS of Durban, South Africa, aims at duplicating the 100 % yield routinely obtained in the "analytical furfural process". As discussed in chapter 7, the "analytical furfural process" gives a proven 100 % yield as it keeps the reaction medium in a state of boiling so that according to the respective phase diagram any furfural generated in the liquid phase is instantly "rejected" into the vapor phase. This cannot be achieved by injecting steam as, at any pressure, condensing steam is thermodynamically incapable of bringing an aqueous pentose solution to boiling, because of the boiling point elevation caused by the pentose. Thus, heating and stripping with steam as

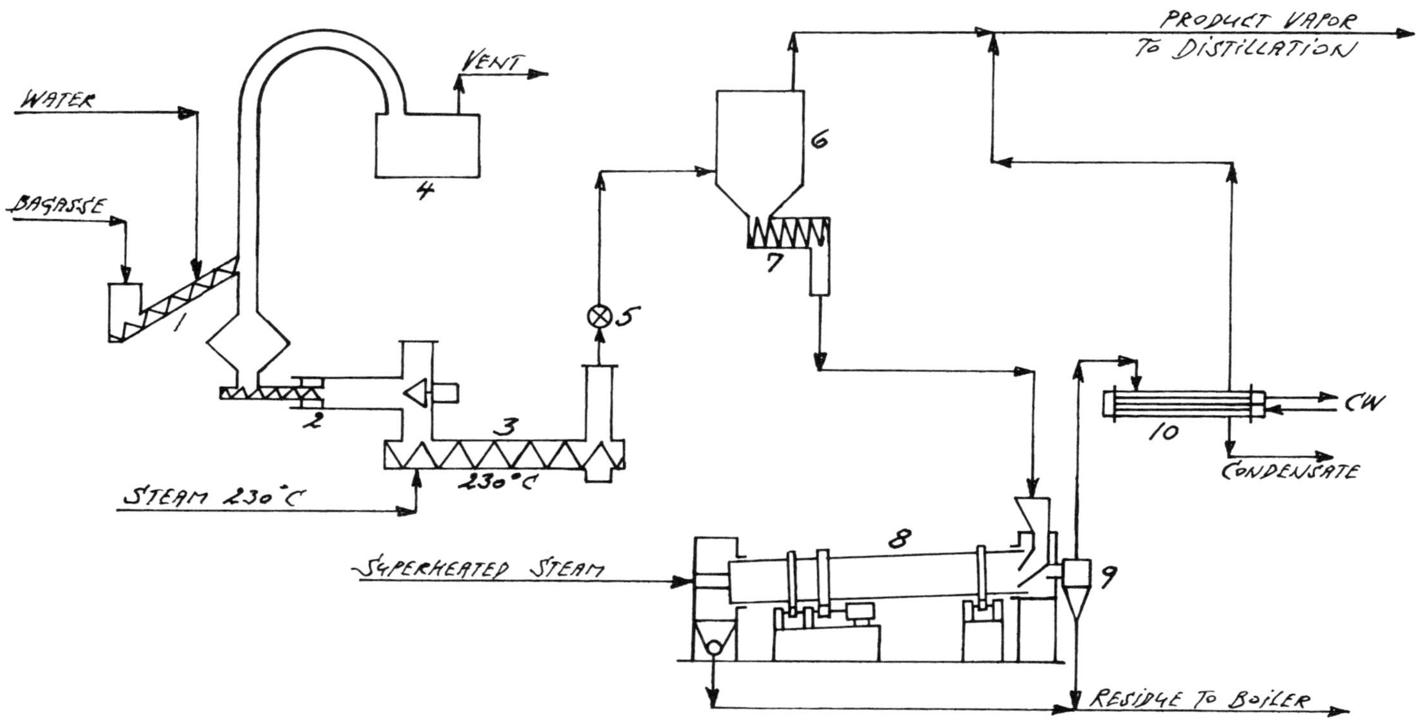

Figure 26. Schematic of the STAKE Process.

used in all industrial furfural processes to date lead to a nonboiling reaction medium, thereby leaving any furfural generated temporarily dissolved in the liquid phase where it can undergo loss reactions with itself and with intermediates of the pentose-to-furfural conversion. These loss reactions could be avoided, as in the "analytical furfural process", by applying a heat source other than steam condensing in the reaction medium, but in furfural production any indirect heating by a heat exchanger must be ruled out as any heat exchange surfaces would tend to undergo fouling. For this reason, in the SUPRAYIELD process the required boiling of the reaction medium is accomplished in a different fashion.

As frequently demonstrated in a well-known high school experiment, water of room temperature can be brought to boiling by exposing it to a pressure below 17 torr. Under conditions of thermal insulation, the boiling will cause the temperature of the water to diminish as the heat of evaporation is taken from the enthalpy of the water ("evaporation cooling"), but the boiling can be maintained for as long as the ambient pressure is kept below the vapor pressure of the water.

At a different thermal level, water heated to a temperature above its atmospheric boiling point, hence under elevated pressure, can be brought to boiling, and can be maintained in a state of boiling, by gradually lowering the pressure so as to keep it below the vapor pressure. For a furfural process aimed at working with a boiling reaction medium, this process, called "delayed decompression", has the advantage that the initial heating can be effected rapidly by steam condensation without any fouling problems. The "price" payed for the state of boiling lies in the acceptance of a gradually decreasing temperature causing an exponentially decreasing reaction rate, but on the other hand the yield should be 100 percent, as in the "analytical furfural process", so that the furfural potential of the raw material is fully utilized.

In the SUPRAYIELD process, the "delayed decompression" is a degenerated flash process slowed down to such an extent that the period of time for going from a high primary pressure to a lower secondary pressure corresponds to the reaction time needed for the desired conversion of pentosan to furfural. To make this a practical proposition, the primary temperature must be high, say 240 °C, and the secondary temperature should not be below 180 °C as in this range the reaction rate would be too slow.

As after heating the reactor to the primary temperature the pressure in the vessel is high, the "delayed decompression" can be readily effected and controlled by a simple leak valve. In special cases, when the primary temperature is limited by the available steam pressure, more than one "delayed decompression" can be applied to ascertain complete conversion of the pentosan to furfural.

Losses can occur only during the short heating period when the reaction medium is not yet boiling. For this reason, the heating to the primary temperature should be an "inverse flash", as fast as possible.

A schematic of the SUPRAYIELD process is shown in Figure 27. A thermally well insulated reactor 1 charged with raw material, preacidified or not, is heated to a temperature T_1 by admitting steam through valve 2 while the valves 3 and 4 are closed. During the very short heating process, the steam condenses, thus increasing the moisture content of the charge. Then, valve 2 is closed and a leak valve 3 is opened so as to produce a steady small flow of product vapor by gradual decompression. This causes a slow drop in temperature. When in this fashion a suitably chosen temperature T_2 is reached, the leak valve 3 is closed to terminate the first "delayed decompression". If at the end of this period no more furfural was obtained, the digestion is completed by opening valve 4 to discharge the residue. If, however, furfural was still obtained, the reactor is reheated and submitted to another "delayed decompression". This procedure can be arbitrarily repeated. All valve operations are governed by an automatic control unit 5.

Depending on the primary temperature, the process can be run with or without a foreign acid. The higher the primary temperature, the smaller is the need for a foreign acid.

If a foreign acid is used, it should not be sulfuric acid as the latter is known to cause some losses by sulfonation. On account of this effect, the "analytical furfural process", having a yield of 100 percent with hydrochloric acid, does not give this theoretical yield when run with sulfuric acid. As in technical operations a use of hydrochloric acid would be inappropriate because of corrosion, and as nitric acid is out of the question because of nitration, the foreign acid of choice is orthophosphoric acid since it does not cause any side reactions [40]. It is not a strong acid, but it is amply strong enough for the given purpose.

10.8. Processes Starting with Sulfite Waste Liquor

The basis of the sulfite pulping process is the discovery by Tilghman in 1866 that white cellulose fibers are obtained by cooking wood under pressure in an aqueous solution of calcium bisulfite $Ca(HSO_3)_2$. In chemical terms, the process is founded on the fact that lignin, the solid binder of wood, reacts with sulfite to form water-soluble sulfonic acids. In this fash-

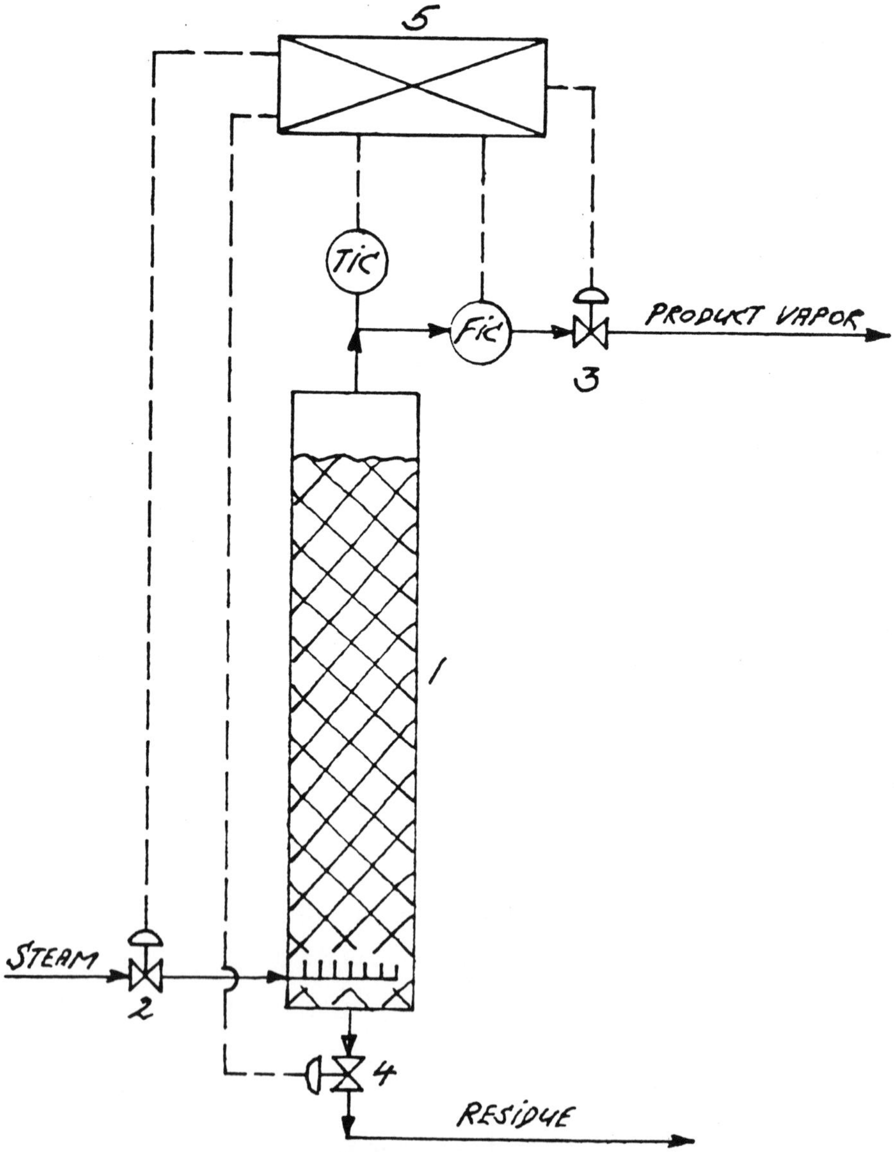

Figure 27. Schematic of the SUPRAYIELD Process.

ion, the fibers of the wood are freed.

The reaction requires a temperature in the order of 140 °C, and as it is acid, pentosan in the wood is hydrolyzed to pentose, and some of this pentose is dehydrated to furfural. Against this background, there has been a persisting interest in converting the pentose of sulfite liquor to furfural, and to recover the furfural already existing in the liquor.

Strictly speaking, the interest for furfural in the pulping industry is essentially limited to the given calcium sulfite process, where disposal of the spent liquor is a problem. In the more modern magnesium process, the spent liquor is readily thickened and then burnt to fully recover MgO and SO_2 used in starting the process:

$$MgO + 2\ SO_2 + H_2O \rightarrow\rightarrow Mg(HSO_3)_2$$

Hence, in the magnesium process the agents required to digest the wood can be used in a closed circle, so that there is no effluent. Thus, there is no interest for doing anything with the pentose and the furfural in the magnesium liquor. They are simply burnt.

By contrast, in the case of the calcium process, the situation is fundamentally different as burning the thickened calcium liquor leads to calcium sulfate (and not to CaO and SO_2). Consequently, the only outlet for the spent calcium liquor is concentrating it to 50 % dry solids content and selling this concentrate as "lignosulfonate", a product usable in various applications from drilling oil bores to building roads in poor dry countries. In these applications, however, the presence of dissolved pentose is detrimental, so that it is desirable to remove the pentose by converting it to furfural. This offers the twofold benefit of an upgraded lignosulfonate and of furfural as a profitable by-product.

10.8.1. Pentose and Furfural in the Sulfite Process

In the sulfite pulping process, there are several streams to be dealt with as illustrated in Figure 28.

The least problematic stream is the "flash condensate" obtained when depressurizing the digesters to 1 ATM. Stemming from a vapor, this stream is, of course, free of pentose as the latter is nonvolatile, and for the same reason it is free of calcium sulfate. Consequently, this small stream containing some furfural is readily submitted to distillation to recover the furfural.

A somewhat different furfural stream is obtained when the total waste water of a sulfite plant is submitted to thickening by a multieffect evaporator. In this case, the vapor

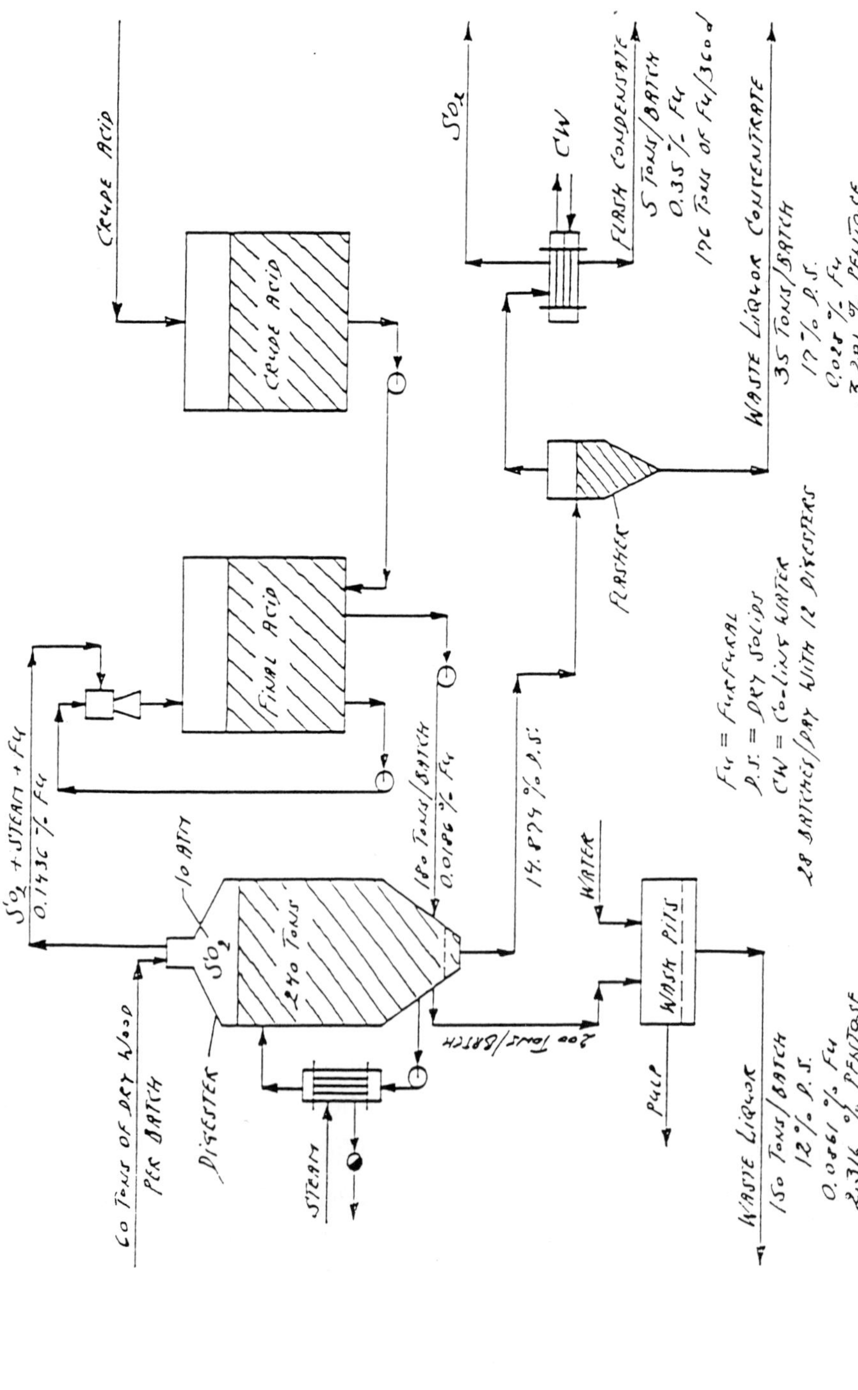

Figure 28. The Output Streams of a Typical Calcium Sulfite Pulping Plant.

condensate of the evaporator contains approximately 2.5 % furfural, 1.2 % methanol, and 4.0 % sulfur dioxide. Processing such a stream is illustrated in Figure 29. At first, the feed stream is preheated in condenser 1 before entering two sulfur dioxide strippers 2A and 2B energized by steam injection. With the feed flowing through these strippers in downward direction, the sulfur dioxide is desorbed from the liquid phase, so that the head vapors of these strippers consist mostly of sulfur dioxide. They pass the partial condensers 3A and 3B, where condensables are liquefied to be refluxed, while the uncondensed sulfur dioxide is separated in cyclones and leaves the system, to be recycled into the pulping process. The sump fractions of the strippers 2A and 2B enter a distillation column 4 energized by steam injection. The purpose of this column is to separate most of the water, which leaves the sump. Column 4 also produces a liquid side stream and a head vapor. The condensate of the head vapor is partly refluxed to effect rectification, and partly fed into a distillation column 5 energized by a reboiler 6. Column 5 also receives, at a lower level, the liquid side stream of column 4. The head fraction of column 5 is methanol, which is liquefied in the condensers 7A and 7B. The condensate of these units is partly refluxed to effect rectification, and partly cooled in heat exchanger 8 before being withdrawn as one of the products. The sump fraction of column 5 consists of water and furfural. This stream is cooled in heat exchanger 9 and then fed into a static decanter 10, where two liquid phases are formed. The light phase consisting mostly of water and little furfural is recycled into column 4, while the heavy phase, consisting mostly of furfural and little water, is collected in a buffer tank 11, from where it enters a distillation column 12 energized by an integrated reboiler 13. The head vapor of this column is liquefied in condenser 14. A part of the condensate is refluxed to effect rectification, while the rest is fed into the decanter 10. Slightly above the sump of column 12, furfural is withdrawn as a side stream vapor, which is liquefied in condenser 15 and cooled in heat exchanger 16 before being collected in the product tank 17. The sump fraction of column 12 is polymeric residue, which is discarded.

Returning to Figure 28, the streams called "waste liquor" and "waste liquor condensate" are seen to contain very little furfural but sizable percentages of pentose. If this pentose were converted to furfural at a good yield, it would upset the furfural market by sheer quantity.

One of the reasons why the pentose in calcium sulfite waste liquors is not being utilized for making furfural lies in the fact that the liquor from the digesters is actually super-

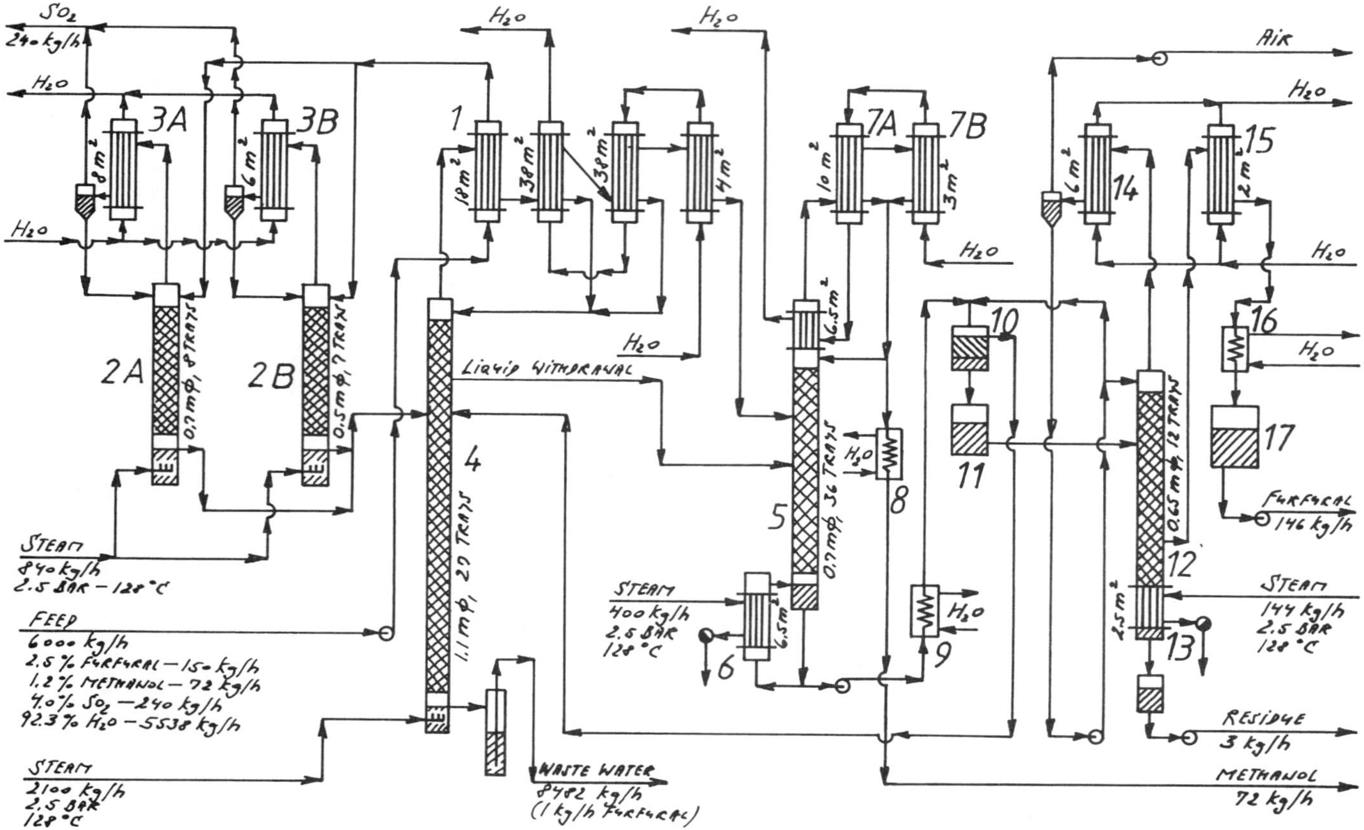

Figure 30. Distillation Plant for the Recovery of Furfural from the Condensate of a Sulfite Liquor Evaporator.

saturated with calcium sulfate. During the digestion, 10 ATM of sulfur dioxide on the charge hugely increase the solubility of calcium sulfate in water, due to a reduction to dithionate:

$$CaSO_4 + SO_2 \rightarrow\rightarrow CaS_2O_6$$

At 30 °C, the solubility of calcium dithionate in water exceeds that of calcium sulfate by a factor of 105. According to the principle of Le Chatelier, under a high sulfur dioxide pressure the equilibrium of the given reaction lies on the right-hand side, while at a low pressure of sulfur dioxide essentially only calcium sulfate exists. This implies that when the digesters are depressurized, the high solubility of calcium sulfate is terminated, thus leading to calcium sulfate precipitation.

In addition, the solubility of calcium sulfate changes with temperature, going through a maximum at 35 °C and then diminishing drastically with increasing temperature as shown in Figure 31.

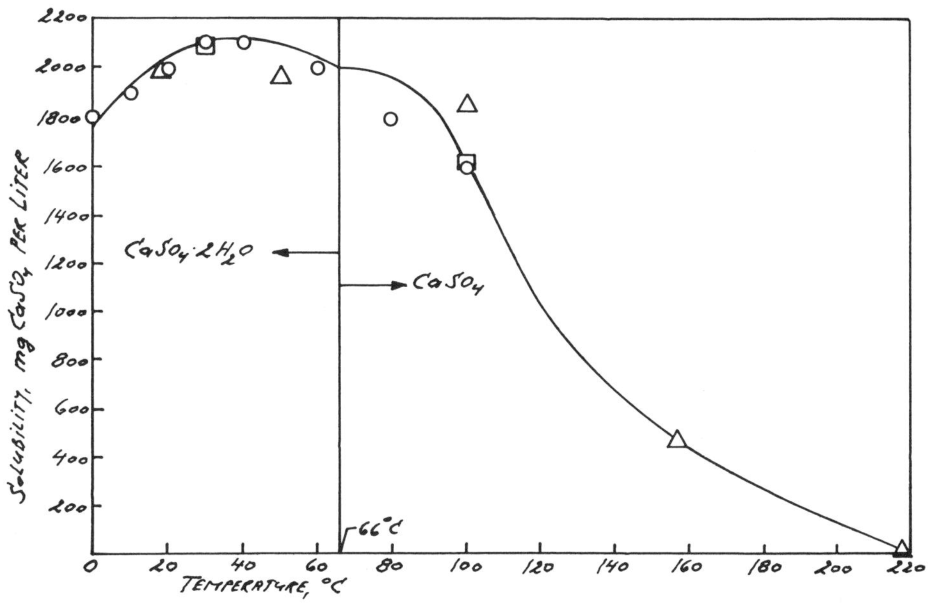

Figure 31. The Solubility of Calcium Sulfate in Water.

○ J. W. Mullin, Crystallization, Butterworths, London, 1972.

□ C. D. Hodgeman et al., Handbook of Chemistry and Physics, 42nd Edition, The Chemical Rubber Publishing Company, Cleveland, 1960.

△ C. Nyman, Svensk Papperstidning 49 (1946) 73-83.

The latter aspect means that any processing requiring surface heaters at high temperatures becomes problematic as calcium sulfate will precipitate on hot surfaces. As can be seen in Figure 31, at 220 °C the solubility of calcium sulfate in water is almost nil. Thus, monstrous fouling problems must be expected at high temperatures unless the heating is accomplished by steam injection.

10.8.2. The VOEST-ALPINE Process

The first process to make furfural from sulfite liquor was offered by VOEST-ALPINE of Austria in 1988. In this process, shown schematically in Figure 32, the sulfite liquor is first thickened to a dry solids content of 30 %. After heating the concentrate to 180 °C, and after holding it at this temperature in a tube reactor for a period of time sufficient to convert some pentose to furfural, the reaction mixture is passed into a distillation column where the furfural is stripped by steam. The treatment of providing residence time at 180 °C in a tube reactor to convert more pentose to furfural, followed by removal of the furfural in a stripping column, is repeated two times. In this fashion, the furfural is removed stepwise soon after its formation, to reduce losses by furfural reacting with itself, with intermediates of the pentose-to-furfural conversion, and with other constituents of the liquor.

Originally, the process was claimed to be applicable to both calcium and magnesium liquor, but later it was realized that use of calcium liquor was prohibitive due to fouling. This left this process without an application since in magnesium sulfite pulping there is no incentive to make furfural since there is no effluent as discussed above.

10.8.3. The Reactive Desorption Process

Lured by the original promises of the VOEST-ALPINE process, a furfural manufacturer built a pilot plant as shown in Figure 33. In this process, the residence time at temperature required for the conversion of the pentose to furfural was provided in a single column energized by steam injection. The feed stock was the calcium liquor of a sulfite pulping mill. The designers called their column a "reactive distillation column", but this was erroneous since steam injection is thermodynamically incapable of boiling sulfite liquor, because of the boiling point elevation caused by the pentose and the lignosulfonate dissolved therein, so that the excess steam bubbled through a nonboiling liquid. Consequently, any furfural formed in this column remained dissolved in the liquid phase until it was picked up by the steam bubbles in a slow diffusion-controlled process. During the time the furfural was

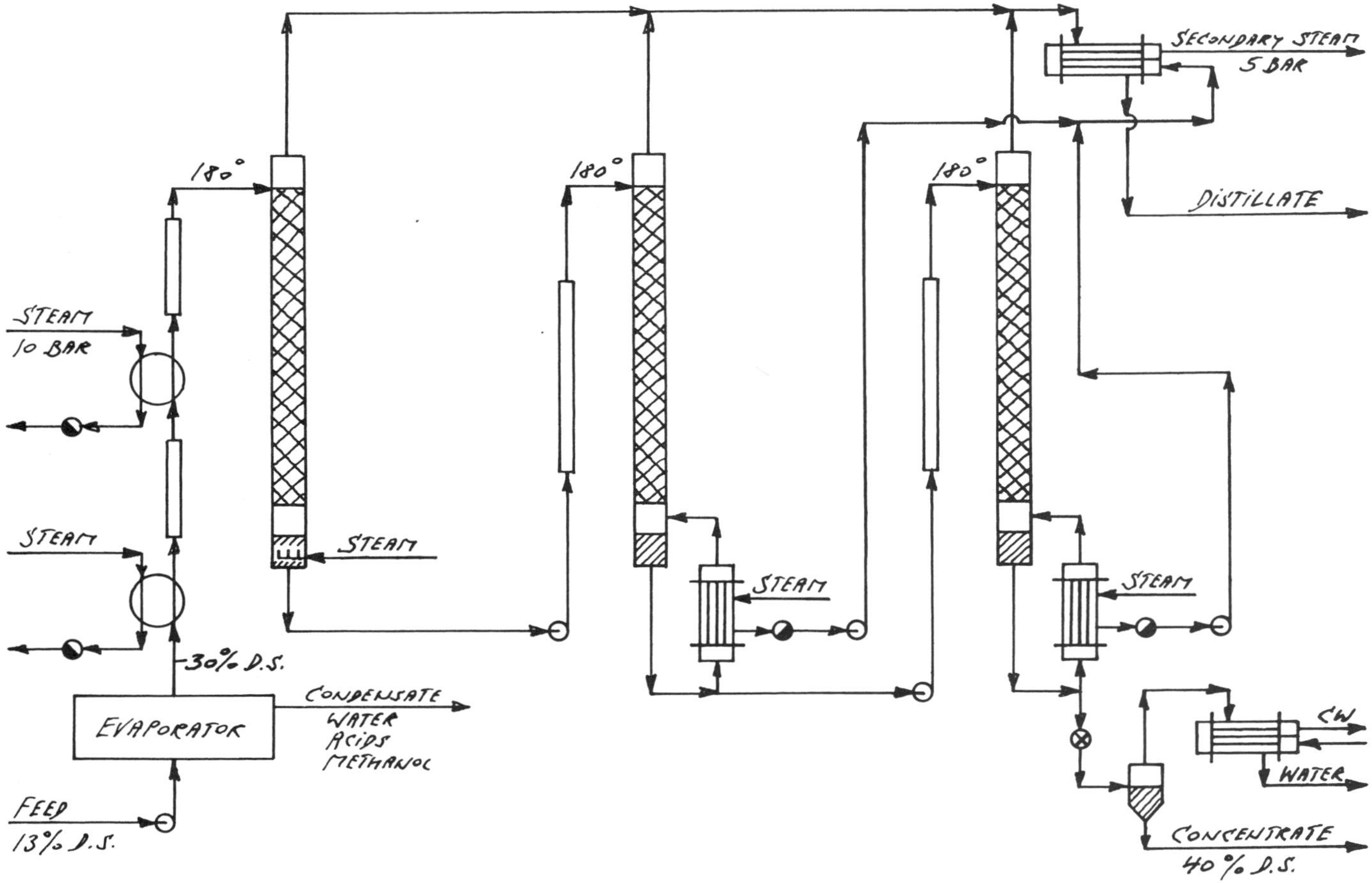

Figure 32. Schematic of the VOEST-ALPINE Process.

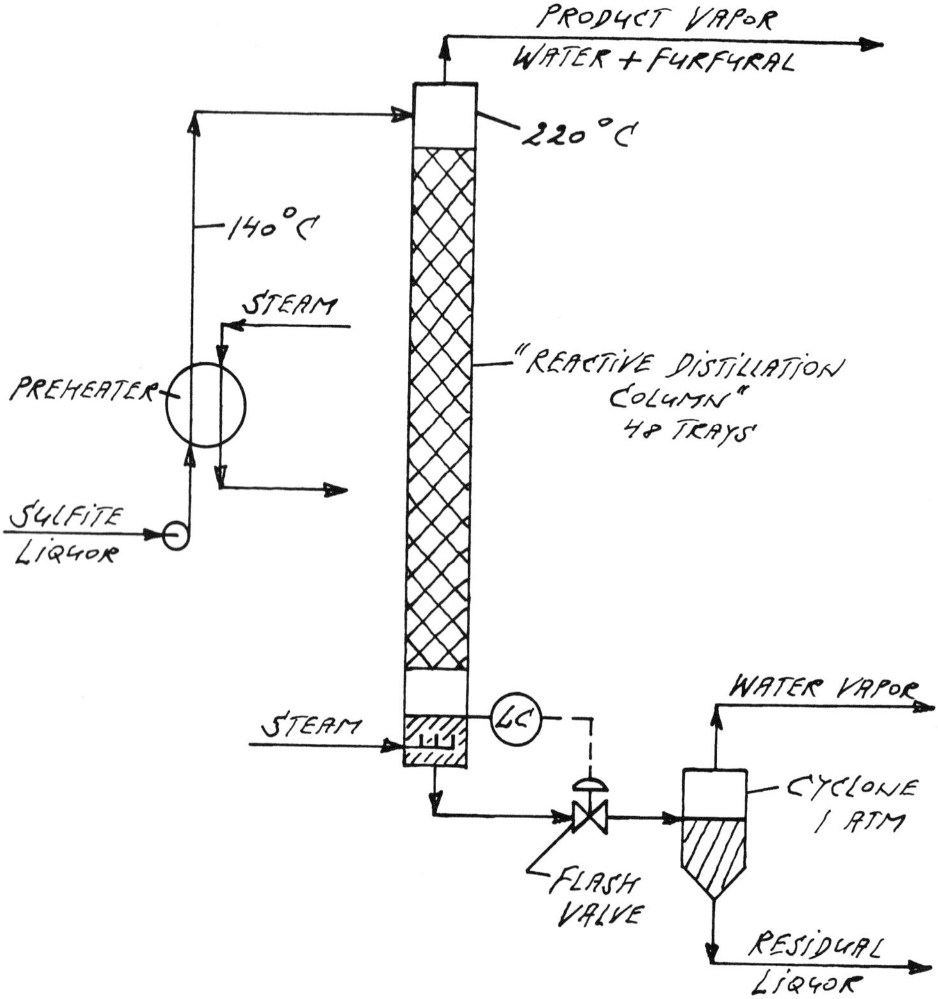

Figure 33. Schematic of the Reactive Desorption Process.

dissolved in the liquid phase, it reacted with itself, with intermediates of the pentose-to-fufural conversion, and with other ingredients of the liquor, thus incurring massive losses.

The "reactive distillation column" was not a distillation column, but a desorption column, as in a distillation column the liquid phase must boil.

In 27 runs at various conditions, the operators of this pilot plant found to their distress that the furfural yields were only in the order of 30 percent, in harsh contrast to their expectations of 85 percent, derived from calculations based on the known kinetics in water. The designers had made a fundamental mistake: They had measured the rate of pentose disappearance in the sulfite liquor, but not the rate of furfural formation, and as the rate of pentose disappearance was far greater than what had to be expected when the liquor's hydrogen ion concentration was used in the known kinetics of xylose disappearance in acid water, they had concluded that the lignosulfonate of the liquor had a special catalytic effect on the pentose-to-furfural conversion. They had overlooked that the fast disappearance of the pentose in the liquor was not due to a mysterious catalysis but caused by loss reactions of the pentose with lignosulfonate and other ingredients of the liquor.

10.8.4. The Enforced Ebullition Process

Based on the yield debacle of the reactive desorption process discussed above, it had become obvious that better yields require measures to bring the liquor to boiling, to the end of causing any furfural generated to be instantly "rejected" into the vapor phase. Of course, this cannot eliminate loss reactions of pentose with lignosulfonate and other ingredients of the liquor, but it can eliminate the loss reactions of furfural with itself and with intermediates of the pentose-to-furfural conversion. In other words, whereas on one hand, with acid water as the reaction medium, the mode of boiling can lead to a 100 % yield, as in the "analytical furfural process", on the other hand, with sulfite liquor as the reaction medium, even boiling cannot produce a 100 % yield, but it can eliminate a sizable portion of the losses otherwise incurred.

This train of thought is the foundation for a new process outlined in Figure 34 [41]. It is termed "enforced ebullition process" as in addition to steam as primary heating agent, a hot gas is used to bring the liquor from the steam condensation temperature to boiling. The process can be described as follows: By means of pump 1, the sulfite liquor is fed through an in-line mixer 2, where it is heated to 240 °C by steam injection. This raises the

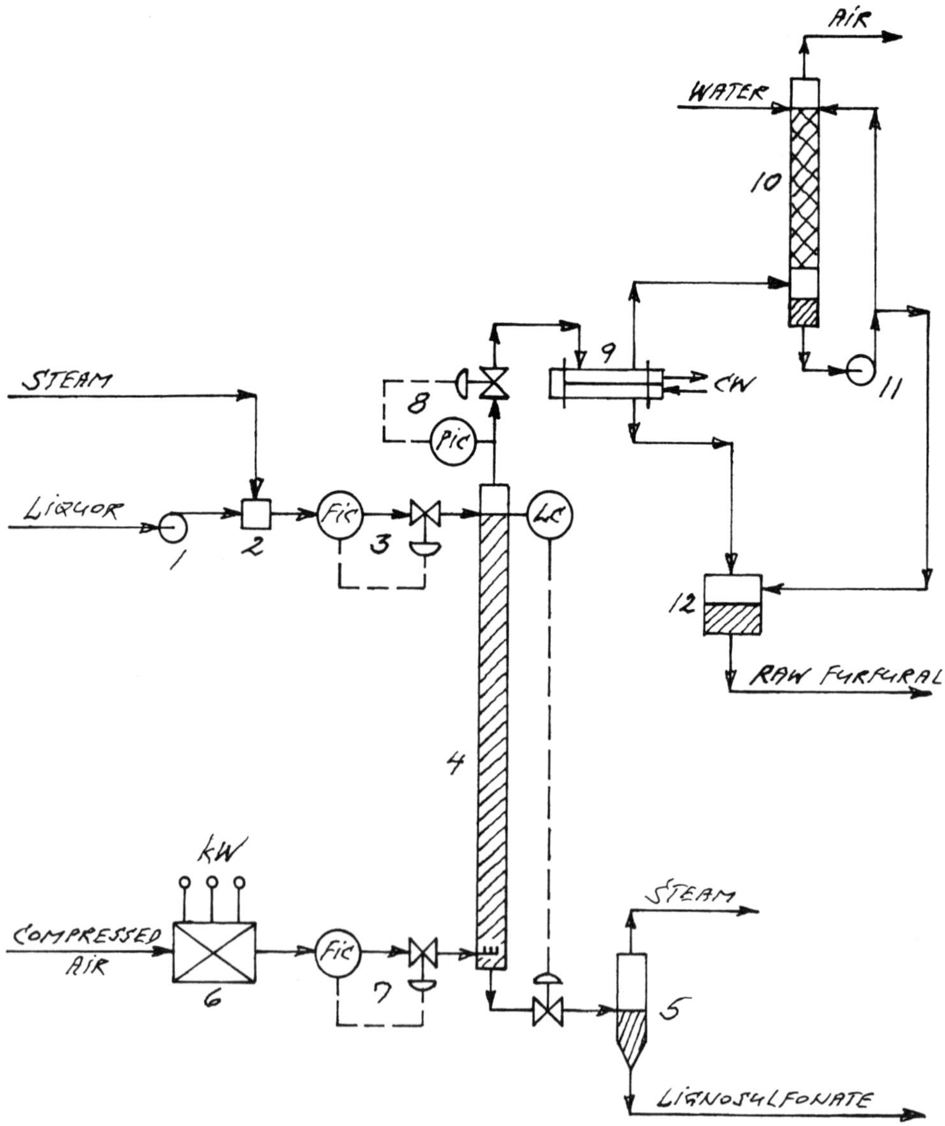

Figure 34. Schematic of the Enforced Ebullition Process.

pressure to 33.116 ATM. Via a throttle valve 3, the hot liquor enters the top of a thermally insulated columnar reactor 4, flows downwards, and is discharged from the bottom by being flashed into a cyclone 5. Compressed air heated electrically to 600 °C in heat exchanger 6 is dosed into the bottom of reactor 4 by a flow control circuit 7. The hot air bubbles through the liquor in reactor 4 and leaves the top via a pressure control circuit 8, a condenser 9, and an atmospheric absorption column 10 equipped with a circulation pump 11. The pressure control circuit 8 maintains the head of reactor 4 at 30 ATM, so that the liquor entering the reactor undergoes flashing, thereby cooling from 240 to 234 °C while the hot air ascertains boiling of the liquor throughout the reactor. Furfural generated from the pentose of the liquor is instantly and completely vaporized, thus joining the water vapor liberated by the boiling, as well as the air stream.

From the gaseous mixture reaching the condenser, most of the furfural vapor and most of the water vapor are liquefied and collected in tank 12, whereas small quantities entrained by the air are recovered in the absorption column.

The liquor input rate and the columnar reactor dimensions are calculated to ascertain that the residence time in the reactor is sufficient for a complete disappearance of the pentose. There is no need for adding acid.

Needless to say, instead of air, other gases can be used if this is desirable.

References

[33] H. J. Brownlee, Ind. Eng. Chem. 19 (1927) 422-424.

[34] M. S. Rushin, Mathematical Modelling and Optimization of a Three Phase Autohydrolysis Reactor, M. Sc. Thesis, University of Natal, Durban, 1992.

[35] K. J. Zeitsch, Process and Apparatus for the Manufacture of Furfural, South African Patent No. 89/0080 (1989).

[36] K. J. Zeitsch, Process for Producing Furfural, US Patent No. 4,912,237 (1990).

[37] K. J. Zeitsch, Verfahren und Vorrichtung zur Herstellung von Furfural, German Patent No. DE 38 42 825 C2 (1991).

[38] K. J. Zeitsch, Verfahren und Vorrichtung zur Herstellung von Furfural, Swiss Patent No. CH 678 183 A5 (1991).

[39] K. J. Zeitsch, Verfahren zur Herstellung von Furfural durch verzögerte Entspannung, German Patent Application No. 199 05 655.2 (1999).

[40] W. M. Dehn and K. E. Jackson, J. Amer. Chem. Soc. 55 (1933) 4284-4287.

[41] K. J. Zeitsch, Verfahren zur Herstellung von Furfural aus Sulfitablauge, German Patent Application No. 199 17 178.5 (1999).

11. Distillation of Furfural

All furfural reactors known so far produce a vapor stream consisting of more than 90 % water, of up to 6 % furfural, and of various by-products. After liquefaction, commonly used to make secondary steam, and sometimes after a filtration or a centrifugal separation of solids, the product stream is fed into a distillation plant.

A typical simple distillation plant is shown in Figure 35. In a first column 1, use is made of the water/furfural azeotrope having an atmospheric boiling point of 97.85 ^{o}C and a water content of 65 percent. Column 1 is commonly called the "azeotropic column", although this is unfortunate as in a subsequent column the same azeotrope is used for the dehydration of furfural, so that the attribute "azeotropic" is not a unique feature of the first column.

From the first column, commonly a tray column, a fraction roughly corresponding to the azeotrope is withdrawn as a liquid side stream and fed into a decanter 2, where it separates to form two liquid phases, a light phase rich in water, and a heavy phase rich in furfural. The light phase is refluxed into the first column, somewhat below the withdrawal point of the azeotrope.

The head fraction of column 1 is mostly a mixture of various low boilers, with methanol as the principal component, but also contains some furfural and some water. This fraction is introduced into a randomly packed column 3 where the furfural and most of the water are separated, to be fed into the decanter 2.

The sump fraction of column 1 is water loaded with some carboxylic acids, mostly acetic acid. This fraction is sometimes discharged into the sea or into a waste water treatment plant, but can be processed further as will be shown later.

The heavy phase of decanter 2 consists typically of 94 % furfural. It is passed through a neutralizer 4 and then fed into a randomly packed vacuum column 5 energized by a reboiler as well as steam injection, where the "raw furfural" from the neutralizer is freed of its water content. The head fraction of column 5 consists roughly of the water/furfural azeotrope. It is condensed and collected in decanter 6, where it is separated into a light and a heavy liquid phase. The light phase is given into the decanter 2, thus removing water from the column, and the heavy phase is refluxed. Below the upper packing, furfural vapor is

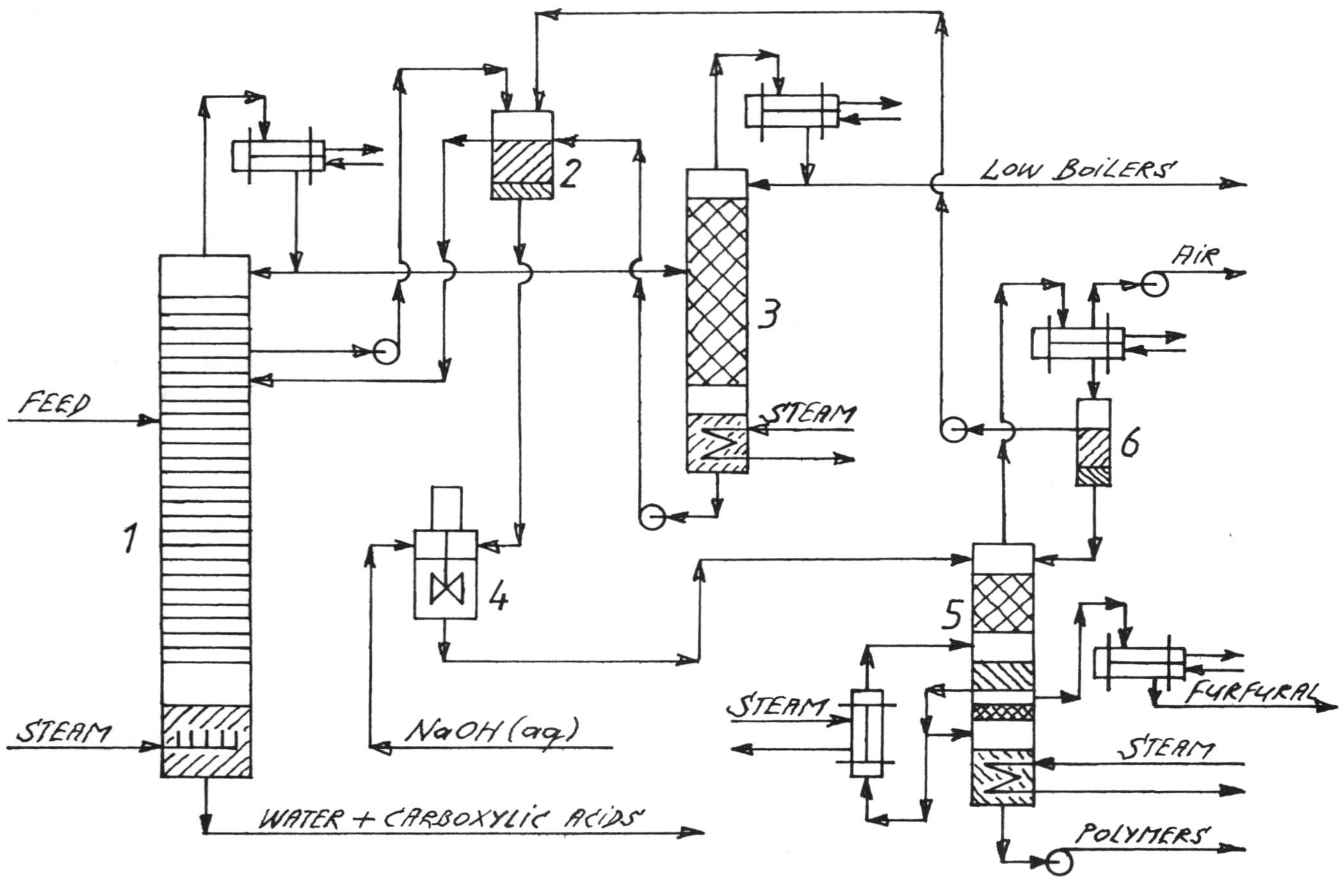

Figure 35. Schematic of a Simple Furfural Distillation Plant.

withdrawn as a side stream. Its condensate represents the desired product. The sump fraction, separated from the furfural withdrawal point by a small packing or merely by a demister, consists of polymers, to be discarded.

A somewhat more elegant furfural distillation plant is depicted in Figure 36. It differs from the scheme of Figure 35 in three points:

(a) The separation of the low boilers is effected in a small tower put on top of the first column.

(b) From the first column, the azeotrope is withdrawn as a vapor, not as a liquid.

(c) The furfural dehydration column is equipped with trays, not with random packings. It features 12 trays between the top and the furfural withdrawal point, and 3 trays of large spacing between the furfural withdrawal point and the sump, to ascertain an improved purity of the product.

Even when properly distilled, commercial furfural always contains sizable impurities of 5-methyl furfural and furyl methyl ketone. A survey of typical impurity levels is given in Figure 37. As can be seen, 5-methyl furfural ranges from 0.32 to 0.95 % by weight, while furyl methyl ketone ranges from 0.04 to 0.35 % by weight.

The explanation for the formation of 5-methyl furfural and furyl methyl ketone, which are isomers, lies in the fact that all raw materials used for the manufacture of furfural contain some methyl pentosan. As the furfural manufacturers are not in a habit to measure the methyl pentosan content of their raw material, it is instructive to look at the pentosan and methyl pentosan contents of various woods as published by Hägglund [42]. These data are compiled in Table 2. The methyl pentosan values were measured by a procedure due to Schorger [43], which is based on the fact that methyl furfural phloroglucide is soluble in hot ethanol while furfural phloroglucide is not.

According to Hägglund, for woods the ratio of methyl pentosan to pentosan ranges from 0.04251 to 0.73256. It is believed that for bagasse, corn cobs, and other raw materials of furfural manufacture this ratio is smaller than the low value of this range for wood, so that levels in the order of 1 percent are considered as reasonable.

In the hydrolysis of the methyl pentosan, the methyl group remains attached to the pentose ring, thus leading to methyl pentose, but it must be noted that the methyl group can be attached at the 5-position or at the 1-position of the pentose ring. Thus, when the methyl pentose is dehydrated by acid catalysis, the result is either 5-methyl furfural or furyl methyl

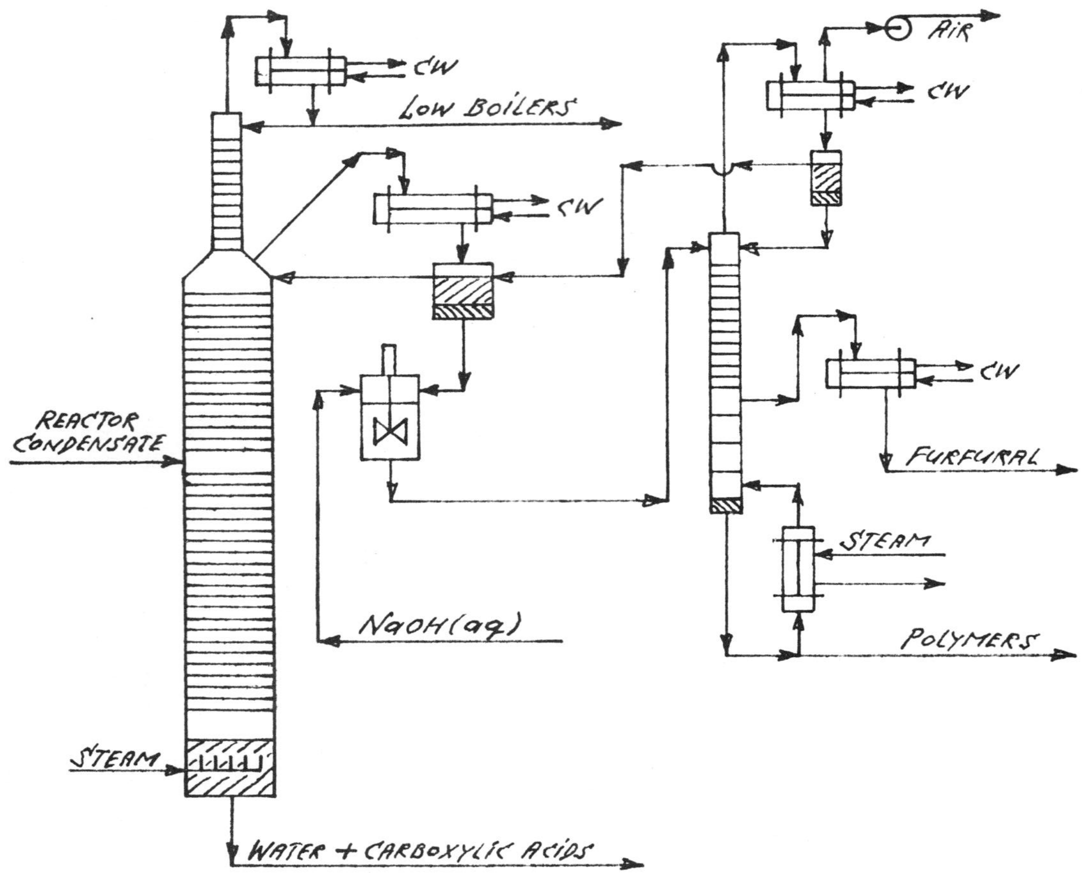

Figure 36. Schematic of an Advanced Furfural Distillation Plant.

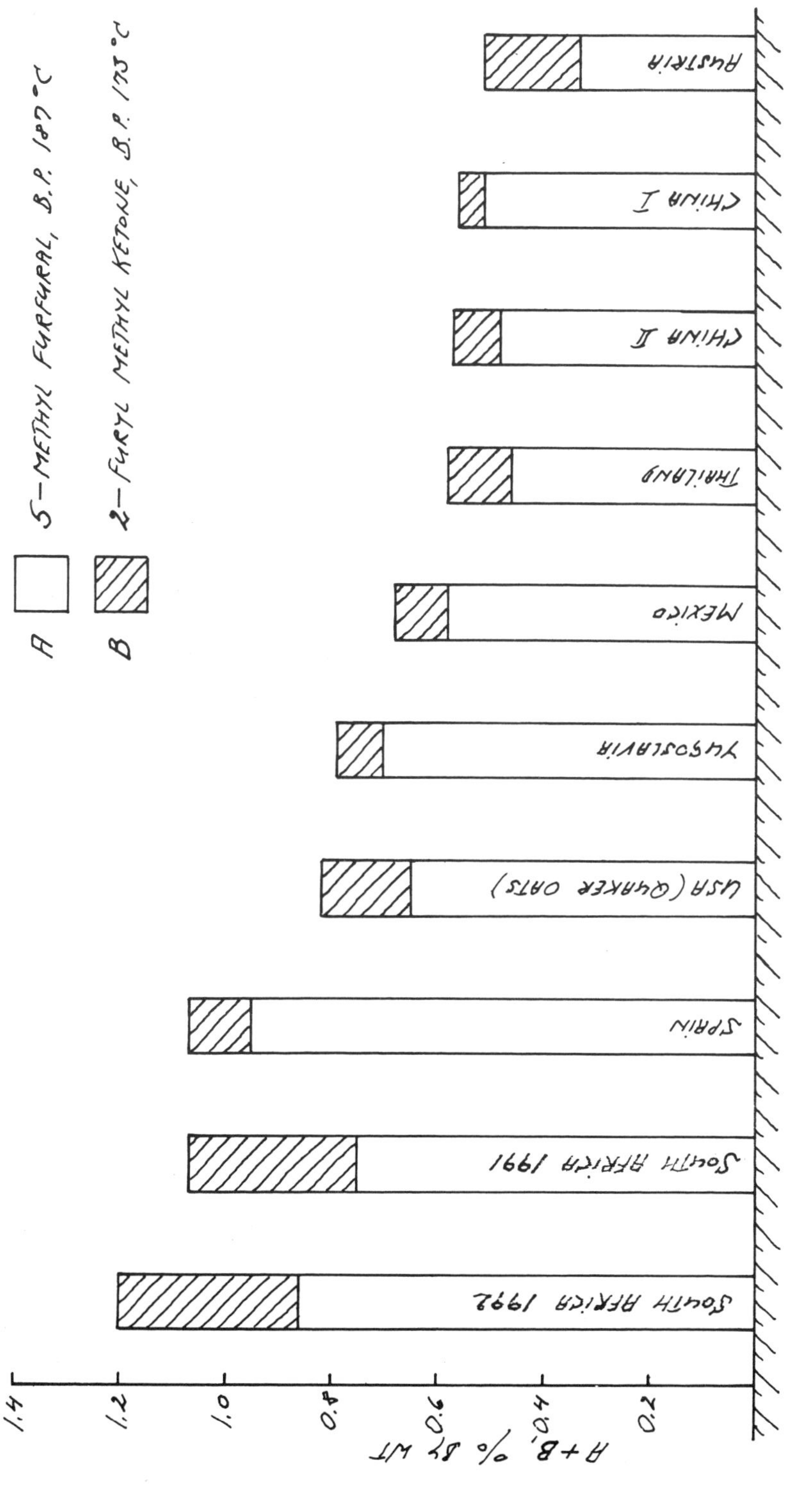

Figure 37. The Principal Impurities of Various Furfural Products.

Table 2. Pentosan and Methyl Pentosan Contents of Various Woods in grams per 100 grams of dry weight [42].

Type of Wood	Pentosan	Methyl Pentosan
Longleaf pine	7.46	3.60
Douglas fir	6.02	4.41
Western larch	10.80	2.81
White spruce	10.39	3.55
Western white pine	6.97	3.22
Western yellow pine	7.35	1.62
Yellow cedar	7.87	3.42
Incense cedar	10.65	1.35
Redwood	7.80	2.75
Mesquite	13.96	0.70
Balsa	17.65	0.86
Hickory	18.82	0.80
Basswood	19.93	3.73
Yellow birch	24.63	2.69
Sugar maple	21.71	2.39
Eucalyptus	20.09	2.33

ketone as shown in the Figures 38 and 39. The observation, seen in Figure 37, that there is more 5-methyl furfural than furyl methyl ketone is simply due to the fact that by a choice of nature the raw material contains more 5-methyl pentosan than 1-methyl pentosan.

If the purity of furfural is to be improved, the split, by distillation, can be effected between furfural (FU) and furyl methyl ketone (FMK), or between FU and 5-methyl furfural (MF). With the atmospheric boiling points being 161.7 °C for FU, 173 °C for FMK, and 187 °C for MF, the choice of the split causes a significant difference in expenditure. If the split is effected between FU and FMK, the FU is obtained at an exceptionally high purity, while a split between FU and MF implies accepting that the FMK will stay with the FU. Although a FU of exceptionally high purity is, of course, desirable, practical considerations speak against this alternative. In the FU/FMK system, the vapor/liquid equilibrium line is very close to the diagonal, which means that separating FMK from FU requires a very high reflux ratio and a

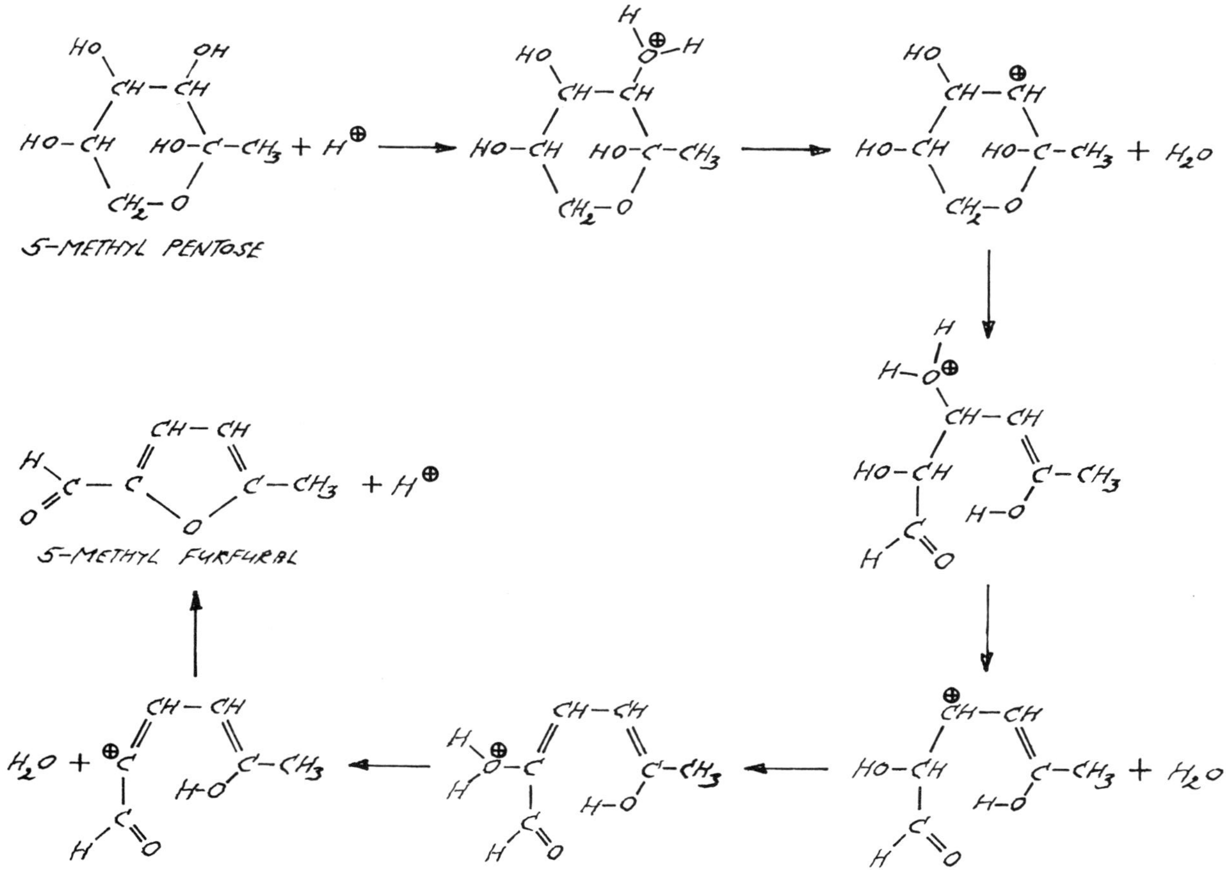

Figure 38. The Formation of 5-Methyl Furfural from 5-Methyl Pentose.

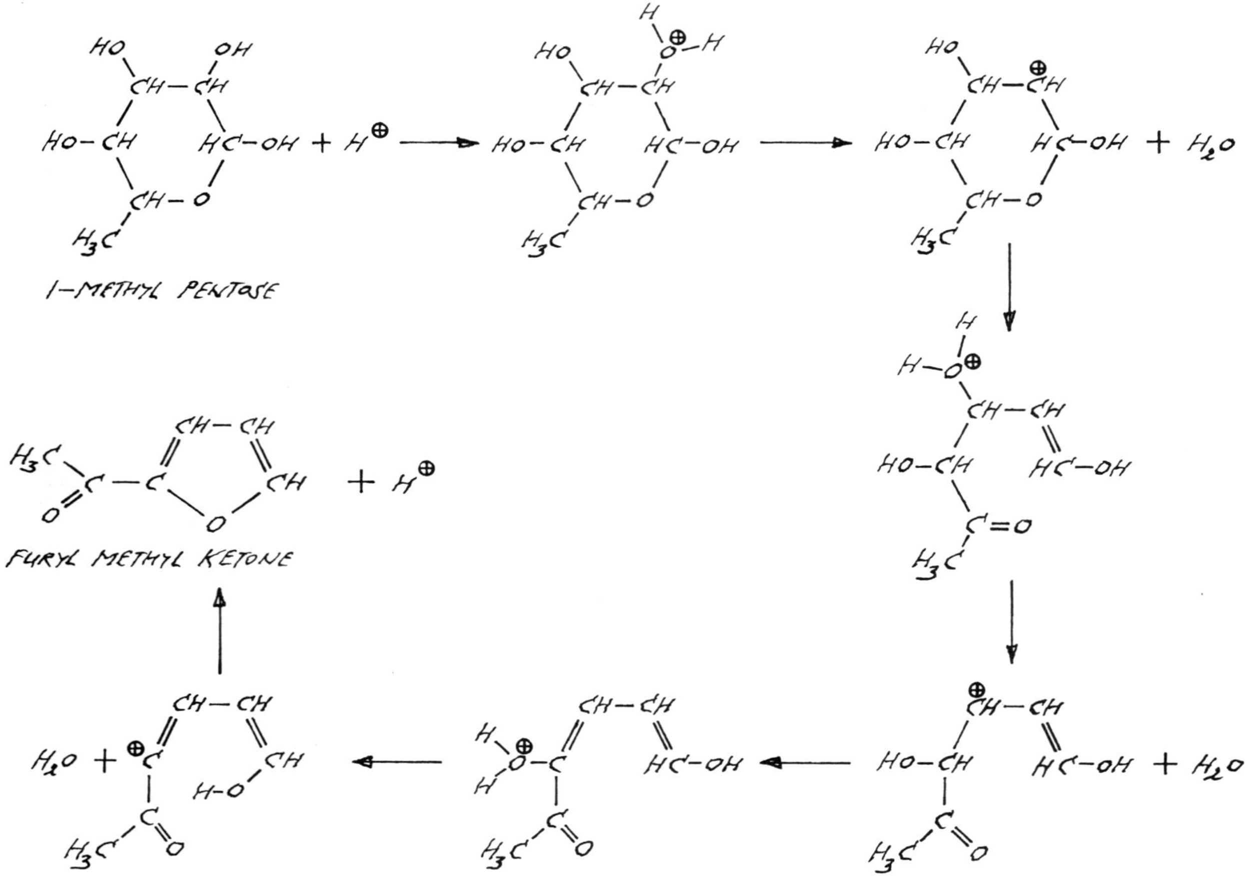

Figure 39. The Formation of Furyl Methyl Ketone from 1-Methyl Pentose.

very high number of theoretical stages, so that a monstrous column would be necessary for this task. It is, therefore, the general consensus in the furfural industry that separating FMK from FU is economically prohibitive.

Against this background, the following treatment of the problem will be limited to separating only MF, the principal impurity.

The boiling curves for FU, FMK, and MF are shown in Figure 40.

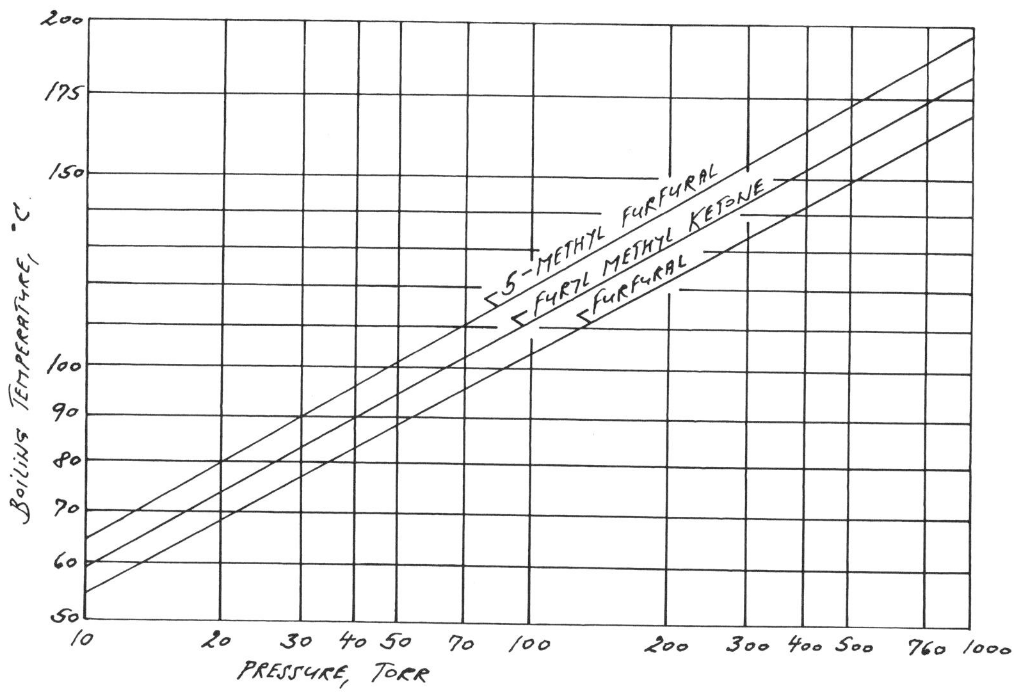

Figure 40. The Boiling Curves for Furfural, Furyl Methyl Ketone, and 5-Methyl Furfural.

To minimize polymerization losses, the purification of furfural must be carried out at greatly reduced pressure, as in the case of dehydrating FU. The separation process is shown schematically in Figure 41. The raw FU is fed into the sump of a rectification column 1 energized by a reboiler 2. The head vapor stream of this column is liquefied in a condenser 3, a portion of the condensate being used as reflux while the remainder is collected in a buffer

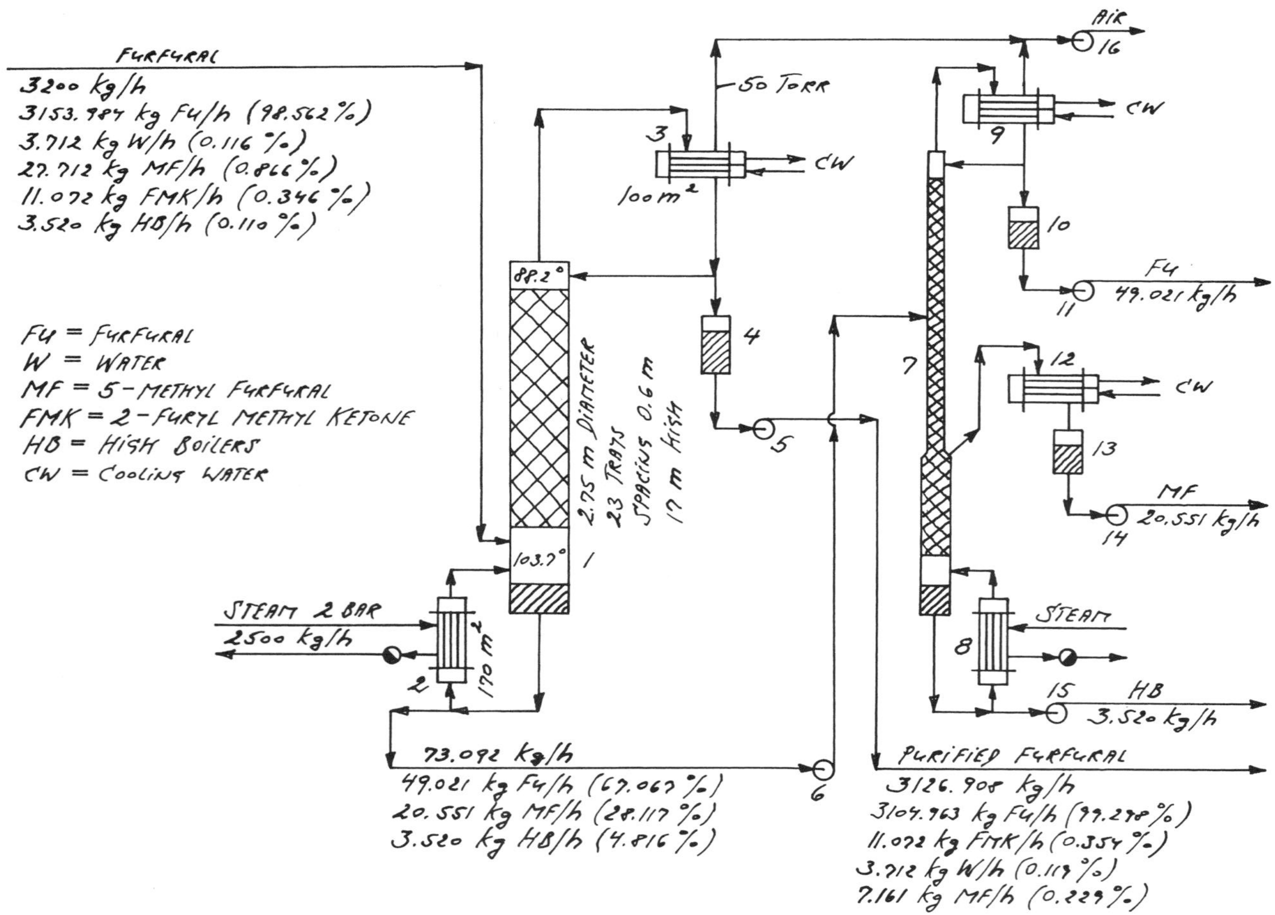

Figure 41. Furfural Purification Plant for a Feed Rate of 3200 kg/h.

tank 4, from where it is withdrawn by pump 5. This fraction represents purified FU greatly depleted of MF. In view of the fact that not all of the MF available can usually be sold, and inasmuch as separating all of the MF would require an infinite number of stages, the given design is based on separating only 74 percent of the MF entering the plant. From column 1, a small sump fraction consisting of FU, MF, and high boilers (HB) is withdrawn by pump 6 and fed into a correspondingly small secondary distillation column 7 energized by reboiler 8. The head vapor consisting of FU is liquefied in condenser 9, a portion of the condensate being used as reflux while the remainder is collected in a buffer tank 10, from where it is withdrawn by pump 11. A side stream consisting of MF is withdrawn as vapor, liquefied in condenser 12, and collected in tank 13, from where it is withdrawn by pump 14. The sump fraction consisting of high boilers is withdrawn by pump 15. A vacuum pump 16 maintains the entire system at a reduced pressure.

Needless to say, if there is no interest in recovering and selling MF, the emphasis being on merely purifying the FU, then obviously the secondary distillation column 7 of the plant shown can be omitted, but this incurs a furfural loss of 1.5 percent of the feed and causes an uncomfortable waste stream to be disposed of.

In this context, it must be noted that it would be unreasonable to use purified FU to make furfuryl alcohol as essentially all of the furfuryl alcohol is used to make foundry resins, where any 5-methyl furfural, hydrogenated to 5-methyl furfuryl alcohol, does not matter at all, and may in fact be beneficial. Thus, only FU used as FU, for instance in extraction processes, is a candidate for the given purification process. For most FU manufacturers, this is only 35 percent of their production.

Finally, it is noted that if there were a demand for a FU free of FMK, the preferred separation process would not be distillation but crystallization as FMK freezes already at +33 ^{o}C whereas the freezing point of FU is –36.5 ^{o}C.

References

[42] E. Hägglund, Chemistry of Wood, Academic Press Inc., New York, 1951.

[43] A. W. Schorger, Ind. Eng. Chem. 9 (1917) 556-560.

12. In-Line Measurement of Furfural

For the processing of furfural in general, and for the control of furfural reactors in particular, it is desirable to have a continuous in-line measurement of the furfural concentration. For batchwise operation, this shows when the charge is exhausted, and for both batchwise and continuous operation it can show the response of the furfural concentration to the steam input, thus permitting an optimization of the furfural output. This is especially important for the ROSENLEW process, where an excess steam input blows out the acetic acid catalyst, so that there is a critical steam input beyond which the furfural output diminishes sharply. In addition, it is profitable to have a continuous record of the reactor operation, to pinpoint mistakes by the operators or other irregularities.

Since 1994, a suitable apparatus for all of these purposes has been marketed by BERNATH ATOMIC GMBH in D-30974 Wennigsen/Germany. Reliably resisting the dirty industrial conditions, this apparatus proved to be a full success. It consists of a continuously operating sampling unit and of a continuously operating process spectrometer.

12.1. The Continuous Sampling Unit

As the measurement is carried out by a process spectrometer requiring a clean gas stream containing the furfural to be determined, the first stage of the measurement process consists of an apparatus producing such a stream. It is shown schematically in Figure 42.

The reactor condensate, usually having a temperature in the order of 100 °C, is withdrawn by a gear pump 1 driven by a pneumatic motor 2 via a magnetic coupling. The driving compressed air is cleaned by a filter 3 and controlled by a pressure regulator 4. The pneumatic motor is lubricated by an oil mist generator 5. The reactor condensate enters a spiral cooler 6 consisting of an inner tube made of TEFLON for the condensate, and of an outer tube made of stainless steel for the cooling water. The cooled reactor condensate flows into a stripper 7 made of glass, from where it is discharged into the condensate tank of the production plant. Compressed air, cleaned by filter 8, controlled by a coarse pressure regulator 9, and a precision pressure regulator 10, passes a flow meter 11 and a one-way valve 12 before entering the stripper 7 through a glass frit 13 acting as a gas distributor. In the

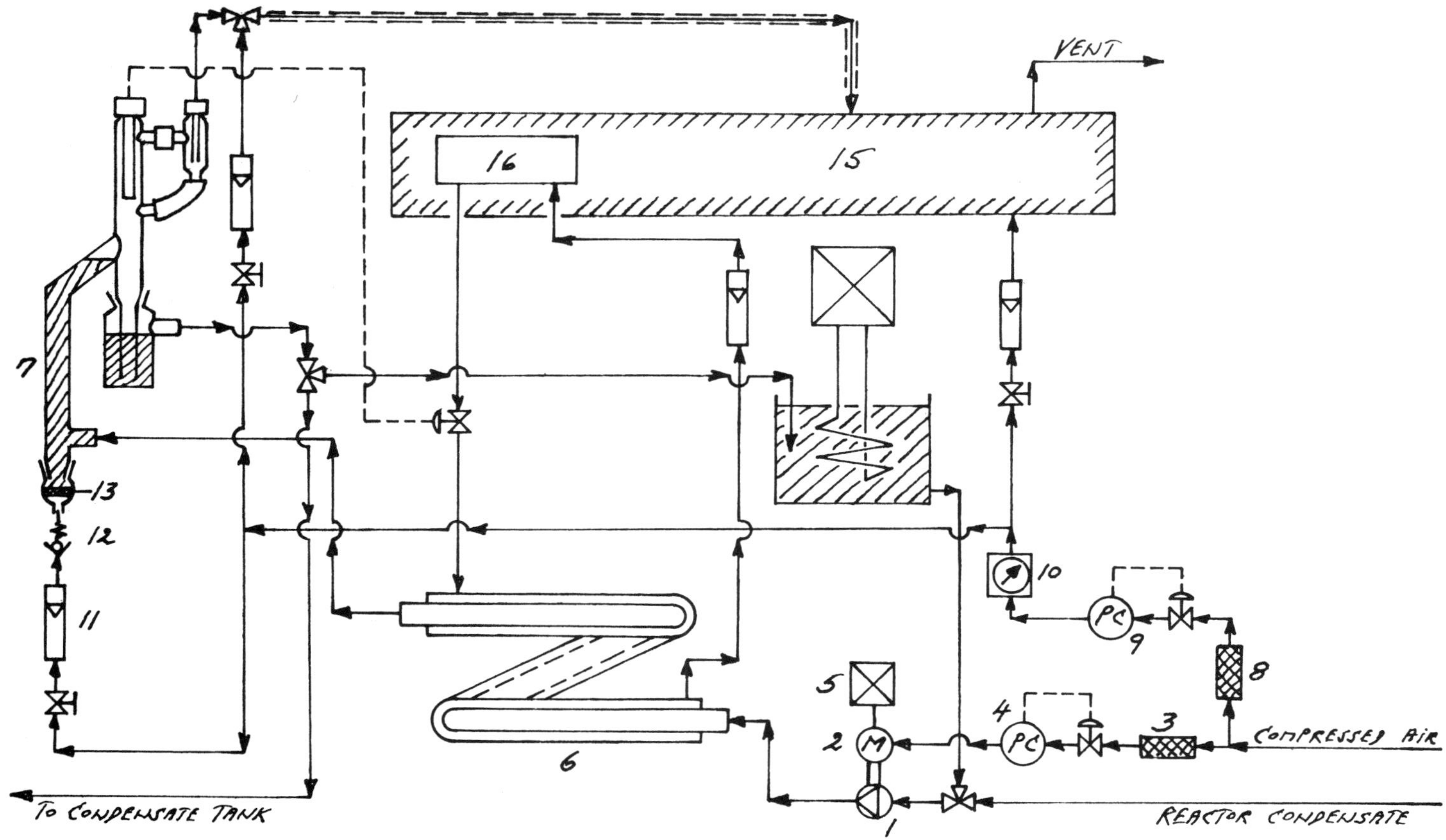

Figure 42. Schematic of the Continuous Sampling Unit.

stripper 7, the small air bubbles produced by the frit pick up an equilibrium quantity of furfural vapor from the reactor condensate, and the air thus loaded with furfural vapor leaves the stripper via a thermostatically heated tube 14 leading the mixture to the spectrometer 15. The spectrometer housing contains a chiller 16 equipped with its own pump which circulates cooling water through the cooler 6. The flow of the cooling water is controlled by a temperature sensor in the head of the stripper 7. A typical stripper temperature maintained in this fashion is 50 °C. The quantity of furfural carried by the air to the spectrometer depends on the percentage of furfural in the reactor condensate and on the stripper temperature. Inasmuch as the latter is kept constant, the concentration of furfural vapor in the air as measured by the spectrometer is proportional to the furfural concentration in the reactor condensate. The other lines and components in Figure 42 are for the calibration and the zero adjustment of the spectrometer.

12.2. The Process Spectrometer

The measurement is based on the fact that furfural vapor exhibits a strong and sharp absorption peak at a wavelength of approximately 256 nm as shown in Figure 43.

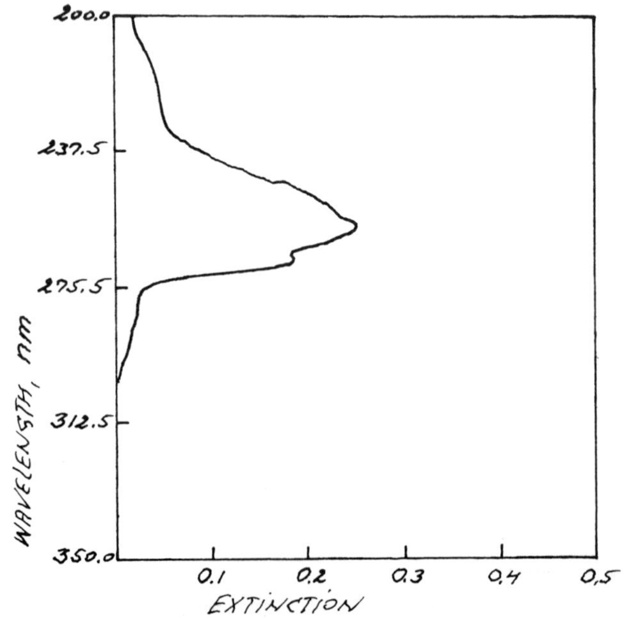

Figure 43. The UV Spectrum of Furfural Vapor in Nitrogen as taken by an Array of 512 Silicon Photodiodes. Courtesy of BERNATH ATOMIC GMBH of Wennigsen/Germany.

The principle of the measurement system is illustrated in Figure 44.

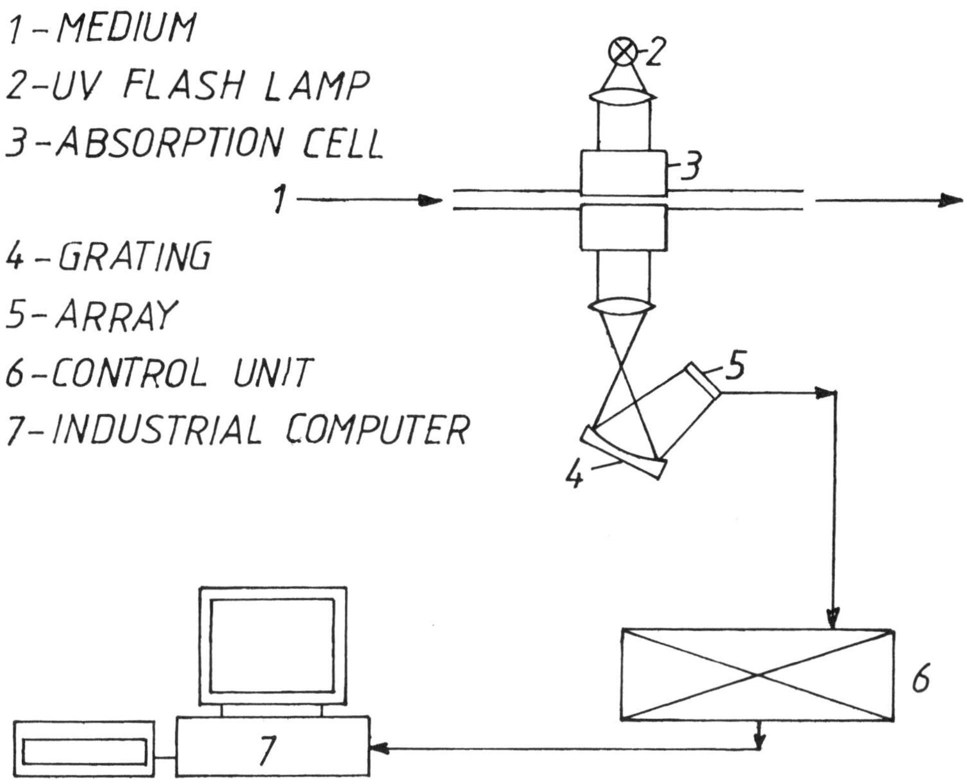

Figure 44. Schematic of the Continuously Operating "Multicomponent Industrial UV Spectrometer" manufactured by BERNATH ATOMIC GMBH of Wennigsen/Germany.

The stream of furfural vapor in air as produced by the continuous sampling unit described above is passed through a UV absorption cell irradiated by a pulsating xenon flash lamp. The nonabsorbed radiation, containing the information on the furfural concentration, is converted to a spectrum by means of a concave holographic grating, and thin "slices" of this spectrum are measured by an array of 512 silicon photodiodes serving as radiation detectors. The spectral signals produced by these photodiodes enter a control unit linked to an industrial computer with screen and keyboard. The computer gives a numerical value of the furfural concentration as well as statistical details, and in addition it can produce a signal to be used for controlling the furfural reactor.

The UV absorption cell is depicted in Figure 45. The stream of furfural vapor in

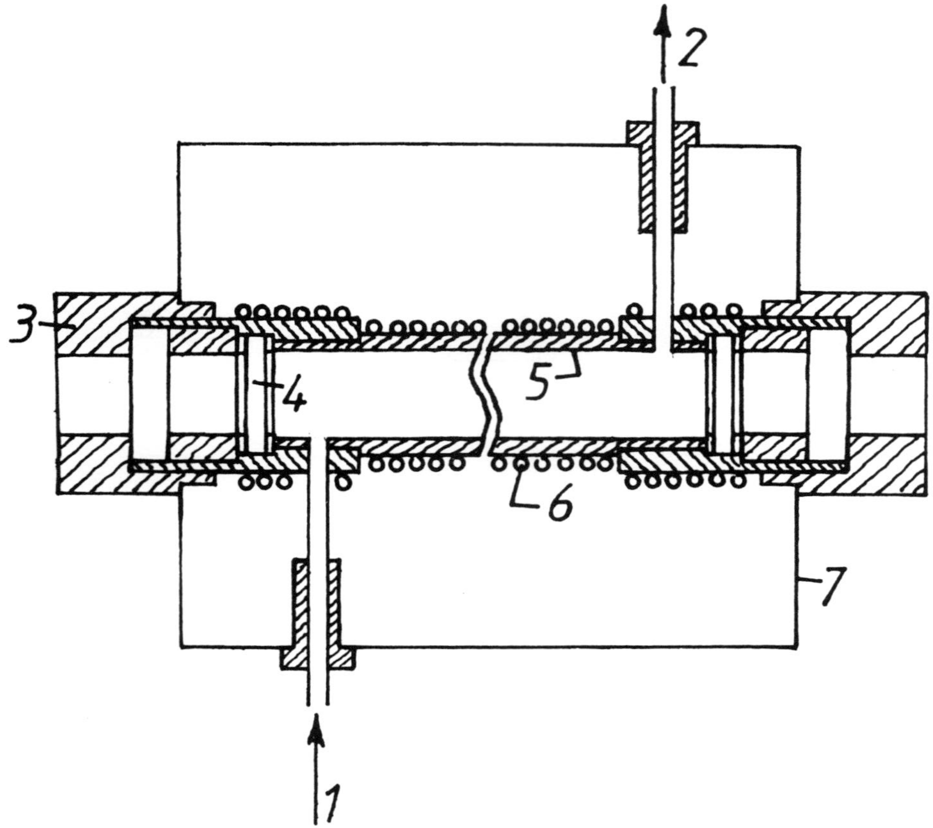

Figure 45. Absorption Cell of the Continuously Operating "Multicomponent Industrial UV Spectrometer" of BERNATH ATOMIC GMBH. (1) Inlet. (2) Outlet. (3) End Cap. (4) Window. (5) Tubular Cell. (6) Electric Heating. (7) Casing.

air produced by the continuous sampling unit flows through a tubular cell passed by the UV radiation in axial direction via appropriate windows. The cell is thermally insulated, and its temperature is closely controlled by a combination of electrical heaters and Peltier panels. (The Peltier effect, also called “electrical cooling”, is the reversal of the thermocouple effect.) The operating temperature of the cell is slightly higher than the temperature of the stream from the sampling unit, so that condensation in the cell is rendered impossible. There are no moving parts, which stands for a rugged design insensitive to vibrations and other harsh treatment.

The spectrometer is calibrated by a known test mixture introduced at regular intervals, and the “background” radiation, established during flushing with air devoid of the product, is deducted automatically.

13. Treatment of Furfural Waste Water

The waste water of all furfural plants contains some carboxylic acids, with acetic acid being the principal load, ranging from 1 to 5 percent by weight, as well as some furfural, in concentrations up to 600 ppm. The furfural concentration depends on the quality of the first distillation column, with good columns going down to 50 ppm.

If the concentration of acetic acid is too low to justify an acid recovery plant, it is common practice, wherever possible, to simply discharge the waste water into a river or the sea. However, with an increasing awareness of environmental issues, and with a tightening legislation in this regard, it is unlikely that such discharges will be permitted for much longer. Against this background, anaerobic digestion of the waste water has become the procedure of choice for such cases.

Application of this process to furfural waste water was studied in detail by Wirtz and Dague [44]. Contrary to widespread erroneous belief, these authors established beyond any doubt that at the concentrations occurring in furfural waste water, the toxicity of furfural for some microorganisms does not apply to methane bacteria. They not only thrive on acetic acid but eat up furfural as well. For the low concentrations of acetic acid in furfural waste water, these bacteria, known to decompose acetic acid to methane and carbon dioxide according to the reaction

$$CH_3COOH \rightarrow\rightarrow CH_4 + CO_2,$$

are employed as "filters" defined as plastic packings on which the methane bacteria are retained.

The study by Wirtz and Dague was made on upflow "filters" of fully and partially packed design as shown in Figure 46. The packing consisted of 1.6 cm FLEXIRINGS manufactured by KOCH ENGINEERING of Wichita, Kansas. The void space of these rings is 94 %. While the right-hand "filter" was fully packed, the left-hand "filter" was packed only in the upper two thirds of the reactor volume. As in start-up both reactors were filled completely with biomass (microorganisms from a municipal waste water treatment plant), the left-hand reactor had only a fluid bed of biomass (but no packing) in the lower third of its volume. For this reason, it was called "Upflow Blanket Filter" (UBF) to indicate that it was a

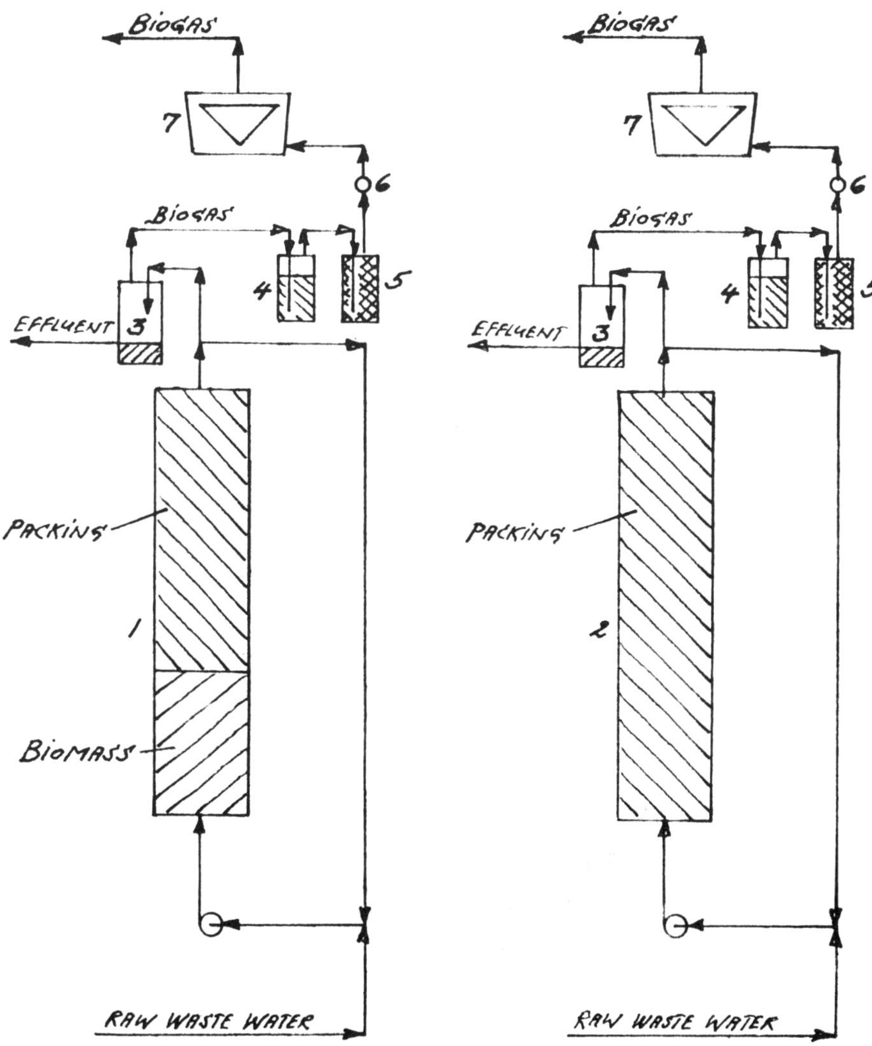

Figure 46. The Experimental Setup used by Wirtz and Dague.

1 – Partially packed reactor (Upflow Blanket Filter)

2 – Fully packed reactor

3 – Separation tank

4 – Bubble flask

5 – Scrubber

6 – Sampling port

7 - Gasometer

hybrid between the normal fully packed "filter" (right-hand reactor) and a now obsolete "Up-flow Anaerobic Sludge Blanket" (UASB) reactor, which had no packing at all, only a "blanket" (fluid bed) of microorganisms.

The furfural waste water (from a QUAKER OATS plant in Cedar Rapids, Iowa) had a load in the order of 13 000 mg COD/liter, an acetic acid content of 1.2 % by weight, a pH of approximately 2.6, and a furfural content of 600 mg/liter.

As shown in Figure 46, some of the waste water, after having passed the "filter", was branched off and added to the incoming stream. The ratio of incoming waste water to recycled waste water was chosen to be 1:1, which means that the stream having passed the "filter" was split in two halves, one half being recycled, and the other half exiting. Due to this recycling, the raw waste water was partially neutralized by the relatively high pH (7.0 to 7.6) of the recycled water, but this was not enough as methanogenic microorganisms work only at pH values above 4.5, so that sodium bicarbonate was fed in to satisfy this condition. Furthermore, essential nutrients and trace minerals had to be added to comply with the requirements of anaerobic processing.

The digestion was carried out at 35 ^{o}C to obtain a reasonable reaction rate. The formation of new microorganisms ("biomass") was extremely small, amounting to less than 0.05 grams per gram of COD removed.

The exiting stream entered a separation tank where the biogas (CH_4 + CO_2) was removed from the liquid phase. The biogas then passed a bubble flask for visual observation, and finally a hydrogen sulfide scrubber before being measured in a gasometer. Ahead of the gasometer, samples could be withdrawn for gas analysis.

Results obtained by Wirtz and Dague with a "systemic residence time" of 12 hours are shown in Figure 47, where their "systemic residence time", also called "hydraulic retention time" (HRT), is defined as the quotient of the reactor volume and the input rate of raw waste water. Due to the 1:1 recycle scheme, the stream entering the reactor is twice as large as the input rate of raw waste water, so that the "true residence time in the reactor" is only 50 percent of the "systemic residence time".

In Figure 47, the percentage of dissolved COD removed is plotted versus the "specific COD load", also called "organic loading rate" (OLR), defined as the COD input (in grams of COD per day) per unit of reactor volume (in liters). As the reactor volume was 9.5 liters (both the residence time and the specific COD load are referred to the empty reactor), a

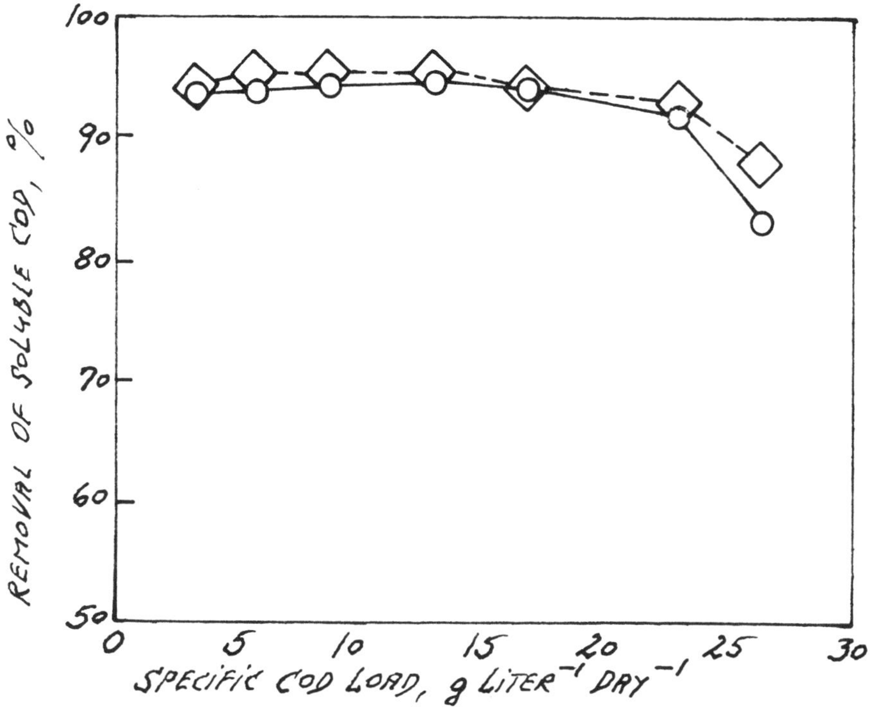

Figure 47. Results of Anaerobic Digestion of Furfural Waste Water as reported by Wirtz and Dague for a Systemic Residence Time of 12 Hours.

Circles: Partially Packed Reactor

Squares: Fully Packed Reactor

systemic residence time of 12 hours = 0.5 days corresponds to a raw waste water feed rate

$$F = 9.5 \text{ liter}/0.5 \text{ days} = 19 \text{ liter/day}.$$

With the raw waste water carrying 13 g of COD per liter, this introduced

$$13 \text{ g of COD/liter} \times 19 \text{ liter/day} = 247 \text{ g of COD/day}$$

so that the specific COD load was 247 g/day per 9.5 liters of reactor volume, i.e. 26 g day^{-1} $liter^{-1}$, as represented by the points on the right-hand side. The points further on the left-hand side were obtained with correspondingly diluted waste water.

In view of the break of the curves in Figure 47, Wirtz and Dague concluded that for a removal of 90 % of the soluble COD, the maximum permissible specific COD load is 23 g day^{-1} $liter^{-1}$.

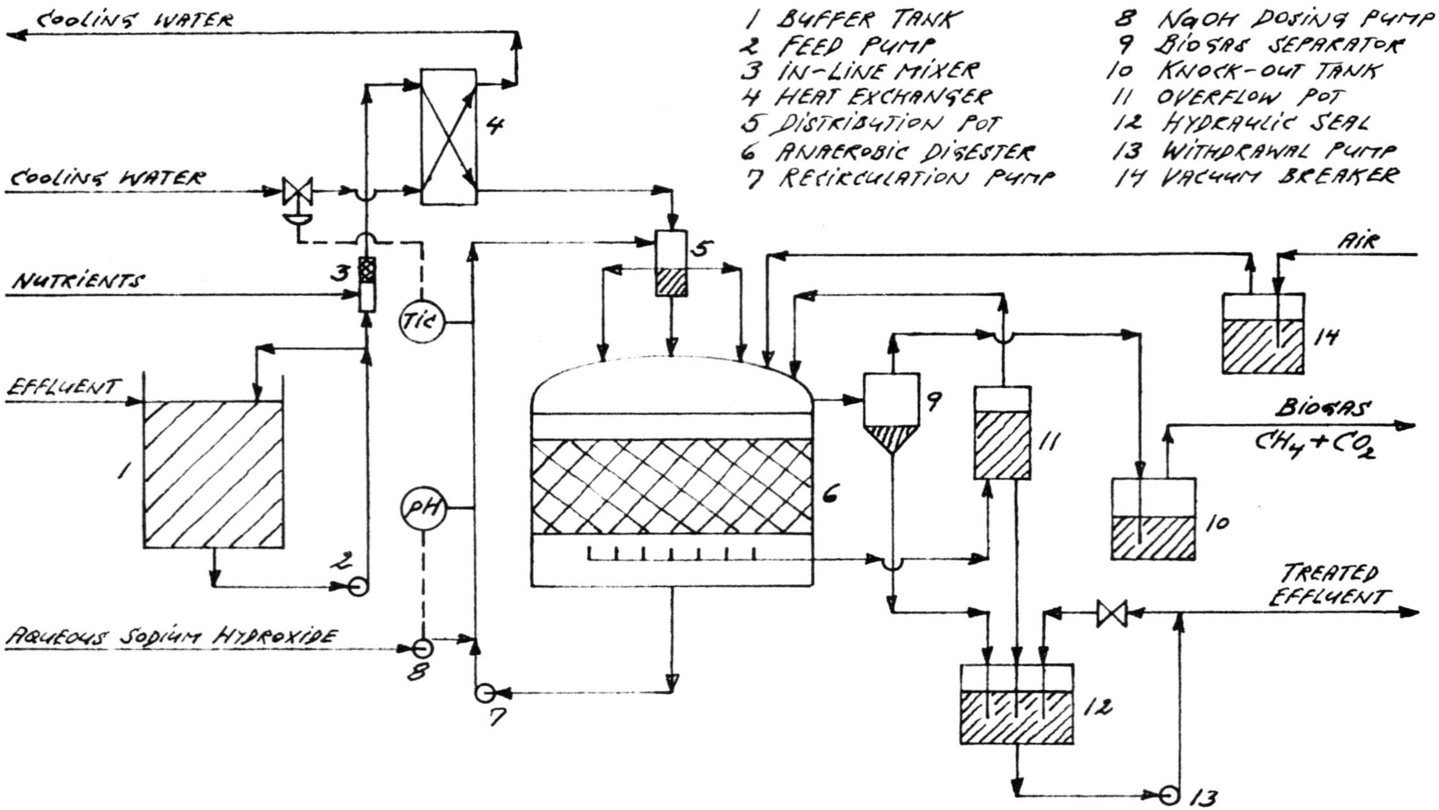

Figure 48. Industrial Anaerobic Treatment of Furfural Waste Water.

In steady state operation, both reactors tested performed equally well, but the fully packed reactor was found to be superior in handling shut-down periods. The methane production was found to be 0.347 liter (STP) per gram of COD destroyed.

Based on these laboratory results, an industrial furfural waste water plant, designed by PROSERPOL of France and UHDE of South Africa, is shown schematically in Figure 48.

After removal of 90 percent of the COD by such anaerobic processing, the remaining 10 percent, down to 100 ppm, can be taken out by a follow-up aerobic treatment. It must be noted, however, that the aerobic removal of the last 10 percent of the COD requires a larger plant than the anaerobic removal of the first 90 percent.

Noteworthily, Figure 47 shows that it is not possible to remove the last 10 percent of the COD by anaerobic processing, not even with a hugely increased residence time. This indicates that these last 10 percent are due to substances which cannot be assimilated by methanogenic bacteria. As furfural waste water is known to contain formic acid at a level amounting to roughly 10 percent of the acetic acid load, it can be concluded that methanogenic bacteria are unable to consume formic acid.

Apart from being painfully large, a follow-up aerobic treatment has two other disadvantages:

(a) A high growth rate of microorganisms, thus posing a sludge problem.

(b) A high consumption of electric energy to get oxygen into the water. Per unit of COD removal, the power requirement of an aerobic plant exceeds that of an anaerobic plant by a factor of 7.7.

Reference

[44] R. A. Wirtz and R. R. Dague, Waste Management 13 (1993) 309-315.

14. Applications of Furfural

Of the many actual and potential applications of furfural, the treatment is limited to a few fields where furfural is used as it is. Cases where furfural is employed as an input material for the synthesis of other chemicals are not discussed in this chapter. It is merely noted that 65 percent of all furfural produced is converted to furfuryl alcohol.

14.1. Furfural as an Extractant

The application of furfural as an extractant is based on a phenomenon called "intermolecular conjugation". This means that when molecules with conjugated double bonds such as furfural come in contact with other molecules containing double bonds, they form an enlarged conjugated double bond system, and this enlargement liberates energy analogous to intramolecular bond formation. Consequently, furfural hooks on to molecules containing double bonds, but "ignores" molecules without double bonds. For this reason, furfural is used

(a) to remove aromatics from lubricating oils to improve the viscosity/temperature relationship,

(b) to remove aromatics from diesel fuels to improve the ignition characteristics, and

(c) to obtain unsaturated compounds (compounds with double bonds) from vegetable oils such as soybean oil to make "drying oils" suitable for paints and varnishes as only the double bond compounds are capable of "drying" by oxidation with air to form cross-linked polymers.

In the applications (a) and (b), the desired product (freed of aromatics) is the raffinate, while in application (c) the desired product (rich in unsaturated compounds) is obtained from the extract.

Of course, in liquid/liquid extraction, the extractant must form two liquid phases with the feed stock. With lubricating oil, diesel fuel, and soybean oil, furfural is capable of satisfying this condition as its oxygen atoms make it a more polar compound. The spectroscopic polarity index E_T^N [45] of furfural is 0.426 whereas the E_T^N values of the oils and of diesel fuel are close to zero.

The application of furfural for the refinement of lubricating oils is illustrated in Figure 49 [46]. In this process, the raw oil is fed into an extraction tower. Furfural enters the top, from where it flows downwards, picking up aromatics and forming an "extract" leaving the bottom, while the nonaromatics flow upwards, forming a "raffinate" leaving the top. Both "extract" and "raffinate" contain furfural (the "extract", of course, much more than the "raffinate") so that this furfural must be recovered. This is done mostly by steam distillation, using the low-boiling water/furfural azeotrope. All of the furfural eventually ends up in a furfural recovery column FRC, where it is freed of water and polymers before it is recycled to the extraction column.

14.2. Furfural as a Fungicide

As early as 1923, it was discovered [47] that furfural is a very effective fungicide. It was found that while even high concentrations of formaldehyde (10 to 15 %) do not prevent the growth of the mold penicillium, one of the most common fungi, as little as 0.5 % furfural is sufficient to entirely prevent the growth of mold even under conditions otherwise most favorable.

It was observed that furfural is particularly effective in inhibiting the growth of wheat smut (Tilletis foetens). This fungus is killed when the wheat is soaked for 3 hours in an 0.05 % aqueous solutuion of furfural, whereas with a formaldehyde solution of the same strength a period of 12 hours of soaking is necessary to destroy the smut.

Most noteworthily, as far as seeds are concerned, it was observed that their treatment with furfural does not diminish their germination power to any significant extent, whereas treatment with the same concentration of formaldehyde proved massively toxic. Wheat can be soaked for 6 hours in an 0.5 % aqueous solution of furfural with a reduction of only 4 % of the germination power of the seeds, whereas the same time of soaking in 0.5 % aqueous formaldehyde destroys the germination power completely.

From the treatment of seeds, the application of furfural as a fungicide was extended to growing plants and to wood.

14.3. Furfural as a Nematocide

Plant-parasitic nematodes, also called eelworms, are small transparent cylindrical organisms usually between 0.5 and 3.0 mm in length. By means of a stylet in the form of a

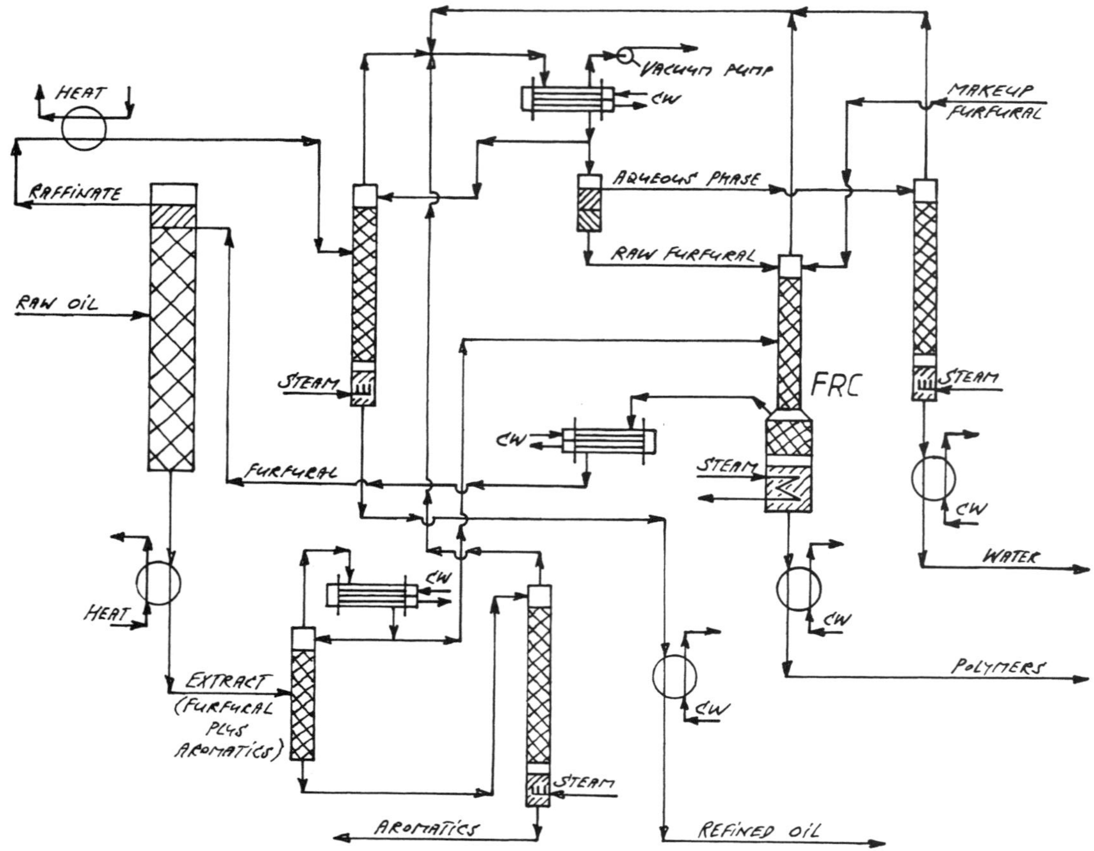

Figure 49. Schematic of the TEXACO Process for the Refinement of Lubricating Oils by Extraction of Aromatics with Furfural. FRC = Furfural Recovery Column.

hollow needle, they puncture plant cells, inject saliva, and suck out part of the cell content. Of greatest importance are nematodes that damage plant parts in the soil. Nematodes considered agricultural pests include root-knot nematodes which cause severe galling on the roots of many crops. Infected plants grow poorly and wilt readily in warm weather.

Worldwide, plant-parasitic nematodes cause an estimated agricultural loss of 60 billion U.S. dollars per annum [48]. Crops particularly threatened by nematodes are potatoes, beets, peanuts, soybeans, tomatoes, bananas, tobacco, berries, citrus fruits, and cotton.

According to various recent publications [49] [50] [51], furfural and other simple aromatic compounds such as benzaldehyde and thymol were found to be very effective nematode control agents. Although they do not kill nematodes, they change the microflora of the soil to an extent that the nematode population decreases to zero. It was discovered that these "indirectly acting nematocides" stimulate a rapid development of bacteria antagonistic to nematodes, so that these agents can be said to breed microorganisms capable of killing nematodes in a biological fashion. The effect of furfural on various types of nematodes is illustrated in Figure 50 [50] showing the nematode counts one week after treatment of the soil with various levels of furfural. For all nematode species investigated, the count was readily reduced to zero. Consequently, planting of soybean in the soils treated with furfural yielded a greatly superior crop, and at the level of 1 ml of furfural per kg of soil, the nematode count was still zero 8 weeks after treatment. Large-scale tests with peanuts in South Africa have confirmed these findings [52]. At an application level of 8 gallons of furfural per acre (74.79 liters per hectare), the peanut quality was converted from unacceptable to perfect, and the crop quantity was significantly increased.

As compared to "directly acting nematocides" (nematocides working by toxicity), furfural has the following advantages:

(1) For equal effect, it costs much less.

(2) It is nontoxic for humans. Furfural is present in fruit juices, beer, and bread. The oral LD_{50} for dogs is 2300 mg/kg.

(3) It is safely and easily applicable (as a saturated aqueous solution, 7.9 ml per 100 ml of water at 20 °C).

(4) It is nonsystemic, which means that it is not taken up by the plant to be protected, so that it can be applied till harvest time. This is prohibited for other nematocides.

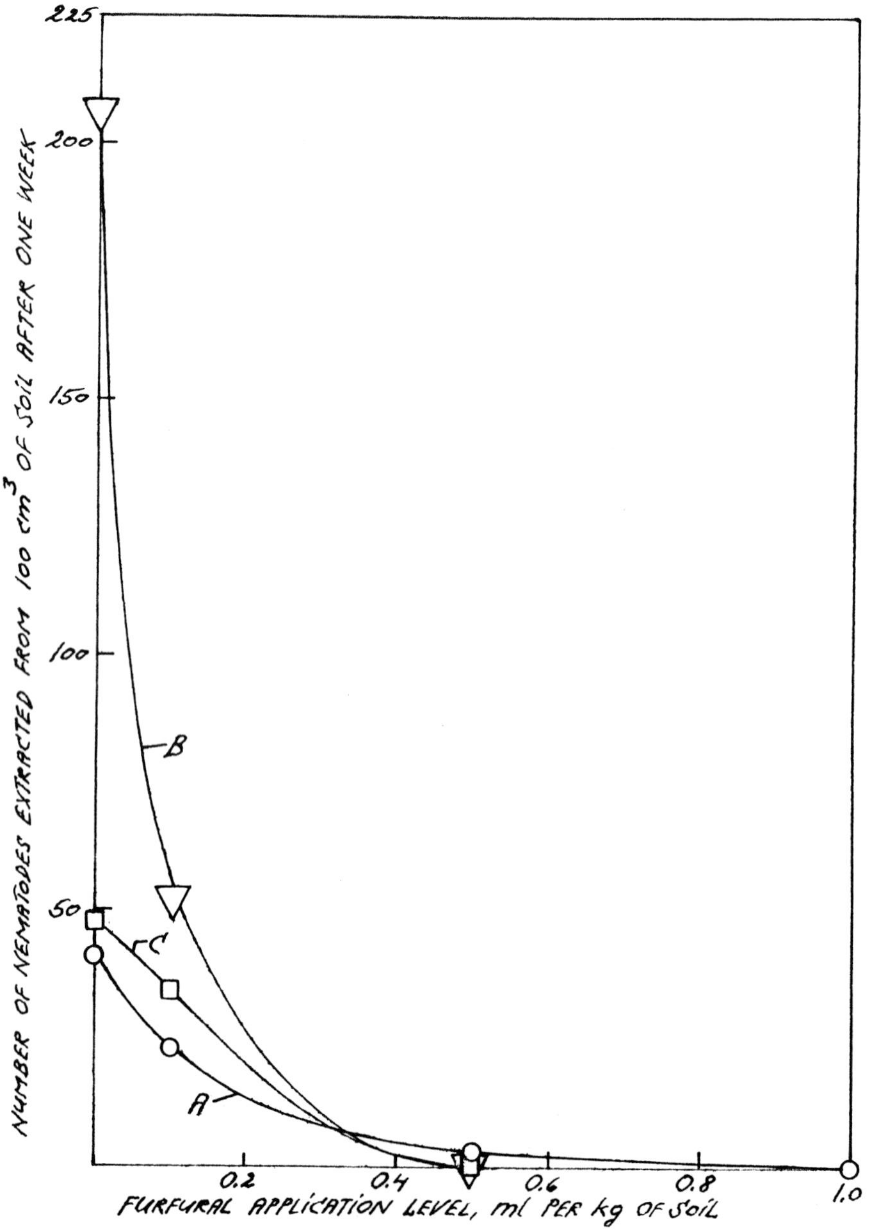

Figure 50. Effect of Furfural on Various Types of Nematodes.

A – Meloidogyne incognita

B – Heterodera glycines

C – Helicotylenchus dihystera

Furfural is not only nontoxic for humans but it is also harmless for the environment. By contrast, methyl bromide, a widely used nematocide having a boiling point of 4.5 °C, is a threat for the ozone layer, and with this being so, according to a directive of the UNITED NATIONS, it must be phased out by the year 2002.

In view of the given advantages, the application of furfural as a nematocide has the potential of becoming the greatest outlet for furfural, requiring the construction of many more furfural plants to satisfy the agricultural demand.

References

[45] C. Reichardt, Solvents and Solvent Effects in Organic Chemistry, VCH Verlagsgesellschaft, Weinheim, 1988.

[46] L. C. Kemp, G. B. Hamilton, and H. H. Gross, Ind. Eng. Chem. 40 (1948) 220-227.

[47] C. S. Miner, US Patent No. 1,592,039 (1926).

[48] C. Venter, Nematode Pests of Agriculture in Southern Africa, Nematological Society of Southern Africa, Potchefstroom, 1996.

[49] G. J. Walter and R. Rodriguez-Kabana, South African Patent No. 91/9037 (1991).

[50] R. Rodriguez-Kabana, J. W. Kloepper, C. F. Weaver, and D. G. Robertson, Nematropica 23 (1993) 63-73.

[51] R. Rodriguez-Kabana, Management of Nematodes by the Use of Botanical Aromatic Compounds, Paper presented at the Nematology Symposium in San Lameer, South Africa, March 9 to 13, 1997.

[52] C. Venter, University of Potchefstroom, South Africa, Private Communication, 1998.

15. Carboxylic Acids

All furfural processes lead not only to furfural but also to carboxylic acids. Depending on the raw material employed, the production of carboxylic acids may exceed the production of furfural.

The principal carboxylic acid produced is acetic acid. Formic acid, propionic acid, butyric acid, and higher carboxylic acids follow in strongly diminishing order.

15.1. Origin of the Carboxylic Acids

To various extents, the pentosan chains of the raw material have acetyl, formyl, and other such groups attached [53]. The ratio of acetyl to formyl groups is usually in the order of 10 : 1.

A typical birch wood pentosan has its acetyl apportioned in such a way that 58 percent of the rings are free from acetyl groups, 24 percent are acetylated at C-3, 12 percent at C-2, and 6 percent have acetyl groups on both C-3 and C-2.

The hydrolysis of a doubly acetylated pentosan to pentosan and acetic acid is depicted below:

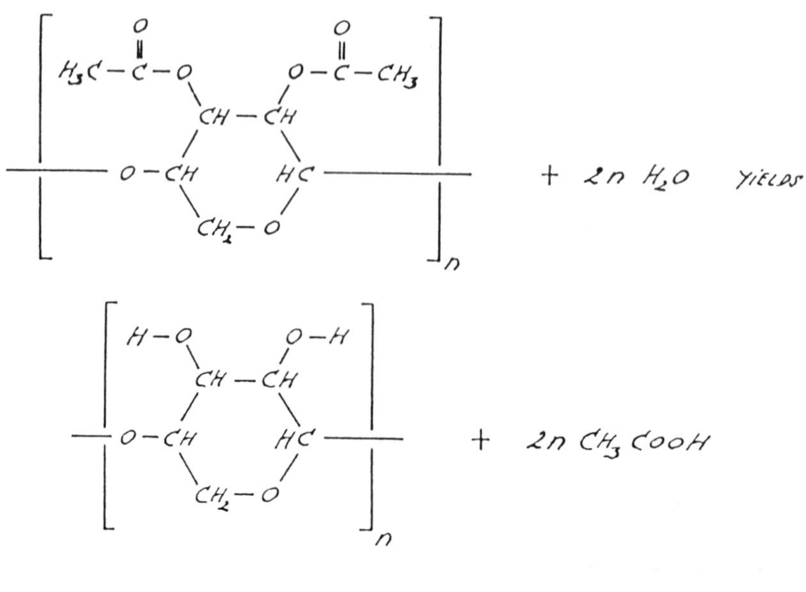

This hydrolysis is the easiest decomposition, and therefore the first to occur. As all hydrolyses, it is catalyzed by acidity, so that the liberation of carboxylic acids is exponential: The first formation of acids catalyzes the liberation of more acids, and so forth. This is of great importance for autocatalytic operation as in the ROSENLEW process.

15.2. Recovery by Extraction

A typical plant for recovering acetic acid and formic acid from the waste water of a furfural process is shown schematically in Figure 51. The waste water, representing the sump fraction of the first distillation column, has a temperature close to its boiling point. By pump 1, this water loaded with carboxylic acids is passed through the heat exchangers 2 and 3 which effect cooling to 30 °C. It then flows from the top to the bottom of an extraction column 4, where ethyl acetate, finely dispersed as a separate phase, forms a countercurrent stream of droplets rising from the bottom to the top. Intermolecular forces cause the ethyl acetate to take up most of the carboxylic acids, thus forming what is called the "extract", while the water takes up some ethyl acetate, thus forming what is called the "raffinate". The latter is passed by pump 5 through heat exchanger 2 into the top of a stripping column 6 energized by reboiler 7. Column 6 serves to recover the ethyl acetate from the water. The latter is withdrawn from the sump by pump 8 and leaves the plant through heat exchanger 9. The head vapor of column 6 has a composition corresponding to the azeotrope of the water/ethyl acetate system, so that when liquefied in condenser 10, two liquid phases are formed which are separated in a static decanter 11. The light phase, being ethyl acetate with very little water, is cooled in heat exchanger 12 and collected in tank 13, from where pump 14 returns it to the bottom of the extraction column 4. The heavy phase of the decanter 11 consists of water and very little ethyl acetate. This phase is used as reflux for column 6. The extract forming a separate phase in the top of column 4 is withdrawn by pump 15 and passed through heat exchanger 9 into an evaporator consisting of a vapor separator 16 and a reboiler 17. Pump 18 passes the concentrate of 16 into the stripping section of column 19 while the vapor of 16 enters the rectifying section. The head vapor of column 19 has a composition very similar to the azeotrope of the water/ethyl acetate system, so that when liquefied in condenser 20 again two phases are formed. These are essentially identical to the products of condenser 10, only the mass ratio of the two phases being different, the products of condenser 20 yielding more of the light phase than the products of condenser 10. The sump of column 19 is energized by

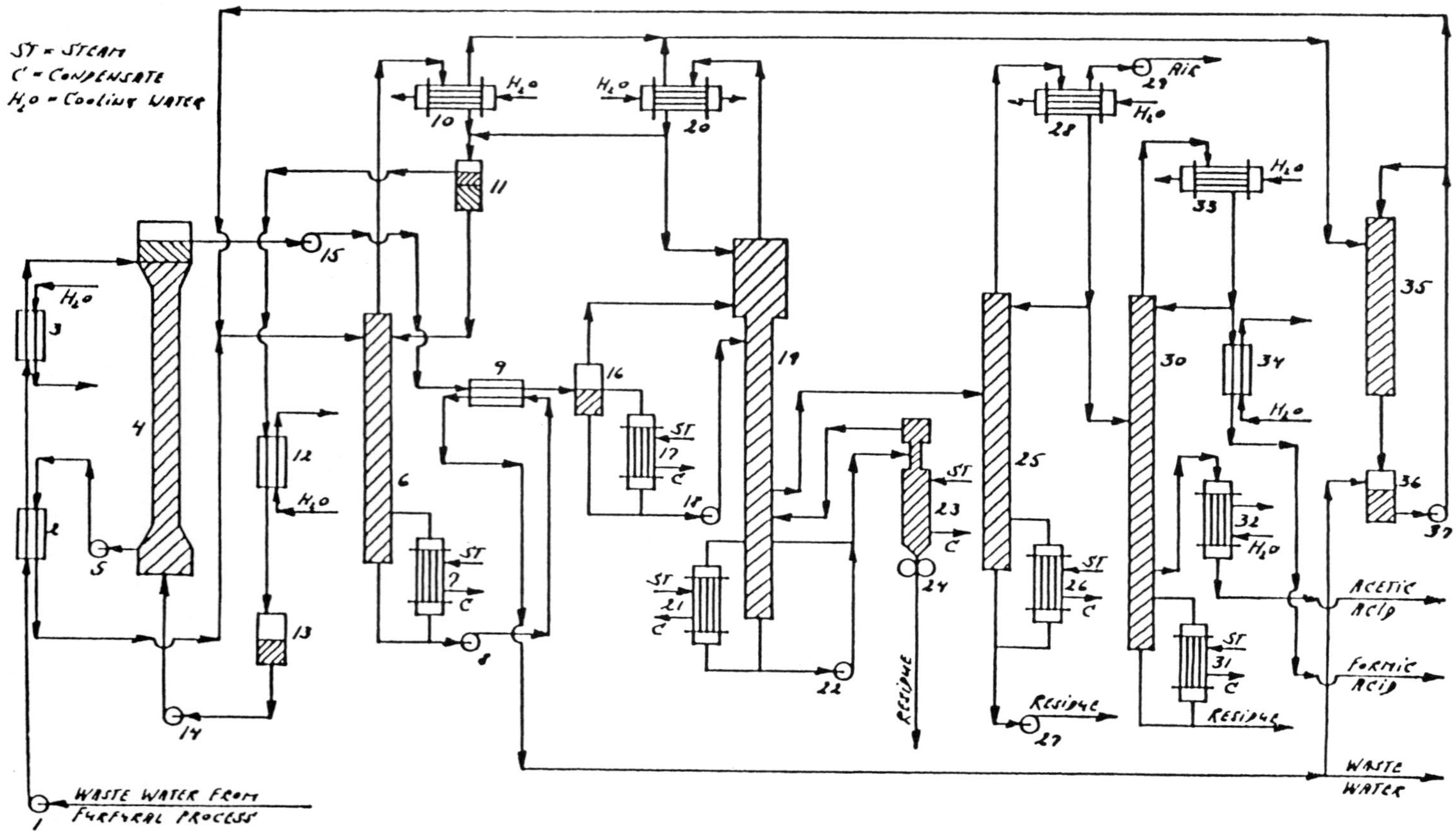

Figure 51. Extraction of Acetic Acid and Formic Acid with Ethyl Acetate.

reboiler 21. The sump fraction consisting of high-boiling components is withdrawn by pump 22 and fed into a thin-film evaporator 23 featuring a mechanical regeneration of the heating surface by rotating scrapers. The concentrate of this apparatus is withdrawn by a gear pump 24 while the vapor is returned to column 19. Several plates above the sump, the components of intermediate boiling points are withdrawn as a vapor phase and introduced into column 25 energized by reboiler 26. The sump fraction, mostly furfural, is withdrawn by pump 27. The head vapor is liquefied in condenser 28. Pump 29 maintains column 25 and condenser 28 under reduced pressure. The product of condenser 28 is partly returned to column 25 as reflux and partly fed into column 30 energized by reboiler 31. The sump fraction of column 30 is discarded, while a few plates above the sump acetic acid is withdrawn as a vapor and liquefied in condenser 32. The head vapor of column 30 consists of formic acid. It is liquefied in condenser 33, the resulting product being partly returned as reflux and partly cooled by heat exchanger 34 for final withdrawal. The condensers 10 and 20 are vented into an absorption tower 35 where any ethyl acetate not liquefied is taken up by water branched off from the waste water line, collected in tank 36, and partly circulated through the tower by pump 37. The remainder of the circulation water is injected into the raffinate stream for recovery of its ethyl acetate in the stripping column 6.

The economic viability of this proven process depends on the concentration of acetic acid in the waste water as the steam consumption increases exponentially with decreasing concentration. This situation is illustrated in Figure 52. While the process is perfectly economic at 5 % acid, it is prohibitive at 1.2 % acid, a level usually encountered with bagasse as the raw material.

Instead of ethyl acetate, other extractants have been used. A furfural plant in Austria employs a mixture of 30 % trioctyl phosphine oxide $(C_8H_{17})_3P{=}O$, known as TOPO, and 70 % hendecane $C_{11}H_{24}$. This plant, associated with a sulfite pulping mill, is shown in Figure 53. The feed stock containing 1.7 % acetic acid and 0.45 % furfural enters a rotary extractor 1 into which the TOPO/hendecane mixture is introduced at the bottom. On its way upwards, this mixture picks up both acetic acid and furfural, and forms an extract layer at the top. The raffinate, containing some residual acetic acid and furfural as well as some TOPO, leaves the bottom of the extractor and flows into a biological waste water treatment plant. The extract enters a distillation column 2 in which the TOPO/hendecane mixture is recovered as the bottom fraction, for reuse in the extractor. The head fraction of column 2 consists of

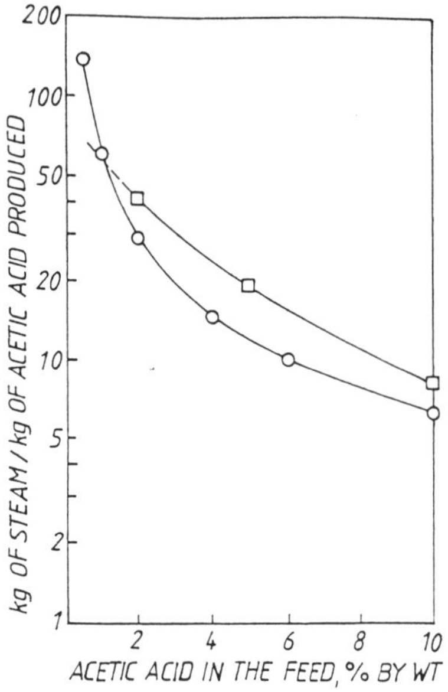

Figure 52. Specific Steam Consumption of Acetic Acid Recovery Plants as based on Extraction with Ethyl Acetate.

○ Data from KRUPP INDUSTRIETECHNIK.

□ Data from W. V. Brown [54].

water, furfural, acetic acid, and formic acid. This stream is liquefied in condenser 3, and partly refluxed to effect rectification. The other part is fed into a distillation column 4 in which formic acid is separated as the head fraction. The sump fraction, consisting of water, acetic acid, and furfural, enters a distillation column 5 where water is removed by forming a low-boiling azeotrope with ether added as entrainer. The azeotropic ether/water head vapor is liquefied in condenser 6. The condensate is collected in a static decanter 7 where two liquid phases are separated. The light phase, consisting mostly of ether and of very little water, is refluxed, while the heavy phase, consisting mostly of water and of very little ether, is introduced into a distillation column 8 producing pure water as the sump fraction and an azeotropic ether/water head vapor. The latter joins the head vapor of column 5 in condenser 6. The sump fraction of column 5 consists of acetic acid and furfural. It enters a distillation column 9 in which raw acetic acid is removed as the head fraction while the sump fraction is furfural. Finally, the raw acetic acid enters a distillation column 10 which separates low boilers as the head fraction, the sump fraction being 99.9 % acetic acid.

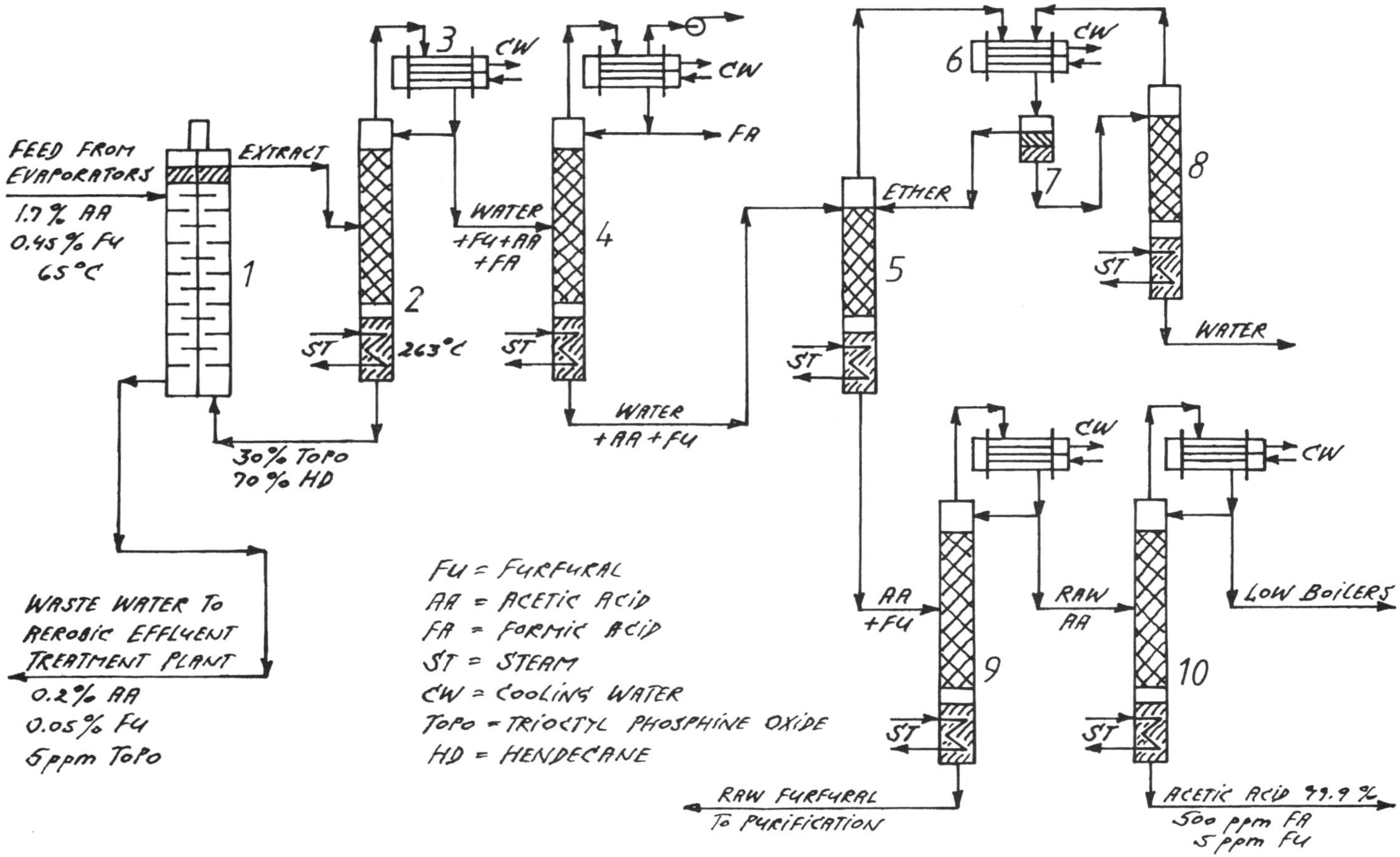

Figure 53. Extraction of Acetic Acid and Formic Acid with TOPO and Hendecane.

This complicated system has the inherent disadvantage that the high-boiling extractant (30 % TOPO and 70 % hendecane) is not recovered as a distillate but as a sump fraction, so that accumulation of high-boiling "crud" necessitates frequent replacement of this expensive extractant. Besides, the high boiling point of the extractant mixture (263 °C) leads to thermal decomposition, thus adding to the "crud" already in the feed.

The same disadvantages of "crud" accumulation and thermal decomposition apply to extracting the carboxylic acids with a mixture of a high-boiling amine (ALAMINE 336 produced by HENKEL) and a high-boiling hydrocarbon fraction (SHELLSOL produced by SHELL).

"Crud" accumulation and thermal decomposition are such grave disadvantages that all high-boiling extractants must be considered inadvisable.

15.3. Recovery by Freezing

According to Heist [55], in the case of low acid concentrations in the order of 1 percent by weight, no process other than freeze crystallization is capable of economic acid recovery. For the acetic acid/water system, the process is based on the phase diagram shown in Figure 54, where the eutectic temperature is seen to be –26.7 °C.

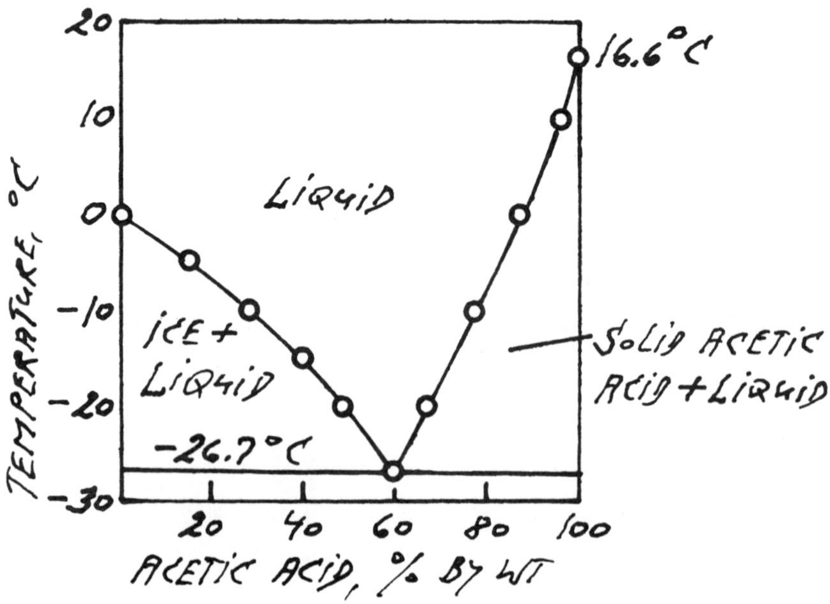

Figure 54. The Phase Diagram of the Acetic Acid/Water System.

Heist gives an example of a plant with an input of 1 million lb/day = 18.875 tons per hour at 1 percent acetic acid in water. He employs two stages for equal ice removal, and a small third stage operating at the eutectic to crystallize water and acetic acid simultaneously. The ice (0.917 g/cc) floats, while the acetic acid crystals (1.05 g/cc) settle to the bottom. The economic data for this plant were summarized as follows:

Operating Cost

(1) Power (1050 hp = 783 kW):.................................195 000 $/a

(2) Manpower:..105 000 $/a

(3) Maintenance:...165 000 $/a

465 000 $/a

Benefit

780 000 $/a for acetic acid at 0.24 $/lb.

Investment

3.72 million $.

15.4. Recovery by Extractive Condensation

In solvent extraction, use is made of weak intermolecular forces selectively attracting some molecules more than others. This concept of selective attraction can be extended to the point of using an "extractant" capable of forming a chemical bond with the molecules to be extracted, but not with other molecules. In this case, the "extractant" can be said to "react" with the partner molecules. In so doing, a new species is formed which will have different physical properties, and in a favorable case these new physical properties will be conducive to facilitating the separation intended.

To the end of extracting acetic acid from water, it is possible to make use of the fact that acetic acid reacts with triethylamine (TEA) to form a high-boiling complex held together by hydrogen bonds. When four molecules of acetic acid (b.p. 118 °C) react with one molecule of TEA (b.p. 89 °C), the resulting complex has a boiling point of 165 °C as shown in Figure 55, thus readily permitting its separation from water by distillation. However, when dealing with a very weak aqueous solution of acetic acid, as in the case of the waste water of furfural plants, a separation of the complex from the water by distillation would require evaporating a huge quantity of water, and although this distillation would be easy because of the great difference in boiling points, in terms of economy it would not be significantly better

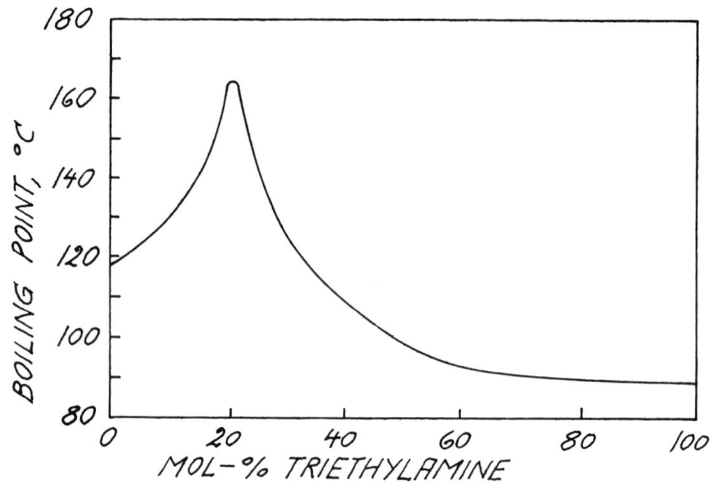

Figure 55. Ebullition Diagram for the Triethylamine/Acetic Acid System at 760 mm Hg.

than distilling the water from the acetic acid without complex formation. Consequently, to make use of the complex in an economic fashion, the separation of the complex from the water must be based on a different principle. In furfural plants, this happens to be possible as the reactors emit a mixture of acetic acid and water in the vapor phase. When TEA vapor is injected into the gaseous product stream of furfural reactors, the ensuing formation of the acetic acid/TEA complex results in the appearance of a fog since the complex has a higher condensation point (dew point) than steam. This effect can be readily demonstrated by a simple laboratory setup as shown in Figure 56.

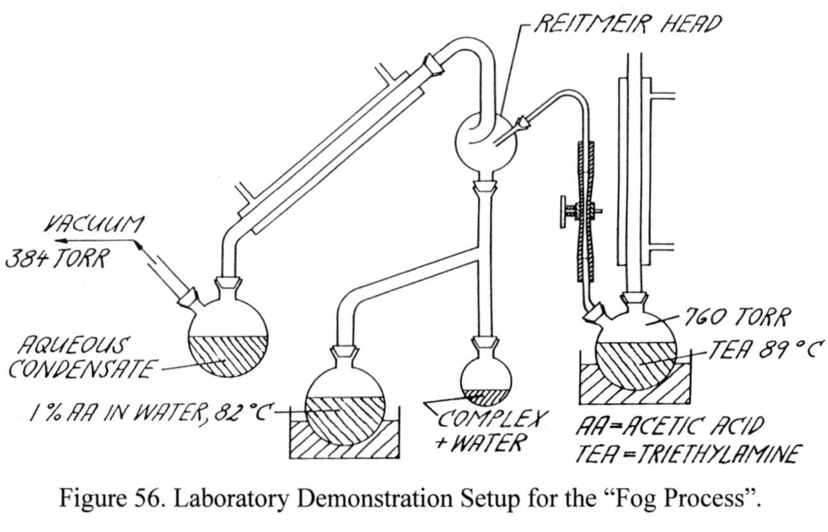

Figure 56. Laboratory Demonstration Setup for the "Fog Process".

TEA vapor is admitted into a vapor stream produced by evaporating a 1 % aqueous solution of acetic acid at slightly reduced pressure. When the two vapor streams meet in a REITMEIR head, a dense fog is observed, and the fog droplets coalesce on the wall of the REITMEIR head to form a small fraction of the complex and water. With such a simple setup, 72 percent of the acetic acid evaporated is found in the "fog fraction".

Of course, in an industrial installation, the REITMEIR head must be replaced by a unit operation capable of separating the fog droplets from the remaining vapor mixture to a high degree of perfection. To this end, use can be made of a coalescence filter (BEGG COUSLAND of Glasgow, England) or of an electrostatic separator (LURGI of Frankfurt, Germany). The latter is based on the fact that fog droplets are electrically charged.

The complex, once recovered, can be readily split by reacting it with ethanol at elevated temperature in the presence of an ion exchange resin. This leads to ethyl acetate, triethylamine, and water. It is noted that many furfural plants are parts of cane sugar mills since they use bagasse as the raw material. Given the fact that such mills make molasses as a by-product, they usually have an ethanol plant to use the molasses, so that the ethanol needed for the esterification is an in-house commodity. The ethyl acetate represents a profitable by-product.

The concentration of acetic acid in the "fog fraction" is essentially identical with the concentration at the maximum boiling point as only the molecules of the TEA/acetic acid complex undergo condensation.

The method of injecting a suitable reagent into a vapor mixture to selectively capture a target compound by forming a fog has been given the name of "extractive condensation" [56] [57]. It is a new separation process of unerring specificity. Although not used so far on an industrial scale, this process deserves attention. There is no problem with the introduction of TEA in a furfural plant for the following reasons:

(a) Furfural plants must be explosion-proof anyway, on account of highly flammable by-products such as acetaldehyde, acetone, methanol, and the like.

(b) There is no problem of furfural contamination as in any furfural plant the low boilers must be separated anyway.

(c) TEA does not react with furfural.

The complex formation described for acetic acid applies to other carboxylic acids as well.

15.5. Recovery by Multieffect Azeotropic Distillation

If the acid concentration is too low for conventional recovery by extraction, and if recovery of water is of interest as well, an elegant industrially proven solution of the problem is a multieffect azeotropic distillation. This process was developed by the BASF in Ludwigshafen/Germany [58].

It was realized that in the water/acetic acid system, in the range of low acid concentrations, the vapor/liquid equilibrium line very nearly coincides with the diagonal, so that obtaining pure water by simple distillation would require a monstrous number of stages (in the order of 150) as well as a huge reflux ratio (in the order of 9). In view of this situation, various considerations have led the BASF to choose azeotropic distillation using butyl acetate as entrainer.

The phase diagram for the water/acetic acid/butyl acetate system at 90 °C is shown in Figure 57.

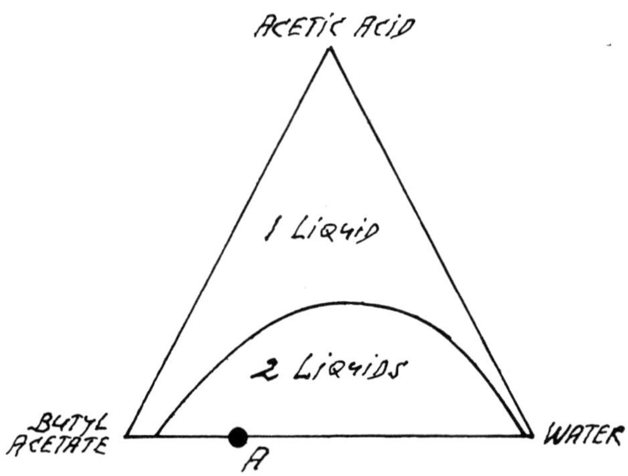

Figure 57. The Phase Diagram for Water/Acetic Acid/Butyl Acetate at 90 °C.

Atmospheric Boiling Points: 100.0 °C for Water

118.1 °C for Acetic Acid

126.5 °C for Butyl Acetate

Point A represents the water/butyl acetate azeotrope at 28.719 % by weight of water, with an atmospheric boiling point of 91.0 °C.

As can be seen, there is a large field with two liquids, and the water/butyl acetate system exhibits an azeotrope at 28.719 % by weight of water, with an atmospheric boiling point of 91.0

^{o}C, which is lower than the boiling points of water, acetic acid, and butyl acetate, so that distillation of any mixture of the ternary system leads to the azeotrope in the head. Against this background, the process is illustrated in Figure 58. Multieffect distillation is employed in that the head vapor of the first column energizes the reboiler of the second column, and the head vapor of the second column energizes the reboiler of the third column. On account of the heat of condensation diminishing with increasing temperature, and because of unavoidable heat losses, the feed stream is split into unequal substreams. Butyl acetate is injected into the heads of the first three columns to produce azeotropism. The pressures in the three effects were chosen to be 2 ATM, 1 ATM, and 0.5 ATM, leading to head temperatures of 112.4 ^{o}C, 91.0 ^{o}C, and 71.2 ^{o}C, respectively.

Condensation of the head vapors leads to two liquid phases: An organic phase consisting essentially of butyl acetate with very little water, and an aqueous phase consisting essentially of water with very little acetate. The organic phases are used as refluxes, while the combined aqueous phases are fed through a cooler into the head of a small fourth column, in which the water is stripped of its acetate by means of a reboiler energized with low-pressure steam. The fourth column is operated at the same pressure as the third column, the head vapors of the third and fourth columns being liquefied in a condenser maintained at 0.5 ATM by means of a vacuum pump.

The compositions of the two liquid phases formed by the azeotrope fix the reflux ratio at 3.22 kg of organic phase per kg of aqueous phase.

As the vapor volume streams in the three effects are large, thus leading to large column diameters, each effect was chosen to consist of two columns operating in parallel, so that the total number of columns is 2 x 3 + 1 = 7. Where applicable, the heat exchange areas given in the diagram represent the sum for two columns operating in parallel.

All columns have 16 trays, but with different spacings at different pressures, so that the column heights range from 7.6 m for the pressure columns to 12.4 m for the vacuum columns. The water produced by this plant has an acid content of less than 50 ppm.

15.6. Recovery by Recirculation

In the normal recovery schemes treated so far, the acid waste water of the furfural process was directly submitted to a separation so as to obtain the carboxylic acids as a by-product, for sale. It is, however, instructive to realize that apart from the small percentage of

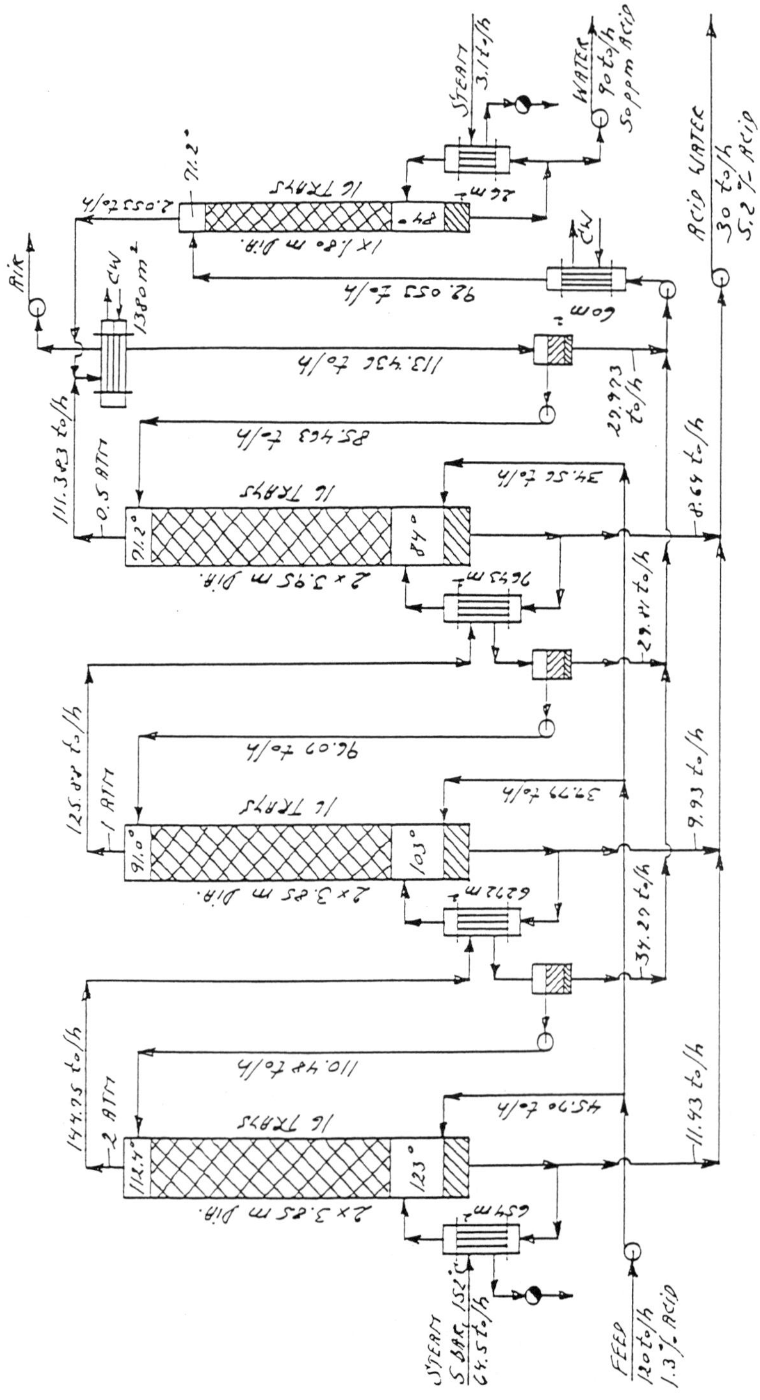

Figurte 58. Multieffect Azeotropic Distillation of Furfural Waste Water.

carboxylic acids liberated from the raw material, the acid waste water consists essentially of the condensate of the steam used in the reactors, so that the most logical and most appealing way of using this waste water would be to reconvert it to steam for the reactors. Needless to say, this would not be normal steam but "acid steam", but this is no drawback as acidity is desirable in the reactors anyway. Of course, in such a recycle scheme, the carboxylic acids generated by the raw material would build up beyond limits unless they are given an outlet, but the buildup as such is perfectly welcome as it improves the economy of any subsequent recovery process and may improve the catalysis in the reactors.

A concept based on this train of thought is illustrated in Figure 59 referring, for the sake of simplicity, to a plant with continuously operating ROSENLEW reactors requiring a temperature of 180 °C normally sustained by superheated steam of 12 bar and 265 °C. The sump fraction of the primary distillation column (i.e. the waste water) is seen to be fed into an evaporator operated at 188 °C, where 75 percent of the stream is vaporized while the remaining 25 percent leave the evaporator as a concentrate. The vaporized portion is passed through a superheater to form an acid vapor of the same temperature and pressure as the steam normally injected into the reactors. As this vapor represents roughly 75 percent of the reactor steam requirement, the latter is satisfied by adding 25 percent of the normal steam input.

As the total acid formation from the raw material would remain unchanged, it follows that in steady state conditions the acid concentration of the evaporator concentrate would be roughly four times as high as in the normal waste water, thereby greatly improving the steam economy of recovering this acid by extraction. In addition, as the steam consumed in the evaporator and in the superheater is recovered as condensate, i.e. boiler feed water, the overall loss of water is reduced to 25 percent of the loss in conventional processing. This is an important benefit in areas of frequent drought, for instance South Africa, where it is a tragic absurdity that in a furfural plant often plagued by water shortage, the effluent containing 98 percent water is discharged into the Indian Ocean.

In summary, at the chosen 3 : 1 split in the evaporator, the "recirculation scheme" is seen to have the following impressive advantages:

(1) The outgoing acid concentration is increased by a factor of 4, thereby greatly reducing the specific steam consumption for an acid recovery by extraction.

(2) The acidity in the reactors is increased, thus leading to improved catalysis.

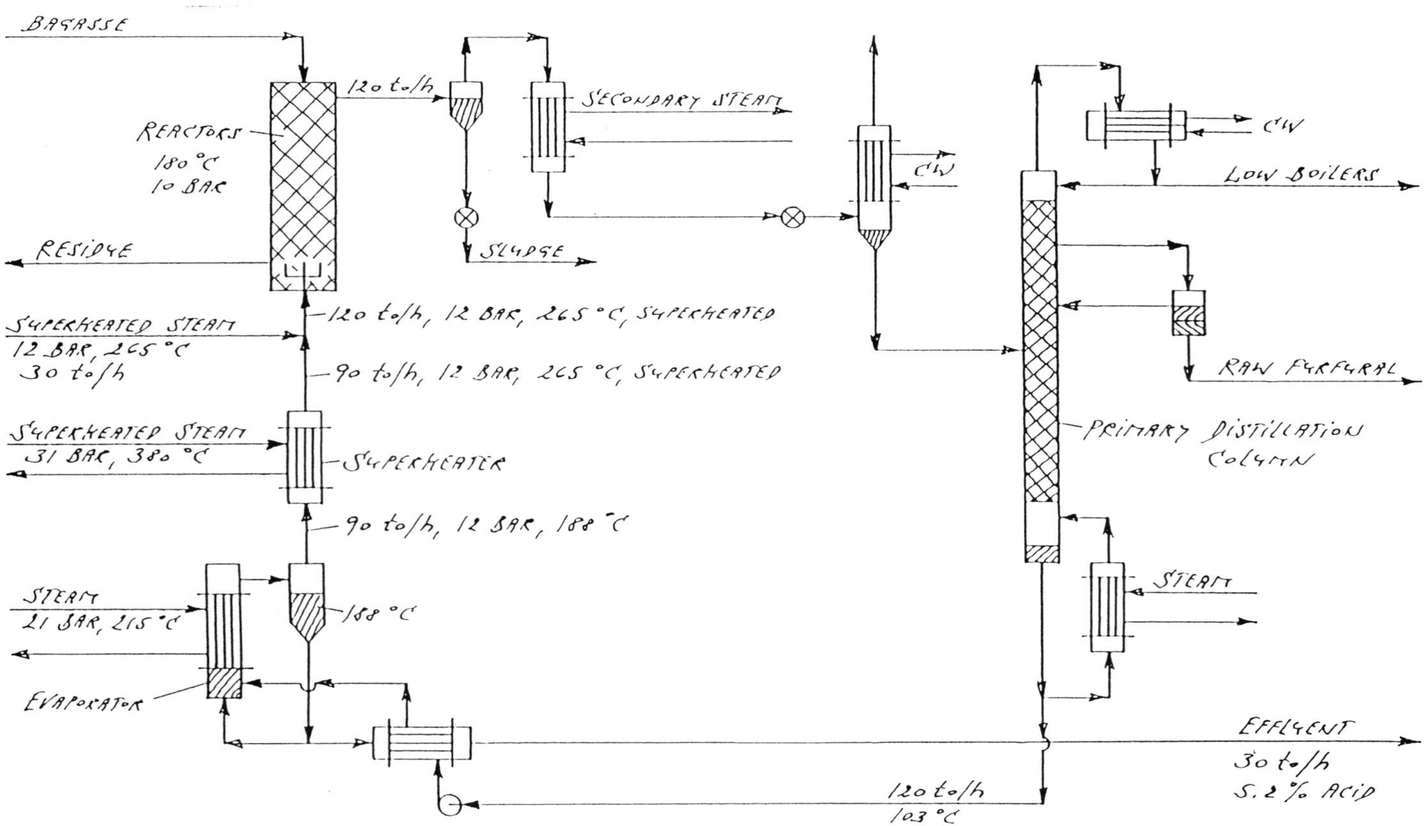

Figure 59. Outline of a Recirculation Scheme.

(3) 75 percent of the process water is recovered.

The only problem with this attractive scheme is the fact that the evaporator and the superheater have to handle an elevated acid concentration at very high temperatures, where corrosion and fouling are severe.

References

[53] E. Hägglund, Chemistry of Wood, Academic Press Inc., New York, 1951.

[54] W. V. Brown, Chem. Eng. Progress 59 (1963) 65-68.

[55] J. A. Heist, Chem. Eng. 86 (1979) 72-82.

[56] K. J. Zeitsch, German Patent Application No. P 40 25 128.4 (1990).

[57] K. J. Zeitsch, Ind. Eng. and Chem. Research, accepted on August 2, 1999 (in press).

[58] B. Hegner, D. Hesse, and D. Wolf, Chemie-Ing.-Techn. 45 (1973) 942-945.

16. Diacetyl and 2,3-Pentanedione

Under special circumstances, serendipitously occurring or produced at will, the production of furfural is accompanied by the formation of diacetyl and 2,3-pentanedione. Used as flavors, these compounds are by-products of enormous value greatly contributing to the profitability of a furfural plant. For this reason, they have been shrouded in secrecy, and what little information on the topic leaked out was erroneous. In volume A 12 on page 123 of the 5th Edition, ULLMANN'S ENCYCLOPEDIA OF INDUSTRIAL CHEMISTRY (VCH Verlagsgesellschaft, Weinheim 1989) presents the following statement on the ROSENLEW process:

> "This milder acid hydrolysis also produces diacetin (glyceryl diacetate) as a saleable coproduct."

The fact that the saleable coproduct is actually not diacetin (glyceryl diacetate) but diacetyl (2,3-butanedione) reveals that the author of the article had somehow misunderstood a word gleaned by poor espionage, and that he had absolutely no idea of the chemical reactions involved in the formation of diacetyl, not to speak of 2,3-pentanedione.

On account of their two conjugated C=O double bonds, both diacetyl and 2,3-pentanedione are intensely yellow compounds, and both are strong flavors. At a level of up to 2 mg/kg, diacetyl is the principal flavor and colorant of butter, and for this reason it is added to margarine to give it the taste and look of butter. Diacetyl is also used as a flavor in ice cream and baked goods. By contrast, 2,3-pentanedione is used as aroma in alcoholic and nonalcoholic beverages.

Diacetyl sells for approximately 14 US $/kg, and 2,3-pentanedione commands a whopping price in the order of 300 US $/kg. Thus, even if the quantities compared to furfural as the main product are small, the profits which can be derived from these by-products are stunning.

For a ROSENLEW reactor 2.5 m in diameter, making 350 kg/h of furfural, the ratio of diacetyl production to furfural production is 0.01396, and the ratio of 2,3-pentanedione production to furfural production is 0.00118. Thus, if furfural sells for 1 US $ per kg, the ratio of the sales value of diacetyl to the sales value of furfural is found to be

0.19544, and for 2,3-pentanedione it is 0.35400, so that the sum of diacetyl and 2,3-pentanedione yields 0.54944. Hence, the sales value of diacetyl and 2,3-pentanedione together amounts to more than half of the sales value of the furfural produced in the same plant. Obviously, this justifies a major effort to recover these diketones.

16.1. The Formation of Diacetyl

Among our forebears, it was common to explain the occasional great reactivity of acetaldehyde by the concept of "nascent acetaldehyde". Today we know that the strange capabilities of acetaldehyde are due to a formation of radicals.

Even at room temperature, acetaldehyde reacts with oxygen to form acetyl and perhydroxyl radicals [59]:

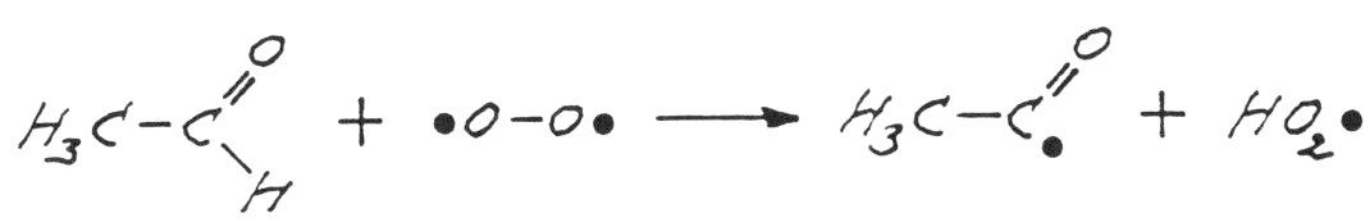

(1)

The acetyl radicals combine with more oxygen to form acetic acid as shown in Figure 60. Thus, any bottle of acetaldehyde that has been opened so that air could enter is known to contain some acetic acid [60].

In all furfural reactors, acetaldehyde is formed from the coniferaldehyde groups of the lignin being part of the raw material [61] as shown below:

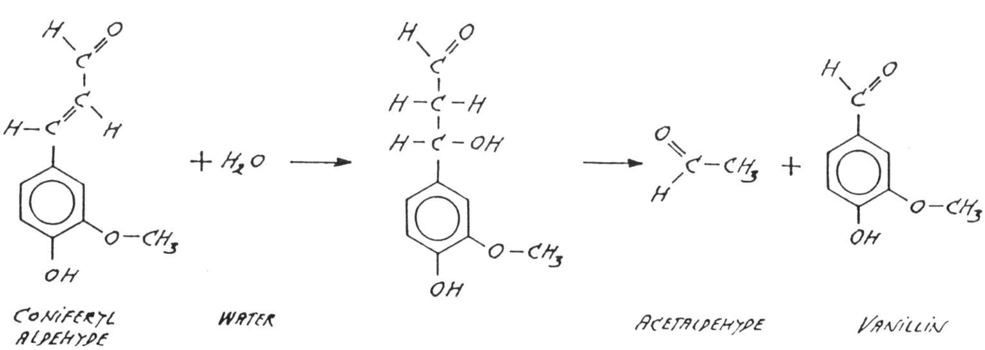

However, there are two different scenarios for the fate of this acetaldehyde. In furfural batch reactors, where a long passage of steam displaces any air in the initial charge, reaction (1) can take place only briefly at the very start, but in continuous furfural reactors, air is admitted all the time with the entering raw material, so that the acetaldehyde liberated from the lignin can

Figure 60. The Formation of Acetic Acid from Acetyl Radicals and Oxygen.

Step 1. Acetyl radicals react with oxygen to form peroxy acid radicals:

Step 2. The peroxy acid radicals abstract hydrogen from acetaldehyde molecules to yield peroxy acetic acid and new acetyl radicals:

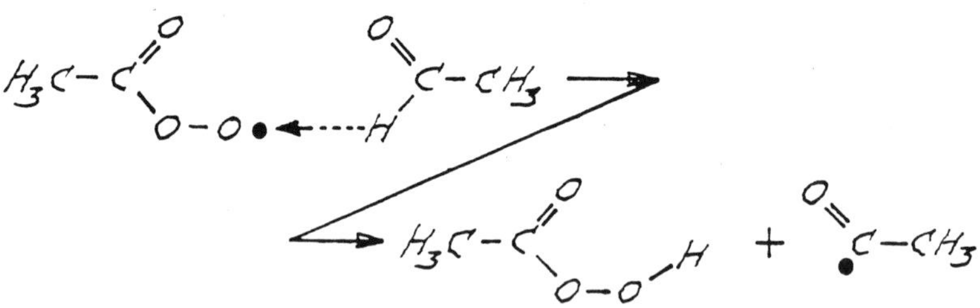

Step 3. The O-O bond in peroxy acetic acid undergoes thermal homolytic rupture to yield carboxyl and hydroxyl radicals:

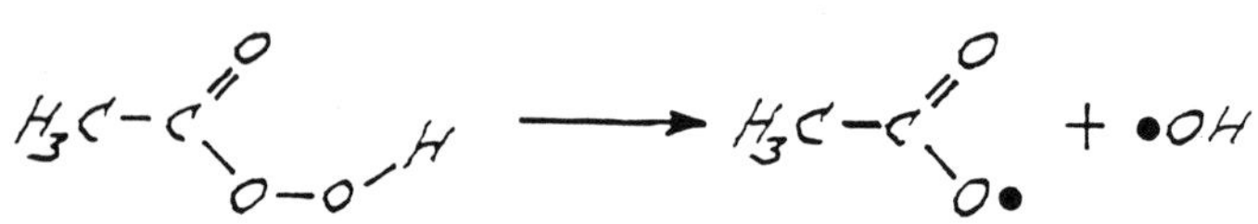

Step 4. The carboxyl radicals abstract hydrogen from acetaldehyde molecules to yield acetic acid and new acetyl radicals:

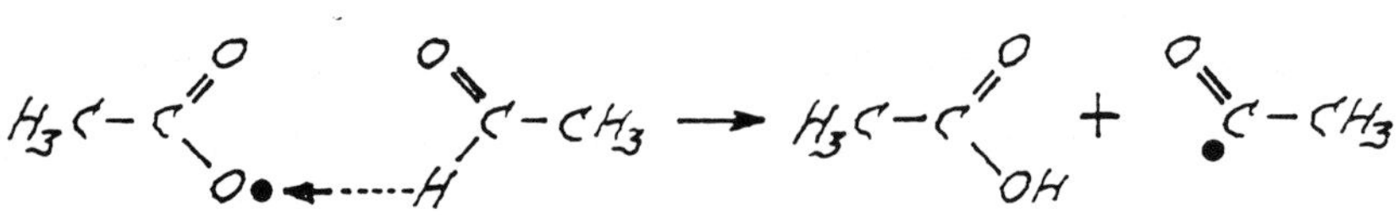

Step 5. The hydroxyl radicals abstract hydrogen from acetaldehyde molecules to yield water and new acetyl radicals:

react with oxygen of the incoming air in an ongoing fashion, thereby producing a continuous supply of acetyl radicals. This is illustrated schematically in Figure 61 showing a ROSENLEW reactor. The steam injected at the bottom splits off acetaldehyde from the lignin and carries this acetaldehyde to the top, where the periodic opening of the shutters introduces air with the feed stock. Thus, in the very top of the reactor, acetaldehyde and oxygen meet and react with each other according to (1). At the elevated temperature in the reactor, (1) is a vicious reaction, and if there were ample oxygen, then according to Figure 60 the acetyl radicals thus formed would react further to yield acetic acid. Inasmuch, however, as the supply of oxygen is limited, the acetyl radicals largely react with each other to form diacetyl [62]:

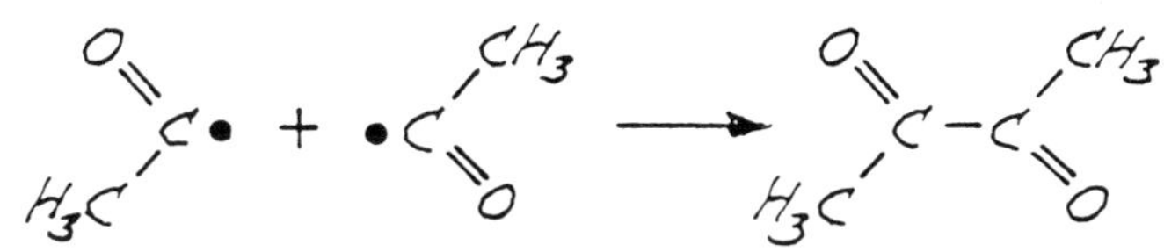
(2)

Apart from the effect of oxygen deficiency, reaction (2) is enhanced by the fact [63] that at the high temperature in a furfural reactor the equilibrium of the reaction

$$R\bullet + O_2 \rightarrow\rightarrow ROO\bullet \qquad (3)$$

lies on the left-hand side, so that the steady state concentration of alkyl radicals is high, thus favoring radical combination. Consequently, in the high-temperature vapor phase oxidation of acetaldehyde, acetic acid is usually not a significant product. By contrast, in a low-temperature liquid phase oxidation of acetaldehyde, acetic acid is overwhelmingly the main product.

Against this background, it is seen that continuously operating furfural reactors, on account of their continuous introduction of air, serendipitously synthesize diacetyl from acetaldehyde and oxygen, and this diacetyl, being more volatile than water, ends up in the low-boiling fraction of the furfural distillation. Being intensely yellow, the diacetyl renders this fraction yellow, and having an intense odor, the diacetyl imparts this odor to the entire plant. On the other hand, with furfural batch reactors, deprived of oxygen and, therefore, incapable of synthesizing diacetyl to any significant extent, the low-boiling fraction of the furfural distillation is colorless, and there is no diacetyl smell in the plant.

Needless to say, what has been described for the continuously operating ROSENLEW reactor also applies to the continuously operating reactor of QUAKER OATS. It, too, generated significant quantities of diacetyl. By contrast, in the continuous SUPRATHERM process, where the air of the raw material is eliminated in the slurry preparation, diacetyl is

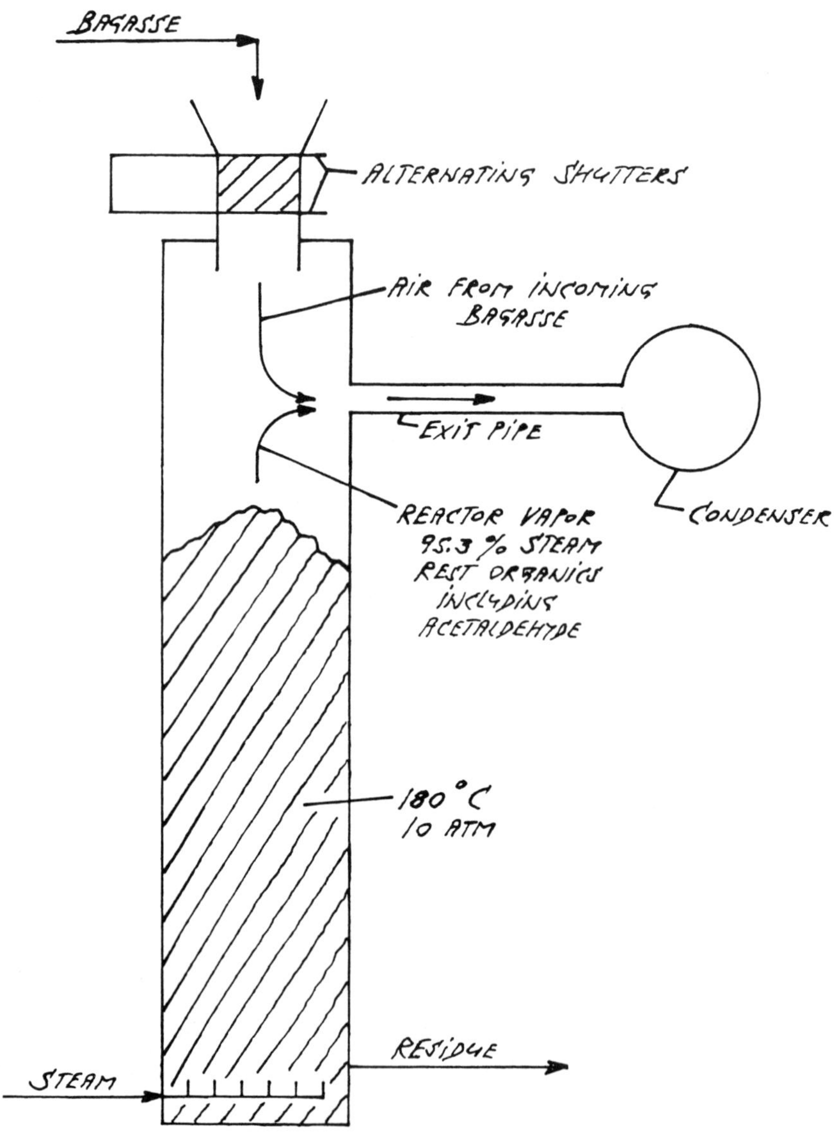

Figure 61. Schematic of a ROSENLEW Reactor.

The exit pipe represents the "free radical reactor" producing diacetyl.

understandably absent.

16.2. Analogy to Charcoal Reactors

A typical charcoal reactor is shown in Figure 62. As in the ROSENLEW reactor, here, too, there are shutters admitting air with the feed stock, and the latter, being wood, contains lignin liberating acetaldehyde at elevated temperatures. Hence, the conditions are analogous to what has been described for the ROSENLEW reactor in that diacetyl is serendipitously synthesized from acetaldehyde and oxygen. For this reason, charcoal producers, like LAMBIOTTE of France, are also major producers of diacetyl.

16.3. Production of Diacetyl in "Free Radical Reactors"

Of course, what is going on unintentionally, by serendipity, in the continuous furfural reactors, can also be carried out on purpose in a reactor specifically designed to make diacetyl as the principal product. According to the arguments brought forward above, this must be a vapor phase reactor, and the reagents should be highly diluted with an inert gas since acetaldehyde and oxygen alone, in the ratio needed, would be explosive. The nominal explosion limits for acetaldehyde in air are 4.0 and 57.0 % by mole [64], but to be on the safe side at all temperatures, rules at HÜLS AG in Marl, Germany, demand that all industrially used mixtures of acetaldehyde and oxygen must have less than 7 % by mole of oxygen [65].

Against this background, an example of a laboratory setup for making diacetyl from acetaldehyde and oxygen by the same "free radical pathway" as in the ROSENLEW reactors is shown in Figure 63. A suitable mixture of nitrogen and oxygen is passed through liquid acetaldehyde held at –20 °C by a cryostat. The resulting mixture of acetaldehyde vapor, nitrogen, and oxygen, containing only 4.382 % by mole of oxygen, is passed through a pipe reactor heated by a tube furnace to 180 °C.

16.4. Modification of Furfural Batch Reactors to Make Diacetyl

As shown above, the formation of diacetyl in furfural reactors requires oxygen as supplied by the air introduced by the feed mechanisms of continuously operating reactors (shutters in the case of the ROSENLEW and ESCHER WYSS reactors, augers in the continuous QUAKER OATS process). By implication, this means that batch reactors,

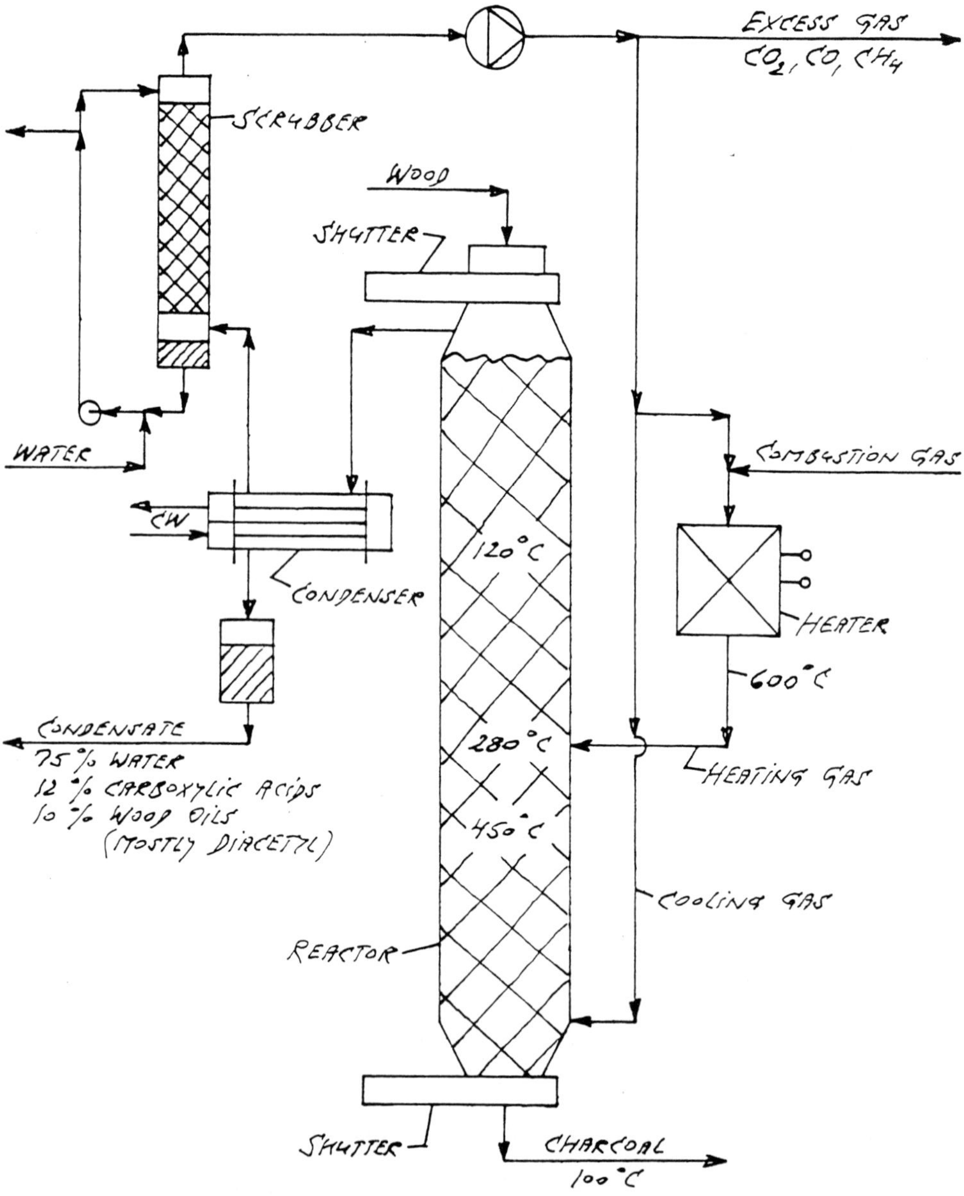

Figure 62. Schematic of a Charcoal Reactor (LAMBIOTTE, France).

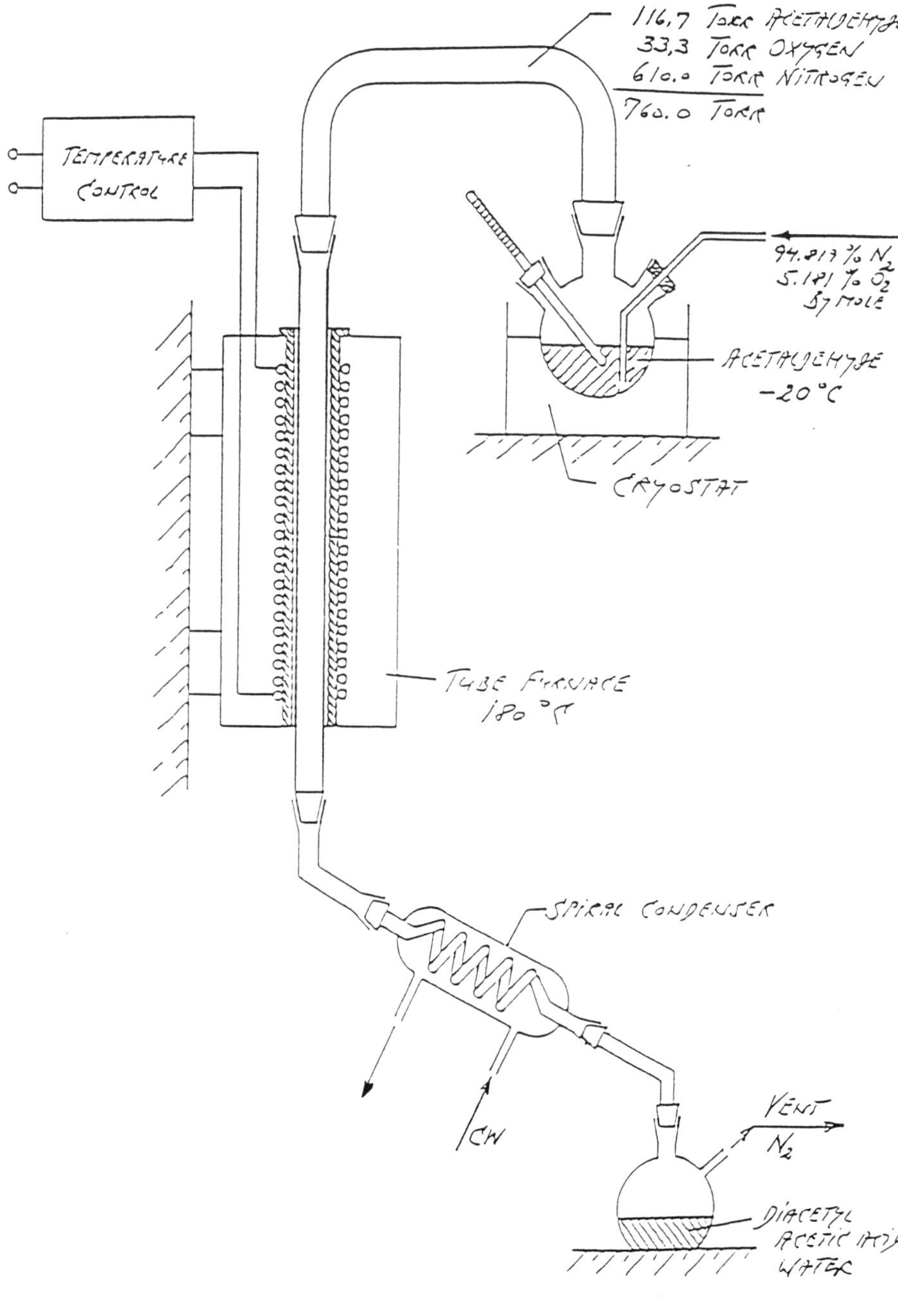

Figure 63. Laboratory Setup for Making Diacetyl from Acetaldehyde and Oxygen.

representing the vast majority of all existing furfural reactors, do not produce any significant quantities of diacetyl as after introduction of the raw material the steam input rapidly displaces all of the air initially in the charge, so that although there is acetaldehyde liberated from the lignin of the raw material as in all furfural reactors, throughout the digestion period there is no oxygen to react with this acetaldehyde to form diacetyl.

Obviously, however, this situation can be readily changed by injecting an appropriate stream of air or oxygen into the gaseous exit stream of such reactors. When this is done, the batch reactors can produce as much diacetyl as the given continuously operating reactors, and the recovery procedures are, of course, the same since the reactants and the process of diacetyl formation are the same. In fact, by injecting not only air or oxygen but also acetaldehyde to supplement the "innate" acetaldehyde actually formed in the reactor from lignin, any quantity of diacetyl can be produced, the furfural reactors providing merely the proper conditions of temperature, pressure, and steam dilution.

16.5. The Formation of 2,3-Pentanedione

The observed formation of 2,3-pentanedione in continuously operating, air admitting furfural and charcoal reactors can be explained by three facts:

(1) A small portion of acetyl radicals decomposes to carbon monoxide and methyl radicals.

(2) A small portion of diacetyl molecules undergoes hydrogen abstraction to yield "diacetyl radicals".

(3) The methyl radicals react with the diacetyl radicals to give 2,3-pentanedione.

This sequence of reactions is shown in Figure 64.

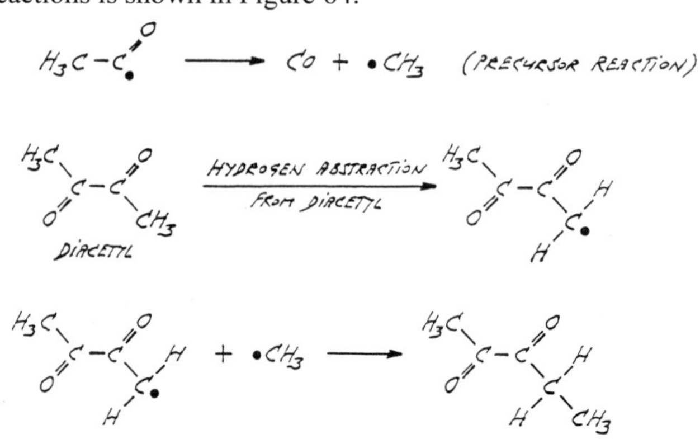

Figure 64. Reaction Sequence resulting in 2,3-Pentanedione.

The quantity of 2,3-pentanedione thus formed amounts to approximately 10 percent of the diacetyl.

The same sequence of reactions is observed when diacetyl is irradiated by UV light [66].

16.6. Recovery Techniques

In the furfural distillation, both diacetyl and 2,3-pentanedione end up in the fraction of low boilers, commonly called "raw solvent". For diacetyl this is clear as it has a boiling point of 88 °C, and for 2,3-pentanedione, having a boiling point of 112 °C, this is due to its azeotrope with water which boils below 100 °C.

The fraction of low boilers obtained in the distillation of furfural is a rather complicated mixture typically consisting of the following components, listed in the order of increasing boiling points:

(1) Acetaldehyde, 2 % by weight, b.p. 21.0 °C.
(2) Acetone, 5 % by weight, b.p. 56.5 °C.
(3) Acetaldehyde dimethyl acetal, 4 % by weight, b.p. 64.5 °C.
(4) Methanol, 42 % by weight, b.p. 64.6 °C .
(5) Ethanol, 18 % by weight, b.p. 78.5 °C.
(6) Diacetyl, 25 % by weight, b.p. 88.0 °C.
(7) Water, 2 % by weight, b.p. 100.0 °C.
(8) 2,3-pentanedione, 2 % by weight, b.p. 112.0 °C.

Of these components, only ethanol is not formed in the furfural reactors. In the typical case of bagasse as the raw material, ethanol results from a partial unwanted fermentation of the cane on the fields and in the sugar mill. In other words, the ethanol is already part of the bagasse before the latter enters the furfural reactors.

16.6.1. Extractive Distillation

It was found that when the given mixture (the "raw solvent") is submitted to a simple distillation, neither the head fraction nor the sump fraction shows any significant increase in diketone concentration. This is synonymous with saying that the diketone concentration in the vapor is roughly equal to the diketone concentration in the liquid, very much as in the case of distilling an azeotropic mixture. This is not surprising as the starting mixture, representing the top fraction of the furfural distillation, is in essence a low-boiling polyazeotrope.

In view of this situation, it was obvious that increasing the diketone concentration by distillation required the addition of a volatility modifier. It was soon discovered that adding water to the original mixture did increase the relative volatility of the diketones as the latter have polarities vastly different from that of water, so that at high water concentrations the diketone molecules are "pushed out of the mixture".

These considerations led to subjecting the initial mixture to an extractive distillation using water as the volatility modifier. In this concept, a high water concentration maintained over most of the column length brought significant quantities of diacetyl and 2,3-pentanedione to the top of the column, while most of the alcohols, because of their high polarity, ended up in the sump fraction. It was discovered, however, that the effectiveness of this distillation was greatly increased by adding water acidified with sulfuric acid rather than plain water. This has two reasons:

(1) Diacetyl and 2,3-pentanedione react with water to form monohydrates:

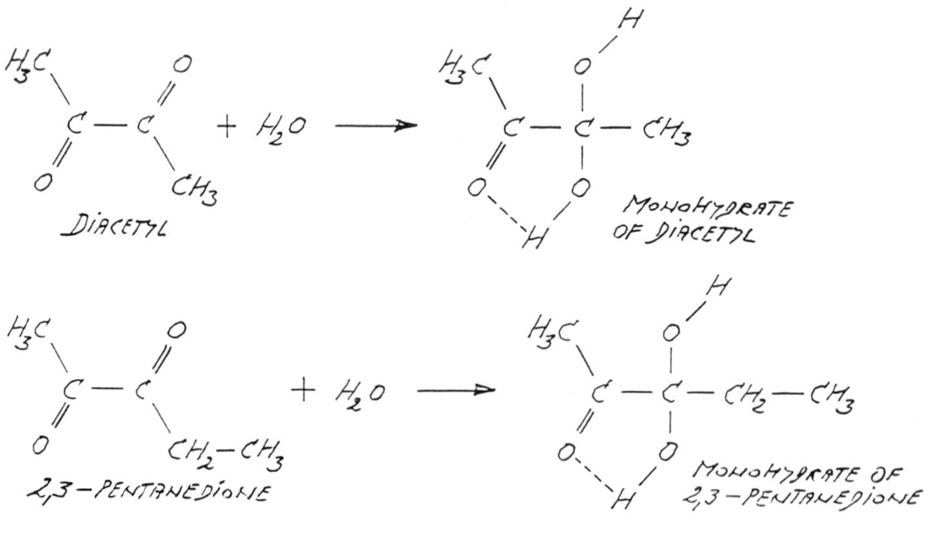

These molecules are stabilized (energetically favored) by intramolecular hydrogen bonds shown as dashed lines. Simple ketones such as acetone cannot form such intramolecular hydrogen bonds as they have only a single carbonyl group, and for this reason simple ketones do not form hydrates.

By NMR spectroscopy [67] it was found that in aqueous solution some 68 percent of all diacetyl molecules are in the high-boiling hydrate form, and a similar figure can be expected for 2,3-pentanedione. In distillation, this equilibrium is upset in that only the low-boiling nonhydrated diketone molecules go to the top of the column. As under normal circumstances the hydration equilibrium is reached very slowly, in a period of many hours, the system in a distillation column does not have sufficient time to reestablish the equilibrium. After a removal of the volatile nonhydrated molecules, there is not enough time for the remaining nonvolatile hydrates to form the equilibrium percentage of volatile nonhydrated molecules, so that most of the hydrate molecules leave the column with the sump fraction. By contrast, when sulfuric acid is added to the "extraction water", the reestablishment of the equilibrium is accelerated (catalyzed) to such an extent that it can proceed in the distillation column, thus permitting the entire diketone input to be transferred into the head fraction.

(2) In equilibrium, approximately 1 percent of the diacetyl molecules are in the enol form [68], and a similar figure can be expected for 2,3-pentanedione:

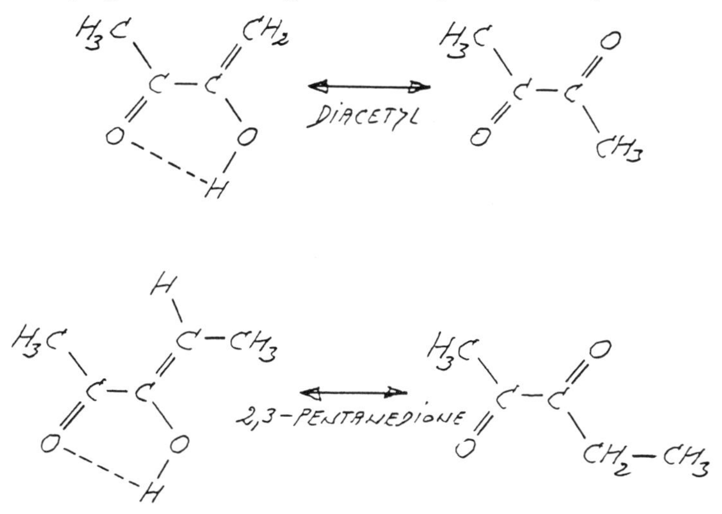

ENOL FORM KETO FORM

The enol forms are more volatile than the keto forms as their intramolecular hydrogen bonds greatly reduce the attractive forces of neighboring molecules. In distillation, the equilibrium is upset in that the more volatile enol molecules will go more readily to the top than the less volatile keto molecules. As reestablishment of the keto/enol equilibrium is also very slow, requiring many hours [69], the time in the distillation column is not sufficient to permit regenerating the equilibrium enol percentage by converting keto molecules to enol molecules. However, when sulfuric acid is added to the "extraction water", the reestablishment of the keto/enol equilibrium is accelerated (catalyzed) to such an extent that essentially all of the diketones are permitted to pass the distillation column in the volatile enol form.

In view of (1) and (2), sulfuric acid acts as a catalyst for a rapid reestablishment of equilibria between a strongly volatile form and a poorly volatile form of the same molecule, so that in combination with the anti-equilibrium effect of distillation, the poorly volatile form is temporarily converted into the strongly volatile form, to the end of greatly improving the separation process. Because of the underlying conversion of a molecule, it is proposed to call this concept a "mutational distillation".

This background led to the development of a distillation scheme as shown in Figure 65. The low boiler fraction of the furfural distillation (the "raw solvent") is mixed with 0.2 % aqueous sulfuric acid in a ratio of 1 : 3 by volume, and this "feed mixture" is introduced into a column of 74 trays comprising a stripping section of 34 trays, a lower rectification section of 34 trays, and an upper rectification section of 6 trays. Between the two rectification sections, 0.2 % aqueous sulfuric acid heated to 81 ^{o}C is added to keep most of the rectification part loaded with acid water as well. The uppermost 6 trays serve to greatly reduce the water content of the head fraction. After condensation, this head fraction forms two liquid phases, a light organic phase and a heavy aqueous phase. The latter is returned into the column as reflux, while the organic phase is the desired diketone concentrate containing typically 62.3 % diacetyl and 9.0 % 2,3-pentanedione. The vastly predominant percentages of the input alcohols (methanol and ethanol) leave the sump of the column together with the "acid extraction water".

This most remarkable column is a drastic example for "volatility manipulation" in that the low-boiling alcohols (with boiling points of 64.6 and 78.5 ^{o}C) end up in the sump, whereas the high-boiling diketones (with boiling points of 88 and 112 ^{o}C) end up in the head, seemingly in defiance of the usual distillation rules demanding the contrary. Because of this situation, the given scheme is called "paradoxical distillation" or "inverse distillation". Its explanation lies in the fact that what counts in multicomponent distillation are the relative volatilities and not the boiling points of the individual components. Due to the presence of the polar "extraction water", the high-boiling diketones of low polarity become more volatile than

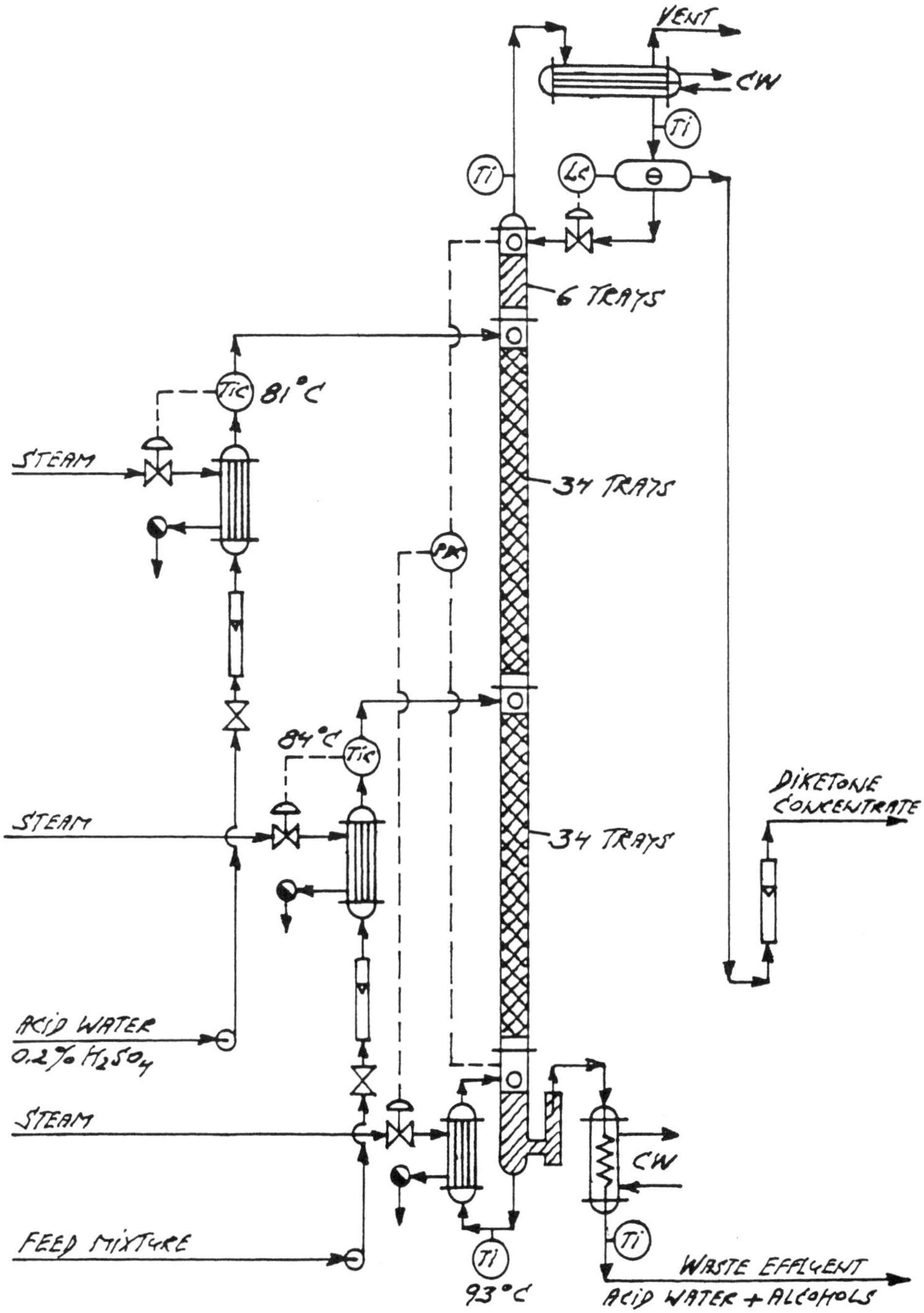

Figure 65. Schematic of an Extractive Distillation Plant to Produce a Diketone Concentrate from the "Raw Solvent".

the low-boiling alcohols of high polarity.

16.6.2. Cryogenic Crystallization

The head fraction of the extractive distillation described above is called "extractive distillate". Historically, the easiest way of getting diacetyl out of this mixture was found to be cystallization as diacetyl has an unusually high freezing point of –2.4 °C. On the other hand, 2,3-pentanedione cannot be conveniently recovered in this fashion as its freezing point is –52 °C. Thus, when crystallization is adopted as the mode of recovery, it is accepted that 2,3-pentanedione is lost completely.

In the recovery by freezing, the "extractive distillate" is submitted to a fourfold cryogenic crystallization using liquid nitrogen as the coolant. The first two crystallization steps are carried out at –30 °C and –15 °C, respectively, the results being as compiled in Figure 66. Subsequently, the sum of residual liquors is reprocessed by another twofold crystallization at –37 °C and –19 °C, respectively. Figure 67 summarizes the corresponding data in the form of a pseudobinary phase diagram, with diacetyl as one component, and all other compounds lumped together as the second component.

The separation of the crystals from the residual mother liquors is carried out by a centrifugal decanter specially designed to be capable of operating at the given low temperatures. This phase separation process is facilitated by a great difference in density between solid diacetyl and the residual mother liquors.

The overall yield of diacetyl in the given four crystallization steps is poor, in the order of 54 percent, due to the inherent equilibrium between liquid and solid phases, leaving large portions of the diacetyl unrecovered in the remaining mother liquors. Besides, there is a high consumption of expensive liquid nitrogen, costing roughly 13 percent of the sales price of the diacetyl produced. In addition, the impossibility of recovering the 2,3-pentanedione by this process is a painful loss.

The crystallization of diacetyl exhibits interesting peculiarities. It was noted that a fresh extractive distillate crystallized much more easily than a distillate "aged" for several days. This can be explained by the fact discussed in chapter 16.6.1 that any distillate of diacetyl produced in acid conditions is rich in the enol form, and that it takes days for the keto/enol equilibrium to be reestablished.

There can be no doubt that the pure enol of diacetyl (so far never prepared) crys-

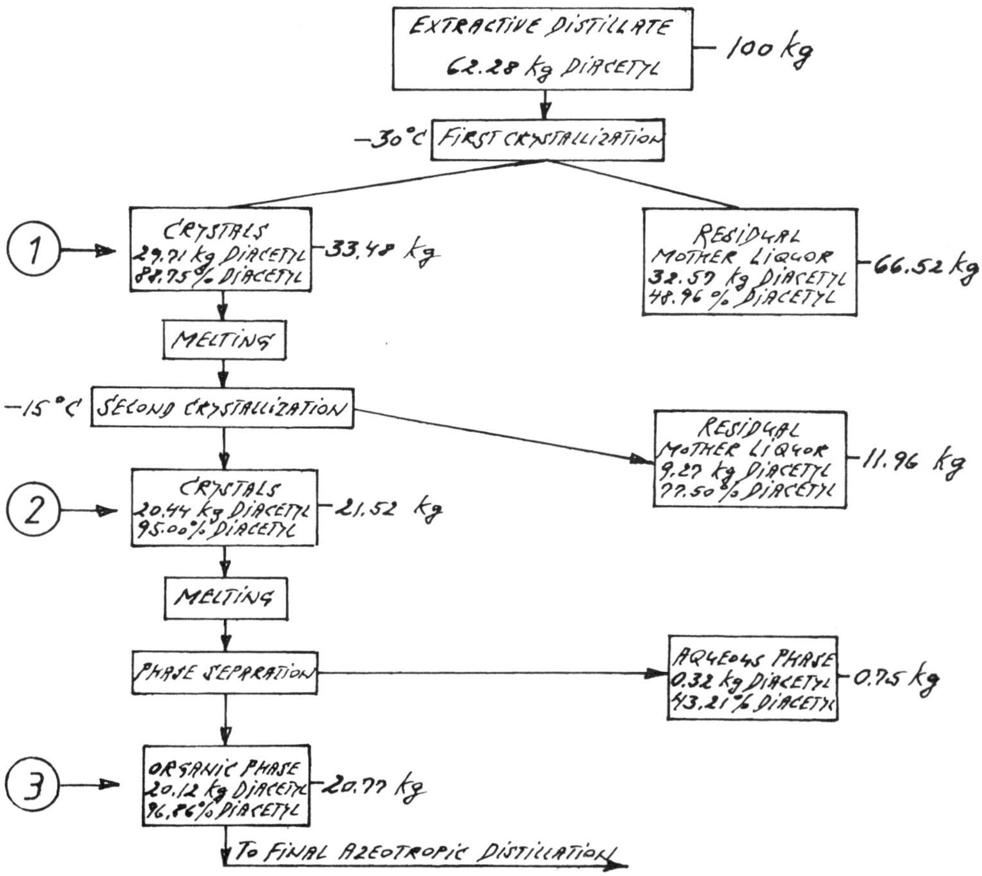

Figure 66. Quantitative Schematic of the first 2 Stages of the Diacetyl Crystallization Process.

Yield 1 = 29.71 kg / 62.28 kg = 0.47704

Yield 2 = 20.44 kg /62.28 kg = 0.32820

Yield 3 = 20.12 kg / 62.28 kg = 0.32306

Thus, the actual yield of two crystallizations is 32.306 percent.

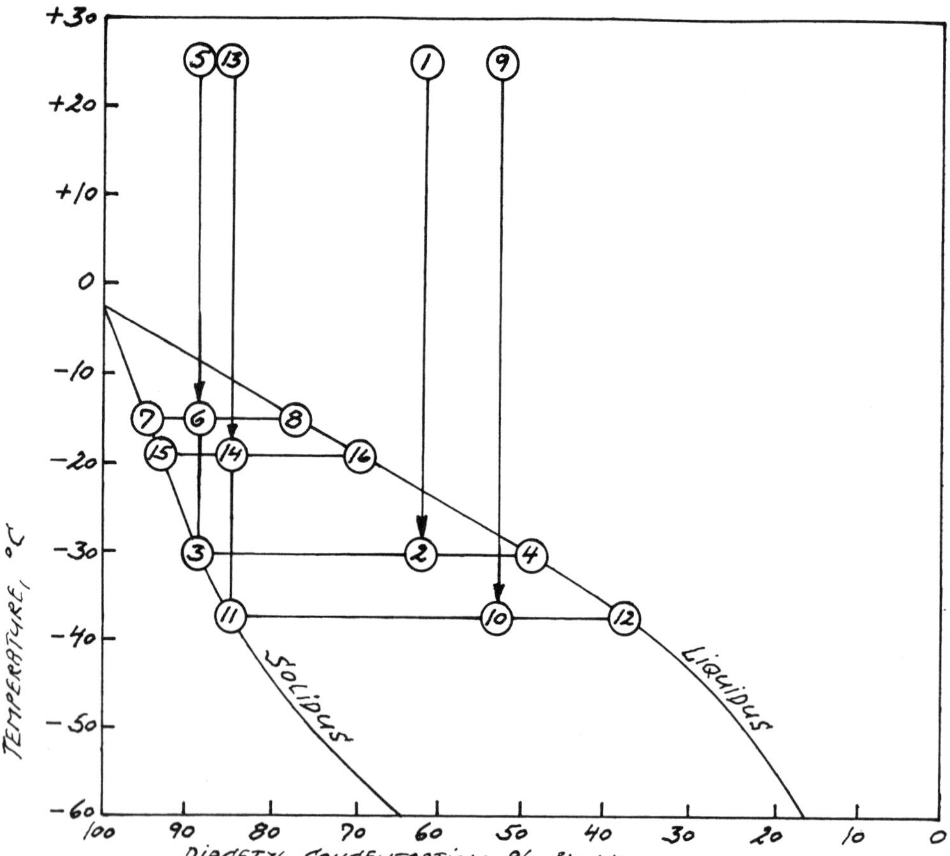

Figure 67. Fourfold Cryogenic Crystallization of Diacetyl as Represented in a Pseudobinary Diagram.

1 – Extractive distillate, 25 °C, 62.28 % diacetyl.

2 – Extractive distillate cooled to –30 °C, leading to crystals (3) and residual liquor (4).

5 – Crystals (3) after melting at room temperature.

6 – Molten crystals (5) cooled to –15 °C, leading to crystals (7) and residual liquor (8).

9 – Sum of residual liquors at room temperature.

10 – Sum of residual liquors cooled to –37 °C, leading to crystals (11) and residual liquor (12).

13 – Crystals (11) after melting at room temperature.

14 – Molten crystals (13) cooled to –19 °C, leading to crystals (15) and residual liquor (16).

tallizes at a higher temperature than the normally predominant keto form since the intramolecular hydrogen bond of the enol fixes the molecule in a planar configuration. whereas the molecules of the keto form permit rotation around the central C-C bond:

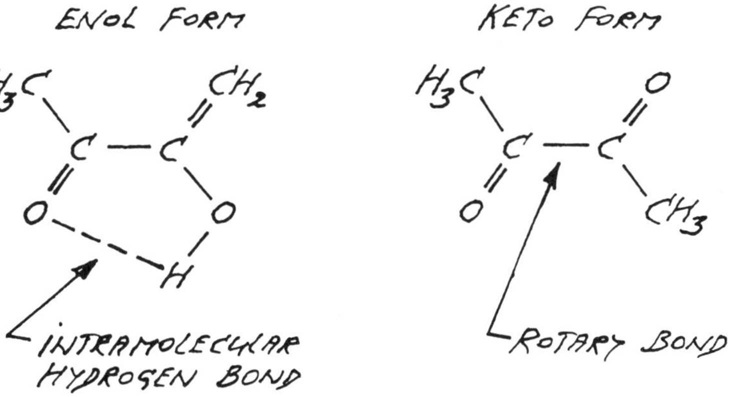

Molecules with a fixed configuration have a higher freezing point than their analogues permitting rotation: Cyclohexane freezes at +6.5 °C while hexane, having five rotary bonds, freezes at –94.3 °C. Against this background, the freezing of diacetyl must be seen as a crystallization in a binary system of enol form and keto form molecules as shown schematically in Figure 68, where it is assumed that the system has a eutectic.

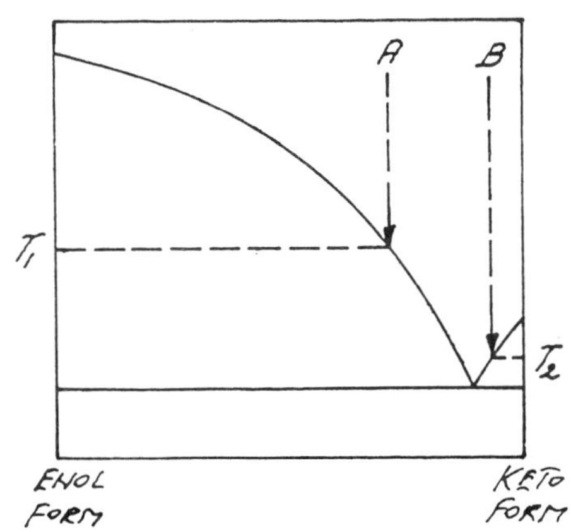

Figure 68. Hypothetical Phase Diagram for the two Tautomers of Diacetyl.

The equilibrium enol content of diacetyl is 1 % [68], and the freezing point of the equilibrium mixture is –2.4 °C. Strictly speaking, normal diacetyl (i.e. 99 % keto form and 1 % enol form) does not have a sharp freezing point unless the enol content happens to coincide with the eutectic composition, which is unlikely. Hence, two cases must be considered:

(1) Crystallization from a liquid A having an enol content greater than the eutectic composition.

(2) Crystallization from a liquid B having an enol content smaller than the eutectic composition.

In case (1), freezing starts at a high temperature T_1, and the crystals will be enol crystals, whereas in case (2) the freezing starts at a low temperature T_2, and the crystals will be keto crystals. If it is assumed that liquid B represents the equilibrium enol concentration of 1 percent, then it would follow that after a few days the liquid A will have been converted to liquid B, so that such a scenario would explain the observed difference in "ease of crystallization" between "fresh" distillate of high enol concentration and "aged" distillate of low enol concentration.

Unfortunately, these interesting phenomena have not been investigated as yet. If the given scenario were correct, it would be possible to prepare enol crystals from a diacetyl antecedently enriched with enol.

16.6.3. Polyazeotropic Distillation

In view of the grave economic disadvantages of cryogenic crystallization (only 54 percent recovery of diacetyl, no recovery of the extremely valuable 2,3-pentanedione, huge cost for liquid nitrogen, and high labor cost due to batch operation), great efforts were made to come up with a superior recovery scheme. In the end, the author invented a polyazeotropic distillation process which satisfied all requirements by recovering essentially all of the diacetyl, by recovering most of the 2,3-pentanedione, by not needing any liquid nitrogen, and by being fully continuous.

Azeotropes between two partner compounds are favored by a decreasing difference of the boiling points and by an increasing difference in polarity, with the latter criterion capable of compensating an unfavorable situation of the former criterion. This can be seen by plotting the known azeotropes of a compound A with different partner compounds B_i in a diagram having the absolute difference in spectroscopic polarity index E_T^N as the abscissa, and the absolute difference in boiling point as the ordinate. In such a diagram, illustrated in Figure 69, the known azeotropes all fall into a bottom field F_b while no azeotropes occur in the top field F_t. A line separating the two fields shows that azeotropes can be produced at high differences of the boiling point provided the difference in polarity is great, or at low

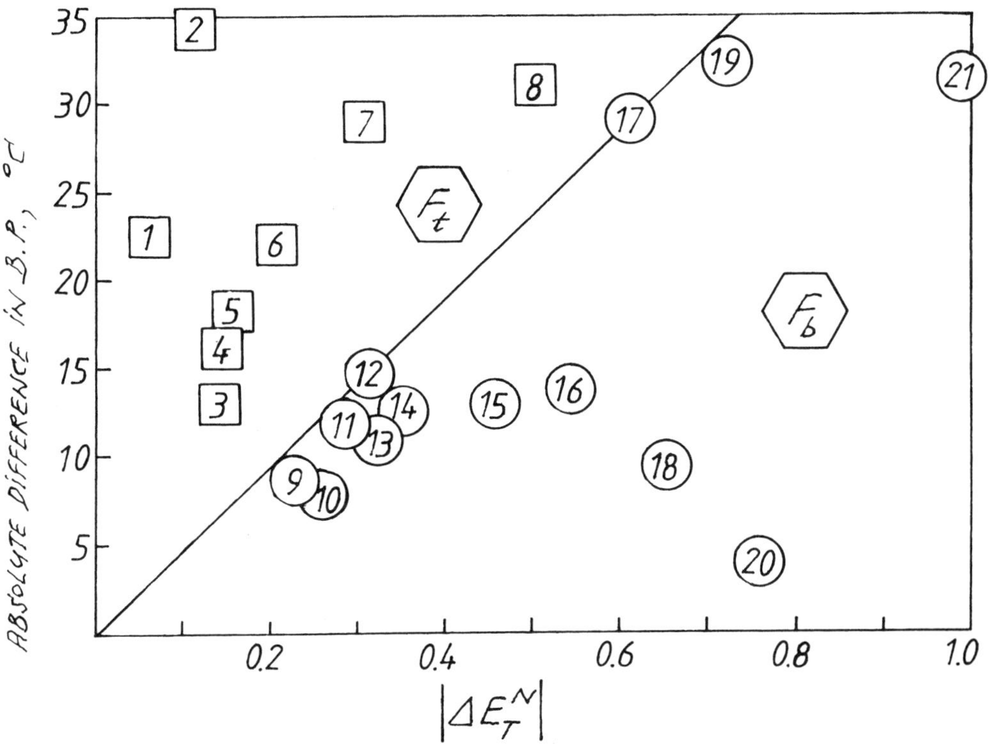

Figure 69. The Azeotrope/Nonazeotrope Diagram for Hexane.

Nonazeotropes: (1) Carbon disulfide. (2) Diethyl ether. (3) Diethylamine. (4)Thiophene. (5) Trichloroethylene. (6) Chloropropane. (7) Dichloromethane. (8) 2-Butyl alcohol.

Azeotropes: (9) Ethyl acetate. (10) Chloroform. (11) Methyl acetate. (12) Ethyl formate. (13) 2-Butanone. (14) Acetone. (15) Acetonitrile. (16) Isopropyl alcohol. (17) Propyl alcohol. (18) Ethanol. (19) Formic acid. (20) Methanol. (21) Water.

differences in polarity provided the difference in boiling point is small. Against this background, from Figure 69, referring to hexane as compound A, it was concluded that diacetyl and 2,3-pentanedione do not form azeotropes with hexane.

The distillate of the extractive distillation described in chapter 16.6.1 had been found to contain 62.3 % diacetyl, 9.0 % 2,3-pentanedione, and minor quantities of water, methanol, ethanol, acetone, and acetaldehyde dimethyl acetal. Contrary to the diketones, the latter more or less polar substances all form low-boiling azeotropes with the extremely nonpolar hexane. The respective data are compiled in Table 3.

Table 3. Atmospheric Binary Azeotropes with Hexane.

Compound	b.p. of partner ^oC	b.p. of azeotrope ^oC	hexane in azeotrope % by weight
Water	100.0	61.5	94.6
Ethanol	78.5	58.7	79.0
Methanol	64.6	49.9	73.6
Acetaldehyde dimethyl acetal	64.5	64.0	30.0
Acetone	56.5	49.7	46.5

It was theorized that what holds for binary systems can be expected to hold in a multicomponent system interpreted as a "superposition of binary systems". In other words, it was predicted that by redistilling the extractive distillate of chapter 16.6.1 with hexane as an "azeotrope former", water, methanol, ethanol, acetone, and acetaldehyde dimethyl acetal would end up in the top as their low-boiling azeotropes with hexane, while diacetyl and 2,3-pentanedione, not forming azeotropes with hexane, would end up in the bottom.

The quantity of hexane necessary to entrain water, methanol, ethanol, acetone, and acetaldehyde dimethyl acetal to the top was estimated as the sum of the hexane quantities required to form the binary azeotropes with the quantities of water, methanol, ethanol, acetone, and acetaldehyde dimethyl acetal in the mixture.

According to this concept, all compounds accompanying diacetyl and 2,3-pentanedione in the extractive distillate of chapter 16.6.1 were expected to arrive in the top condenser of the proposed "hexane column", and on account of the hexane entrainer also arriving there, two liquid phases were expected to form: A nonpolar phase rich in hexane, and a polar phase rich in water, methanol, and ethanol. Head vapor compounds of intermediate polarity, such as acetone, acetaldehyde dimethyl acetal, and diacetyl, were expected to be found primarily in the nonpolar phase.

Where known, the position of the various compounds on a ladder of spectroscopic polarity indices E_T^N is illustrated in Figure 70,

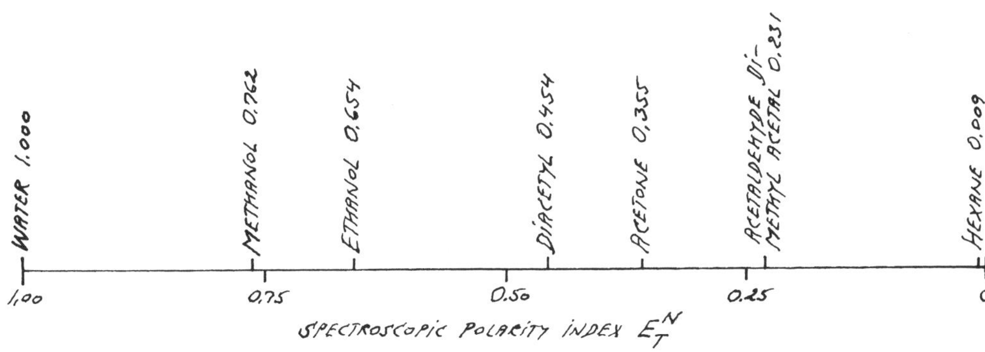

Figure 70. Ladder of Spectroscopic Polarity Indices E_T^N accorduing to C. Reichardt [70].

with the understanding that in a static decanter the polar compounds from water to ethanol would make up the heavy phase, and the other compounds the light phase.

The laboratory setup used to study this concept is shown in Figure 71. The distillation column, 30 mm I.D., consisted of a tray section in the bottom and four packed sections, each of the latter loaded with nine cartridges of type SULZER EX. The column was energized by an oil-heated falling film evaporator featuring a hermetic circulation pump. The distillate of the extractive distillation column of chapter 16.6.1 was introduced below the two uppermost SULZER columns, while hexane was introduced one SULZER section below the feed inlet. Between the bottom tray section and the lowermost SULZER section, the diketones were withdrawn as a vapor.

The head vapor was liquefied in a condenser cooled by a chilled glycol solution. The condensate, forming two liquid phases as expected, was collected in a static decanter separating a light phase rich in hexane and a heavy phase rich in water and alcohols. A vacuum pump connected to the condenser maintained the system under a reduced pressure in the order of 200 torr to lower all temperatures to the end of reducing polymerization in the falling film evaporator.

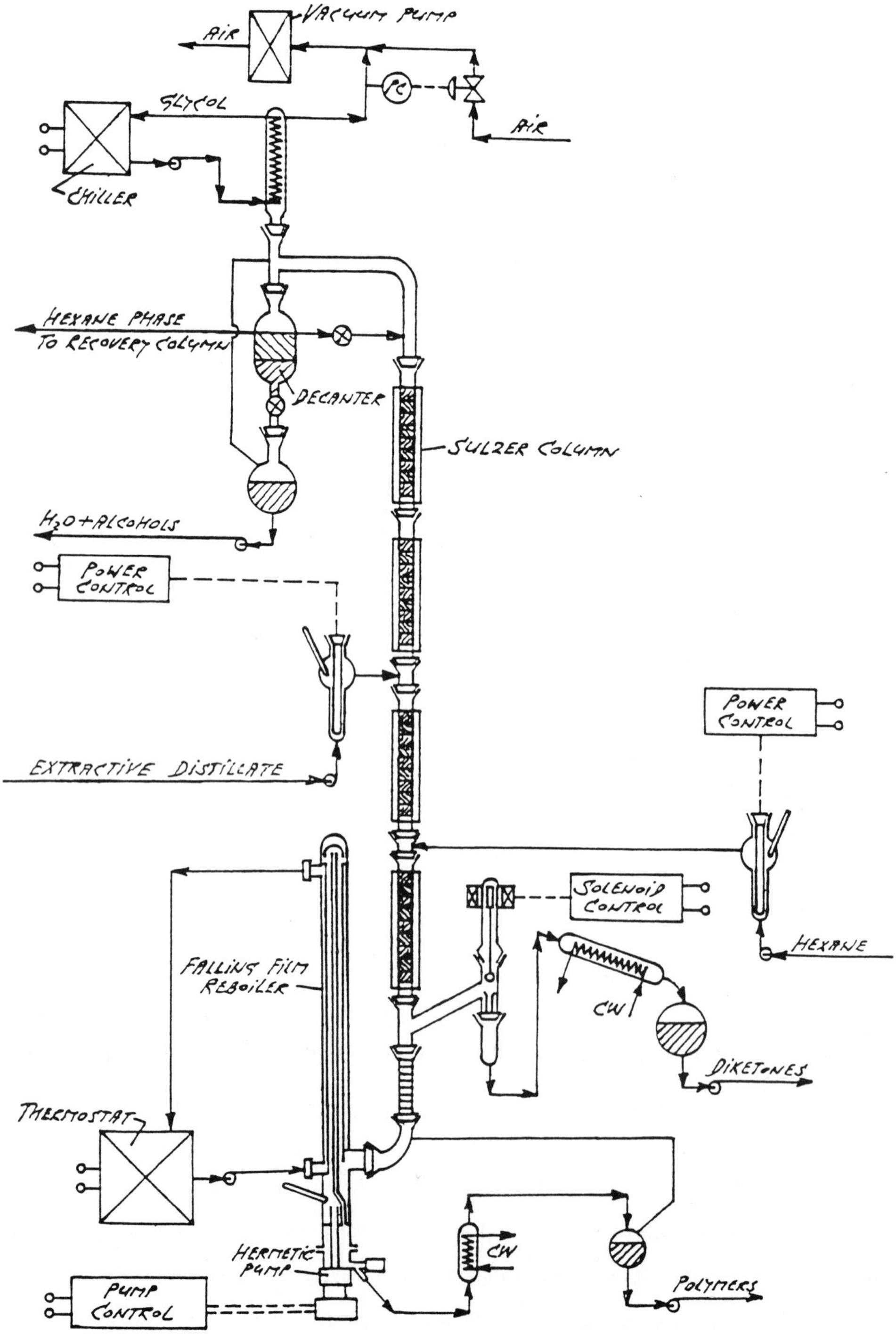

Figure 71. Experimental Setup for Testing the Polyazeotropic Concept.

The concentration profiles in this setup are illustrated schematically in Figure 72. As can be seen, there is a small diketone concentration at the top, and a diketone maximum at the withdrawal point.

The measurements carried out with this apparatus confirmed the expectations. From the head vapor condenser, water, methanol, and ethanol formed a heavy polar phase to be discarded, while essentially all of the acetone and the acetaldehyde dimethyl acetal as well as some diacetyl ended up in the light hexane phase. From the latter, in an auxiliary distillation column, the hexane and the diacetyl were readily recovered as entrainer to replace the pure hexane, whereas acetone and acetaldehyde dimethyl acetal were separated as an impurity fraction to be discarded.

Based on the experimental results obtained with the laboratory setup, a full-scale plant was designed as shown in Figure 73. In addition to the polyazeotropic column on the left-hand side, it is seen to feature a hexane recovery column in the middle, and a finishing column on the right-hand side. In the latter, the raw diketones produced in the polyazeotropic column are separated into diacetyl and 2,3-pentanedione, both of these products being withdrawn as side streams. Their separation is straightforward since they exhibit ideal behavior as shown in Figure 74.

The overall diacetyl yield of this plant is 93.3 % as compared to 54 % for the cryogenic crystallization, and in recovering the 2,3-pentanedione, completely lost in the cryogenic process, it offers an extremely valuable fringe benefit.

16.6.4. Final Distillation

For both the cryogenic crystallization or the polyazeotropic distillation, the diacetyl obtained by these routes may be good enough for most applications, with typical purities in excess of 96.8 % by weight (86.358 % by mole), but to satisfy market demands it is customary to "polish" the product by a final batch distillation. This step can readily remove small quantities of water and methanol as both of these possible impurities form low-boiling azeotropes with diacetyl.

As an example, the vapor/liquid equilibrium diagram of the diacetyl/water system is shown in Figure 75. At the high diacetyl concentrations obtained by both the cryogenic crystallization or the polyazeotropic distillation, the water is more volatile than the diacetyl, and there is a low-boiling azeotrope at 63.57 % diacetyl by mole corresponding to 89.29 % by

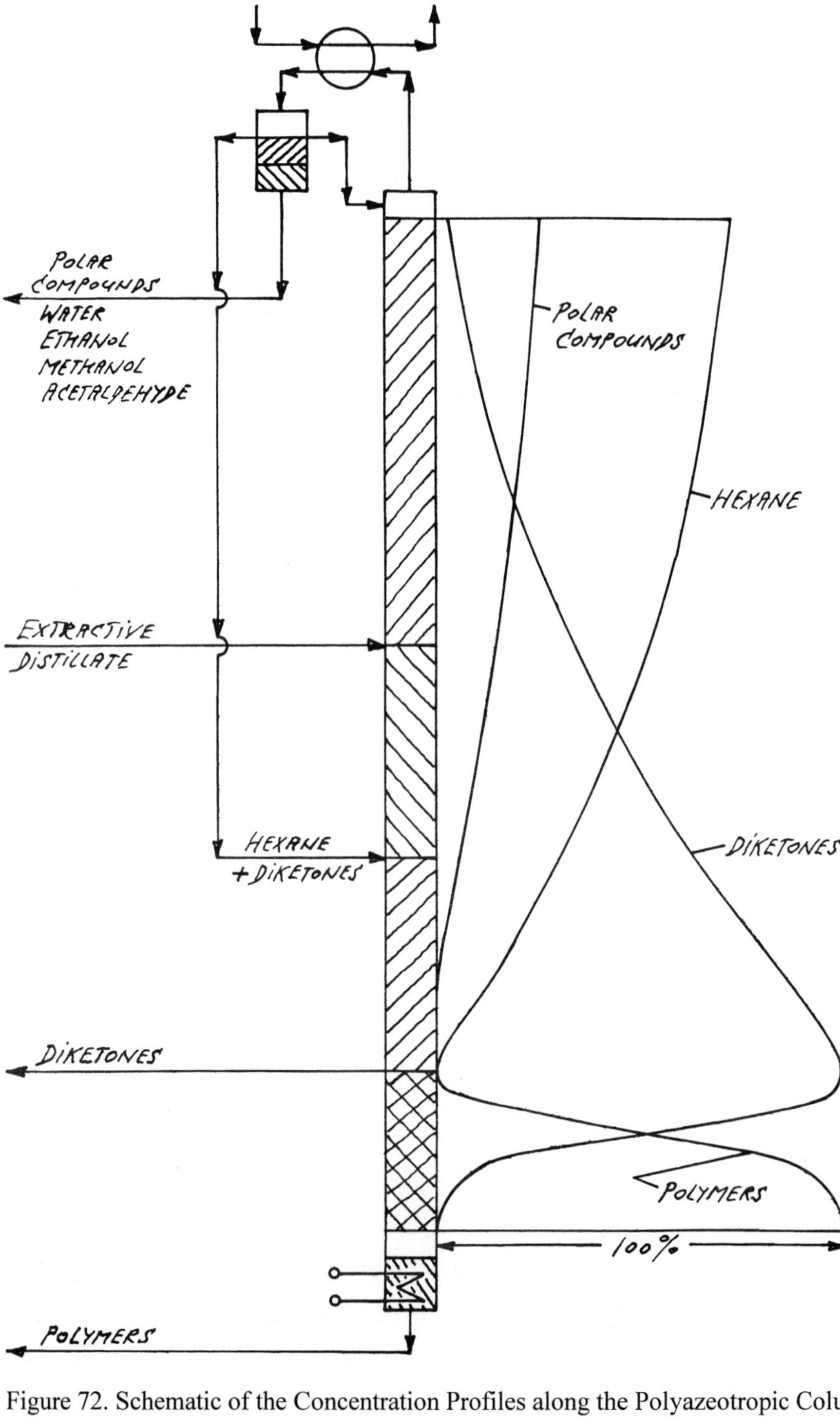

Figure 72. Schematic of the Concentration Profiles along the Polyazeotropic Column.

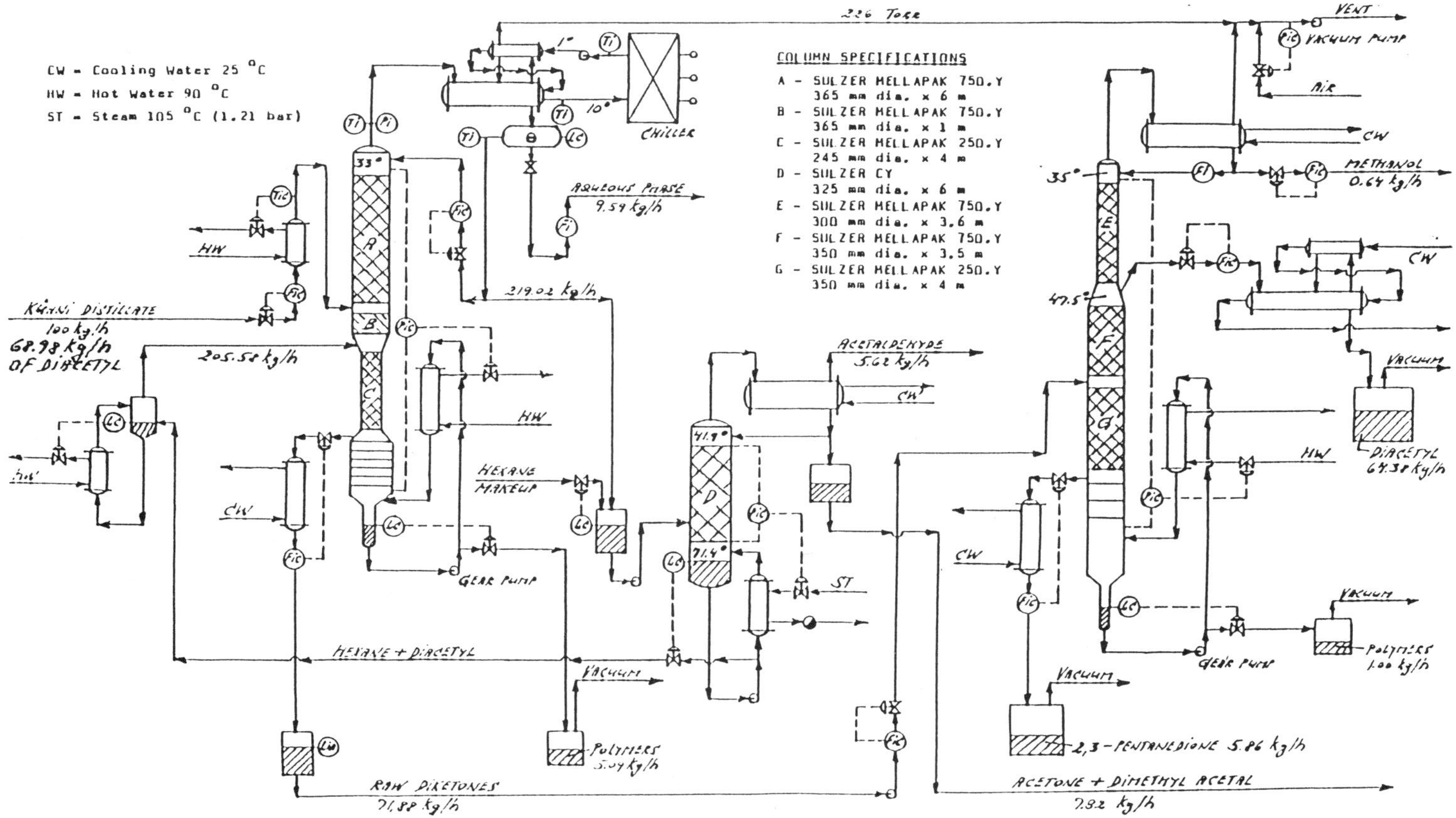

Figure 73. Full-Scale Plant for the Recovery of Diacetyl and 2,3-Pentanedione by Polyazeotropic Distillation.

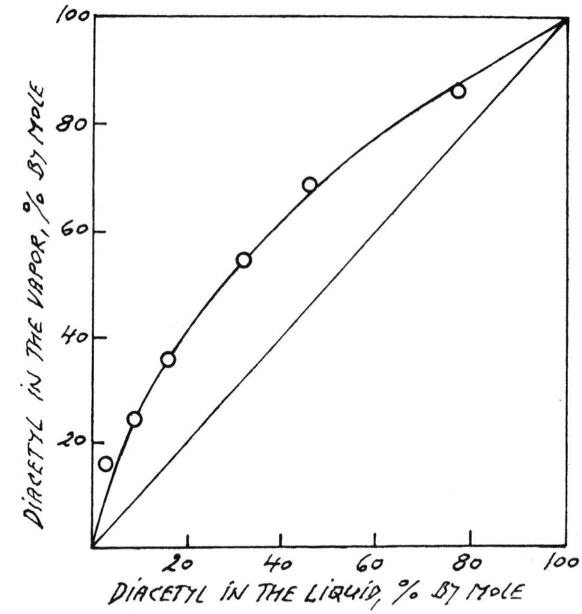

Figure 74. The Vapor/Liquid Equilibrium of the Diacetyl/2,3-Pentanedione System at 60 °C.

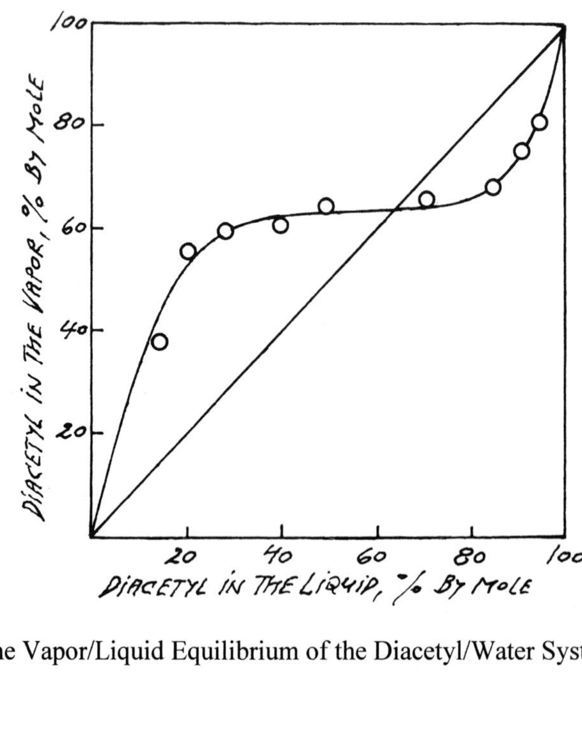

Figure 75. The Vapor/Liquid Equilibrium of the Diacetyl/Water System at 1 ATM.

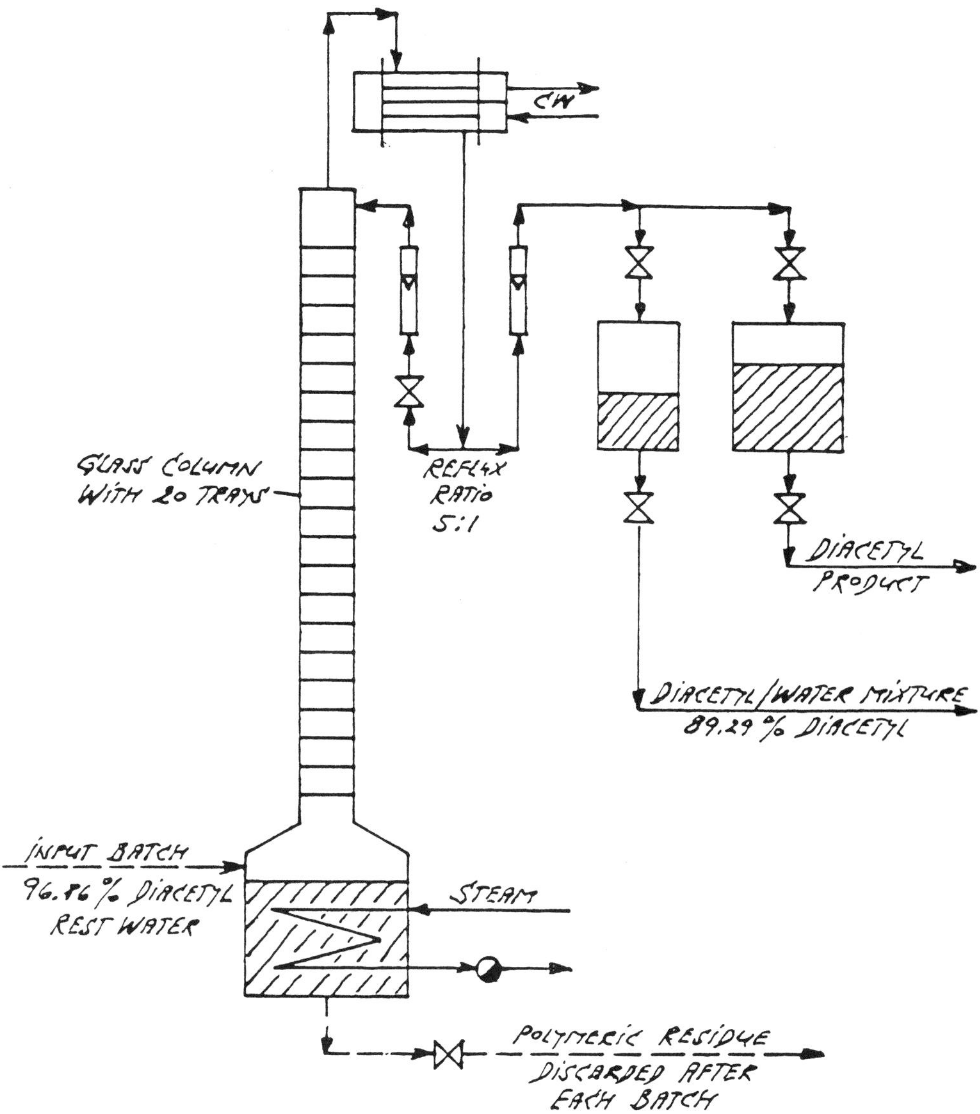

Figure 76. Final Batch Process for the Purification of Diacetyl by Azeotropic Distillation. Shown for the removal of water.

weight. Thus, submitting the products of cryogenic crystallization or polyazeotropic distillation to a final batch distillation leads at first to a distillate of azeotropic composition, and only after all of the water is removed by this head fraction, pure diacetyl is obtained. The respective process is illustrated in Figure 76. Of course, the azeotropic head fraction is not discarded but recycled into the front end of the recovery process. At the end of each batch, a small quantity of polymeric residue amounting to less than 3 percent of the diacetyl input remains in the still. This residue is discarded. Preferentially, but not necessarily, this final distillation is carried out under reduced pressure to lower the temperatures for the purpose of reducing the polymer formation.

Noteworthily, the diacetyl/water azeotrope at 89.29 % diacetyl by weight implies that the preceding process (cryogenic crystallization or polyazeotropic distillation) must provide a diacetyl concentration exceeding this value as otherwise concentrating the diacetyl to 100 % purity by distillation would be impossible.

What has been said for the elimination of impurities of water applies analogously for the elimination of impurities of methanol. As both form low-boiling azeotropes with diacetyl, in actual fact water and methanol are eliminated together in the same final batch distillation.

References

[59] C. A. McDowell and J. H. Thomas, J. Chem. Soc. 1949, 2208-2216.

[60] D. R. Larkin, J. Org. Chem. 55 (1990) 1563-1568.

[61] E. Hägglund, Chemistry of Wood, Academic Press Inc., New York, 1951.

[62] G. Franz, Volume A 18 of Ullmann's Encyclopedia of Industrial Chemistry, 5th Edition, VCH Verlagsgesellschaft, Weinheim, 1991, 261-311.

[63] C. C. Hobbs, Volume 13 of Kirk-Othmer' Encyclopedia of Chemical Technology, 4th Edition, John Wiley & Sons, New York, 1998, 682-717.

[64] K. Nabert and G. Schön, Sicherheitstechnische Kennzahlen brennbarer Gase und Dämpfe, 2nd Edition, Deutscher Eichverlag, Braunschweig, 1978.

[65] G. Franz, HÜLS AG, Private Communication, 1998.

[66] F. E. Blacet and W. E. Bell, Discuss. Faraday Soc. 14 (1953) 70-76.

[67] D. L. Hooper, J. Chem Soc. B 1967, 169-170.

[68] A. Gero, J. Org. Chem. 19 (1954) 1960-1970.

[69] N. Sleszynski and P. Zuman, J. Org. Chem. 52 (1987) 2622-2623.
[70] C. Reichardt, Solvents and Solvent Effects in Organic Chemistry, VCH Verlagsgesellschaft, Weinheim, 1988.

17. Furfuryl Alcohol

Furfuryl alcohol is the most important derivative of furfural. At present, approximately 65 percent of all furfural produced is converted to furfuryl alcohol as there is a good demand for this product in the manufacture of foundry resins.

The manufacture of furfuryl alcohol is a simple hydrogenation, with copper chromite used as catalyst:

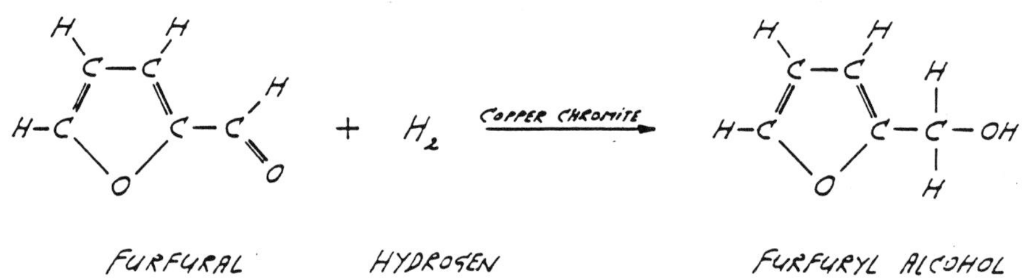

17.1. The Vapor Phase Process

The process employed almost universally is shown schematically in Figure 77. Furfural is fed into an evaporator system comprising a packed column 1, a circulation pump 2, and a heater 3 energized by steam to maintain the furfural temperature at 120 °C. Below the packing of column 1, a dosed quantity of hydrogen is introduced. In the countercurrent system of hydrogen flowing upwards and liquid furfural flowing downwards, the hydrogen gets saturated with the vapor pressure of furfural at 120 °C. The resulting mixture of hydrogen and furfural vapor passes a demister pad 4 and a superheater 5 before it enters a tubular catalytic reactor 6 maintained at a temperature in the order of 135 °C by means of hot oil. The tubes are filled with copper chromite pellets catalyzing the desired reaction of furfural with hydrogen to form furfuryl alcohol. This reaction is slightly exothermic, releasing 14.5 kcal/mole (60.668 kJ/mole), so that the oil flowing around the reactor tubes must withdraw heat from the system. The gaseous mixture of reaction products enters a condensation system comprising a packed column 7, a pump 8, and a cooler 9. The pump circulates raw furfuryl alcohol through the cooler 9 and onto the packing of column 7 where it meets a countercurrent of the gaseous products. From the latter stream, most of the condensables are liquefied. The remaining

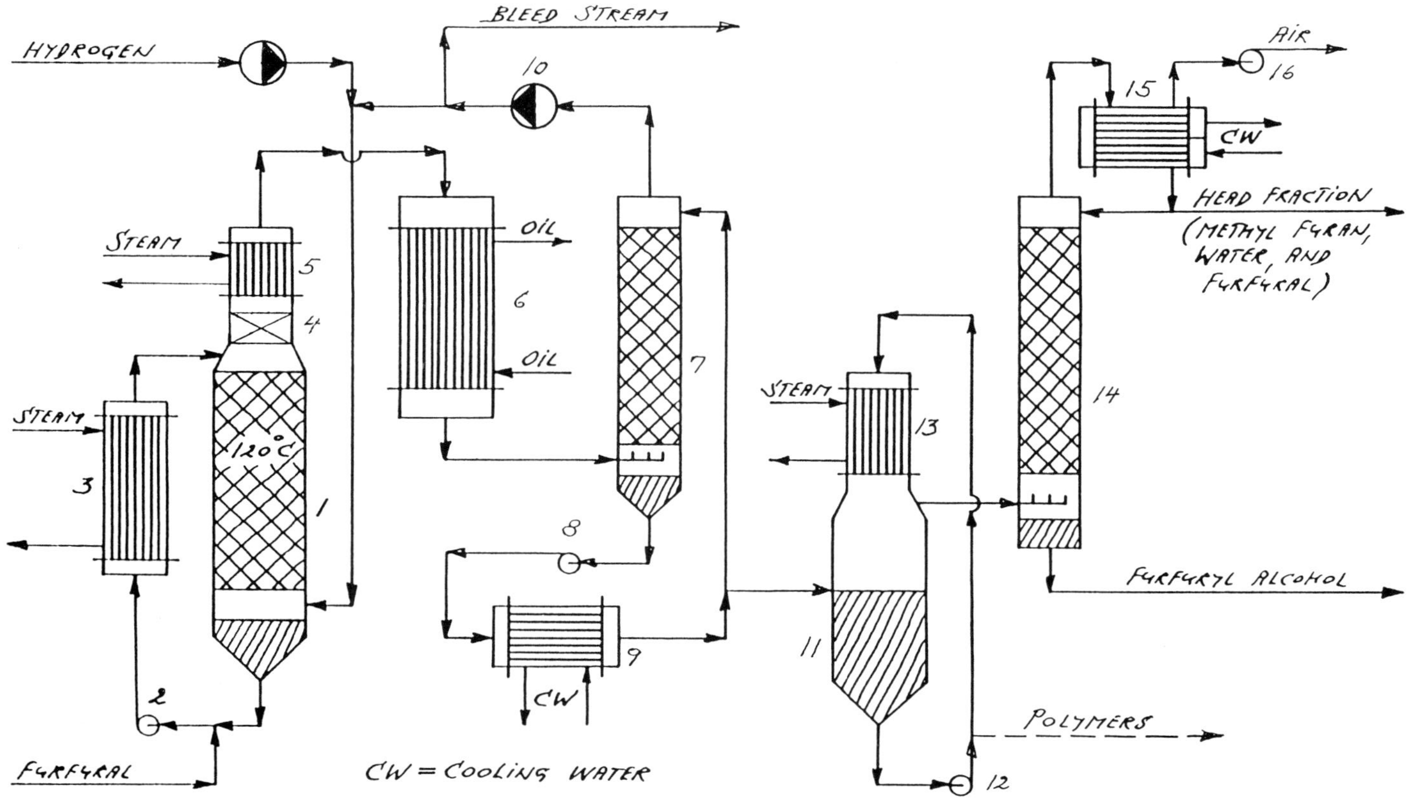

Figure 77. Schematic of the Vapor Phase Hydrogenation of Furfural to Furfuryl Alcohol.

portion, consisting of unreacted hydrogen and the saturation quantities of the condensables at the column temperature, is recompressed by a ROOTS pump 10 and added to the hydrogen feed to prevent losses. A small bleed stream prevents a buildup of impurities. The condensed portion is fed into a reboiler system comprising a tank 11, a circulation pump 12, and a heater 13 energized by steam. The vapor produced by this system enters a packed vacuum distillation column 14. The head vapor of this column is liquefied by a condenser 15 maintained at a reduced pressure by a vacuum pump 16. Most of the condensate is returned to the column as reflux, while the rest represents a small head fraction consisting of 2-methyl furan, unreacted furfural, and reaction water from the 2-methyl furan formation and polymerization effects. The sump fraction is the purified furfuryl alcohol. High-boiling polymers remaining in the reboiler are withdrawn intermittently.

In actual practice, the temperature of reactor 6 is gradually increased from 122 °C to 152 °C to compensate a progressing decrease in catalyst activity due to carbonaceous deposits. A typical rate of temperature increase is 3 °C/day. After ten days, when 152 °C are reached, the feed is turned off, and the reactor is heated to 220 °C to remove the deposits by oxidation. Then the catalyst is reactivated by hydrogenation at 160 °C before a new production cycle is started at 122 °C.

The yield of the process is in excess of 92 percent. The principal by-product is 2-methyl furan. Its formation increases when the reactor temperature is raised to compensate decreasing catalyst activity. At high temperatures, commercial quantities of 2-methyl furan can be produced.

17.2. The Liquid Phase Process

An older, less elegant process for making furfuryl alcohol is shown in Figure 78. In this process, the catalyst is used as a slurry, and the hydrogenation is carried out at a pressure of 200 ATM and at a temperature of 120 °C. Furfural and a copper chromite catalyst are mixed in tank 1 by means of a circulation pump 2. Pump 3 feeds the slurry continuously through a preheater 4 into a tubular bubble reactor 5. Hydrogen, usually from a water electrolysis plant, is injected by compressor 6. The mixture leaving the reactor flows through a cooler 7 into a cyclone 8 where excess hydrogen is separated from the slurry and reinjected into the reactor feed stream by means of compressor 9. The slurry is depressurized in tank 10, a relatively small quantity of hydrogen thereby released being vented into the ambient air.

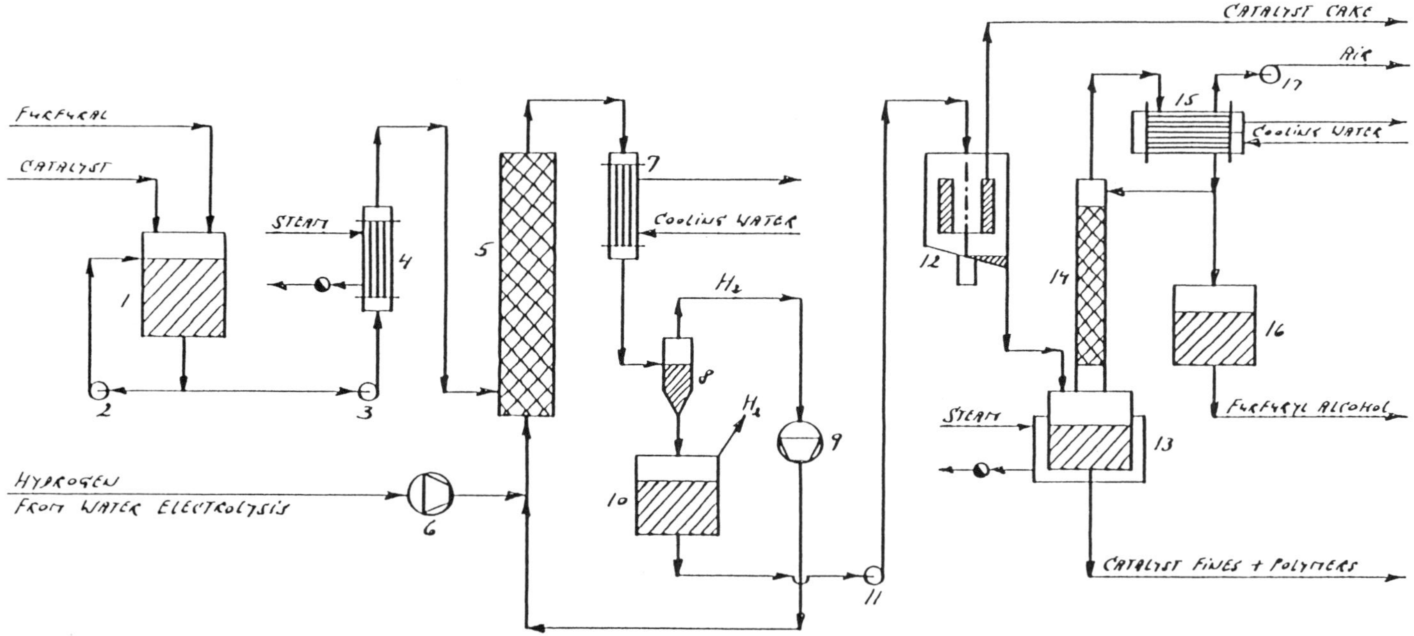

Figure 78. Schematic of the Liquid Phase Hydrogenation of Furfural to Furfuryl Alcohol.

Pump 11 takes the depressurized slurry into an overflow sedimentation centrifuge 12 where most of the catalyst particles are separated from the liquid phase. Removal of the solids from the bowl is effected manually at appropriate intervals. The liquid phase flows into a still 13 topped by a rectification column 14. The head vapors of the column are liquefied in condenser 15, the resulting distillate being partly returned to the column to effect rectification and partly collected in tank 16. This distillate is pure furfuryl alcohol. Vacuum pump 17 maintains a reduced pressure to permit distillation at moderate temperatures. Catalyst fines and high-boiling polymers inevitably formed in the reactor remain in the still and are discarded.

17.3. Comparison of Different Catalysts

All catalysts used for the hydrogenation of furfural are made of copper chromite, but there are significant differences between various grades. For the customary vapor phase process, this is illustrated in Figure 79, where the temperature required to maintain 99 % furfural conversion is plotted versus the time of operation [71].

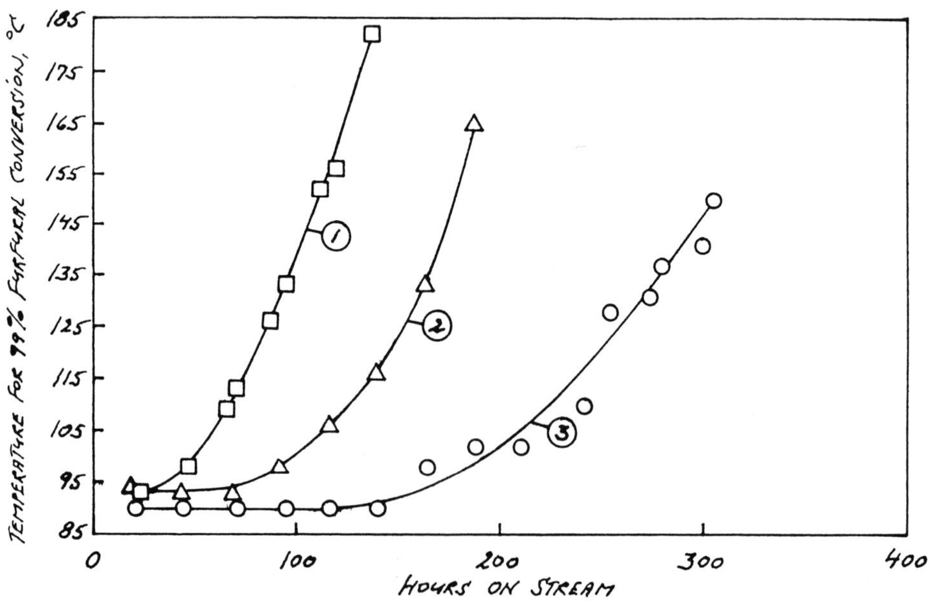

Figure 79. Comparison of Three Different Catalysts.

1 – HARSHAW Cu-1132

2 – CALSICAT X-407 TU

3 – Modified CALSICAT X-407 TU

All three catalysts compared are seen to gradually lose their activity so that maintaining a constant conversion of 99 % requires increasing the temperature, but catalyst 3 loses its activity much more slowly than catalyst 1, so that with catalyst 3 the 99 % conversion can be obtained at relatively low temperatures for a relatively long period of time. This is advantageous as with all catalysts the formation of 2-methyl furan, an unwanted by-product, increases with increasing temperature as shown in Figure 80. Thus, catalyst 3 leads to a product of greater purity which in the end is synonymous with a higher yield.

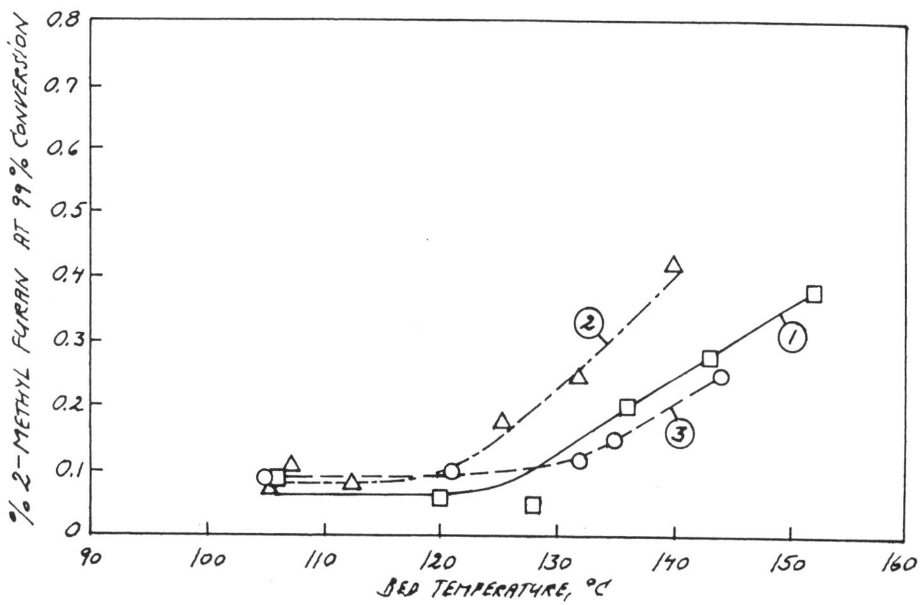

Figure 80. Formation of 2-Methyl Furan as a Function of Temperature.

1 – HARSHAW Cu-1132

2 – CALSICAT X-407 TU

3 – Modified CALSICAT X-407 TU

Reference

[71] D. G. Rodgers, Comparison of CALSICAT Modified X-407 TU, X-407 TU, and HARSHAW Cu-1132 for Vapor Phase Hydrogenation of Furfural, Publication of the MALLINCKRODT SPECIALTY CHEMICALS COMPANY, 1990.

18. Furan

A catalytic decarbonylation of furfural leads to furan:

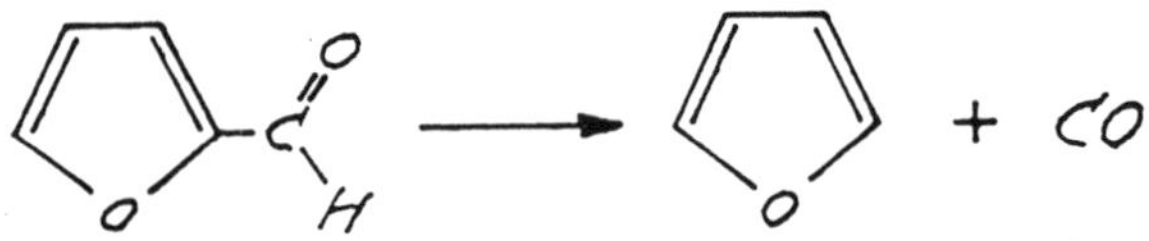

The most efficient way of doing this is heating furfural at atmospheric pressure to 158 °C (3.7 °C below its boiling point) while stirring it in the presence of a catalyst consisting of 5 % palladium on microporous carbon and a promoter consisting of potassium carbonate [72] [73]. The stirring is important to assist the desorption of furan and carbon monoxide from the catalyst surface. It is recommended to employ 170 liters of furfural per kilogram of catalyst, and a promoter to catalyst ratio of 1 : 1 by weight. Up to 36000 g of furan can be obtained per gram of palladium before the catalyst must be regenerated after 240 hours of operation.

The process is illustrated in Figure 81. Furfural freshly distilled under vacuum to remove polymers is fed into a stirred reactor 1 heated to 158 °C. Then the catalyst and the promoter are added to meet the conditions specified above. The gaseous phase emanating from the reactor consists of carbon monoxide, furan vapor, furfural vapor, and some water vapor, the latter stemming from the usual small water content of industrial furfural (in the order of 0.2 % by weight). This gaseous phase enters a distillation column 2 from where a small side stream of furfural/water azeotrope is withdrawn and liquefied in condenser 3, the resulting condensate being returned to the furfural plant. From the head stream of column 2, furan is liquefied in condenser 4 cooled to –30 °C by a chiller 5, whereas carbon monoxide remains of course gaseous. Failure to remove the water as the furfural/water azeotrope would lead to troublesome ice formation in condenser 4. As the raw furan, because of its low temperature, contains a sizable quantity of dissolved carbon monoxide, it is fed into a degassing column 6 energized by hot water to produce furan free of carbon monoxide in the sump. Pump 7 takes this furan through a heat exchanger 8 cooled by a chiller 9 for storage in tank 10. To recover entrained furan vapor, the combined carbon monoxide streams are introduced into an absorption tower 11 through which furfural is recirculated by pump 12. A small portion of the absorber stream consisting of furan dissolved in furfural is fed into the reactor 1.

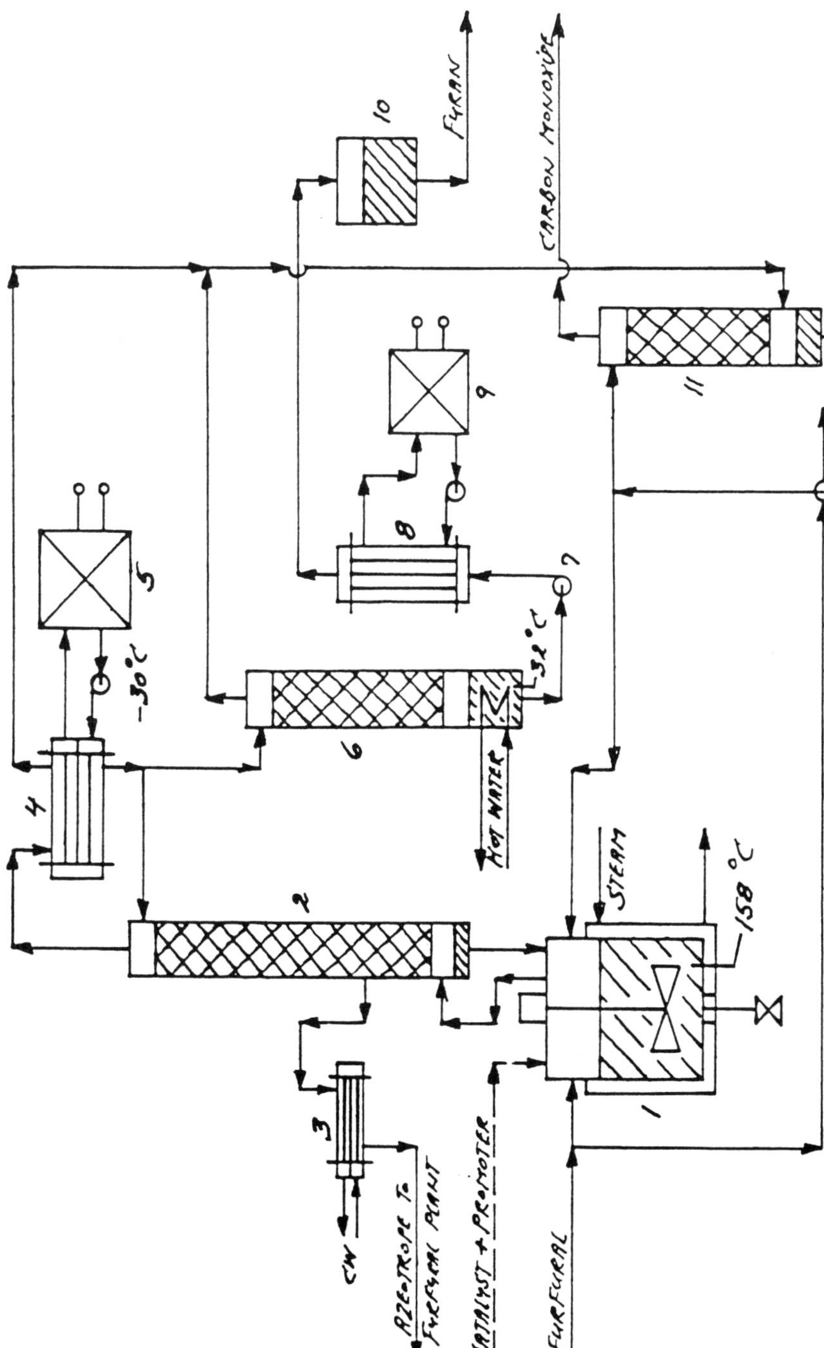

Figure 81. Plant for the Production of Furan from Furfural.

The yield of this process was found to be in excess of 98 percent.

References

[72] K. J. Jung, P. Lejemble, A. Gaset, P. Kalck, and Y. Maire, Bulletin de la Société Chimique de France 1986, 459-464.

[73] K. J. Jung and A. Gaset, Biomass 16 (1988) 89-96.

19. Furoic Acid

Furoic acid is the first down-line oxidation derivative of furfural. It has a market in the pharmaceutical and agrochemical field, where it is frequently converted to furoyl chloride to be used in the production of drugs and insecticides.

The customary process for making furoic acid starts with a Cannizzaro reaction between furfural and aqueous sodium hydroxide [74] to yield furfuryl alcohol and sodium 2-furancarboxylate:

$$2\ C_5H_4O_2 + NaOH \rightarrow\rightarrow C_5H_6O_2 + C_5H_3O_3Na \qquad (1)$$

After removal of the furfuryl alcohol, the sodium 2-furancarboxylate is usually acidified with sulfuric acid to yield furoic acid and sodium hydrogen sulfate:

$$C_5H_3O_3Na + H_2SO_4 \rightarrow\rightarrow C_5H_4O_3 + NaHSO_4 \qquad (2)$$

The crude furoic acid crystals thus precipitated are then submitted to a purification.

Alternative processes are not viable on an industrial scale. For instance, a straight oxidation of furfural with oxygen over a catalyst is completely unacceptable as the furfural undergoes not only oxidation to furoic acid but also a competitive nucleus oxidation resulting in ring cleavage and unwanted by-products. An electrochemical process was found to result in a competitive decomposition of furoic acid as the oxidation potentials of furoic acid and furfural are extremely close, so that it is not possible to selectively produce furoic acid by this route. Treatment of furfural with sodium chlorite in the presence of hydrogen peroxide is extremely sensitive to temperature fluctuations which result in the formation of maleic acid. Besides, the given oxidizing agents represent a safety risk.

As reaction (1) is strongly exothermic, the furfural is cooled to 2 ^{o}C before the sodium hydroxide solution is added, and the reactor content is stirred vigorously while this is done. Typically, a 28 % sodium hydroxide solution is used. The addition of the sodium hydroxide must be carried out slowly to make sure that the temperature does not exceed 20 ^{o}C. In other words, the rate of sodium hydroxide addition depends on the efficiency of cooling. Customarily, this addition requires a period in the order of half an hour.

Since the Cannizzaro reaction is slow, after completion of the sodium hydroxide addition the stirring must be continued for at least one hour. In the course of the reaction, considerable quantities of sodium 2-furancarboxylate precipitate in the form of scale-like crystals. These are separated by filtration and then washed with acetone. The dry furoate salt is dissolved in a minimum amount of water and then treated with sulfuric acid to form raw furoic acid and sodium hydrogen sulfate.

The purification of the raw furoic acid is best carried out by sublimation as the triple point pressure of furoic acid is high (10.3 torr) and as the impurities (polymers of furfuryl alcohol) are essentially nonvolatile. Thus, passing a hot carrier gas over the raw furoic acid selectively vaporizes the desired compound while leaving the nonvolatile impurities behind. If, by a conservative estimate, the vapor pressure of furoic acid is 100 times greater than the vapor pressure of the polymeric impurities, then the quantity of furoic acid vaporized by a hot carrier gas exceeds the quantity of vaporized polymer by a factor of 100 as the rate of vaporization is proportional to the vapor pressure. Hence, if in the initial raw furoic acid the ratio of furoic acid to polymer were 100 : 1, then in the carrier gas the ratio of furoic acid to polymer would be 10^4 to 1, so that desublimation yields an enormously purified product, and contrary to recrystallization from a solution, where huge losses are incurred, this effect is obtained at essentially no loss at all as the desublimation temperature can be chosen so low that practically no furoic acid is retained in the carrier gas. This is shown graphically in Figure 82.

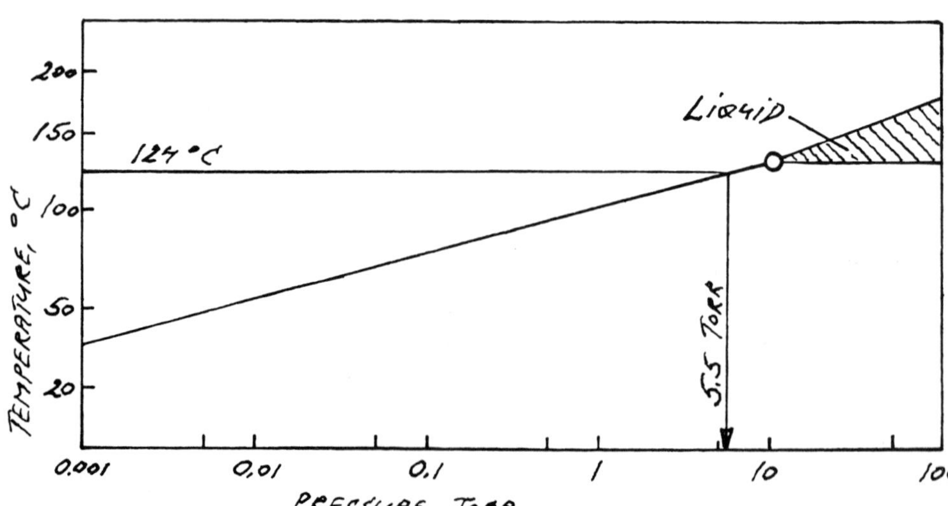

Figure 82. Sublimation and Desublimation of Furoic Acid in the Phase Diagram.

The triple point of furoic acid is at 10.3 torr and 133 °C. A sublimation at 124 °C corresponding to a furoic acid vapor pressure of 5.5 torr, and a desublimation at 36 °C corresponding to a furoic acid vapor pressure of 0.001 torr, result in a relative loss of only 0.001/5.5 = 0.00018 (0.018 %).

One possible mode of carrying out the purification on an industrial scale is shown in Figure 83. The process consists in feeding the raw furoic acid into a rotary drier operating as a "sublimer". The fuel and air inputs to the burner must be controlled in such a way that the combustion gas heats the charge to a temperature slightly below the melting point. The vapor pressure of furoic acid at the temperature in the "sublimer" will prevail in the gas stream. After leaving the "sublimer", the gas stream is passed through a cyclone to remove entrained particles, and then fed into a "desublimation chamber" cooled by the ambient atmosphere and by some air injection. Due to the resulting transformation from the gas phase to the solid phase, the furoic acid is obtained as a fluffy powder resembling snow which does not stick to any walls and is very easy to handle. (Natural snow is also produced by desublimation in cooling the water vapor of air at a partial pressure below the pressure of the triple point. By contrast, artificial snow is made by freezing tiny water droplets obtained from a spray nozzle. For this reason, artificial snow is not fluffy, thus being less pleasant for skiing.)

From the "desublimation chamber", the gaseous phase is vented while the product is retained by a filter. Furfuryl alcohol polymers, not being sufficiently volatile, remain in the "sublimer" and are discarded as residue. The entrained particles separated in the cyclone are returned into the feed stream.

For the conditions shown in Figure 83, the sublimation rate is 26.7 kg of furoic acid per 100 kg of combustion gas. The customary design value for the "desublimation chamber" is 3.33 m^3 of chamber volume per kg/h of desublimate [75].

According to the reactions (1) and (2), if there were no losses due to unwanted by-products, 192.164 g of furfural, 39.999 g of sodium hydroxide, and 98.082 g of sulfuric acid would yield 112.082 g of furoic acid, 98.098 g of furfuryl alcohol, and 120.065 g of sodium hydrogen sulfate. Thus, the theoretical yield of furoic acid amounts to 58.326 percent of the furfural input. However, the actual yield of furoic acid is only between 45 and 53 percent of the theoretical yield, so that each ton of furfural gives a mere 262 to 312 kg of furoic acid.

According to the reactions (1) and (2), stoichiometrically each ton of furoic acid produced would be accompanied by 1.07122 tons of sodium hydrogen sulfate, but when the

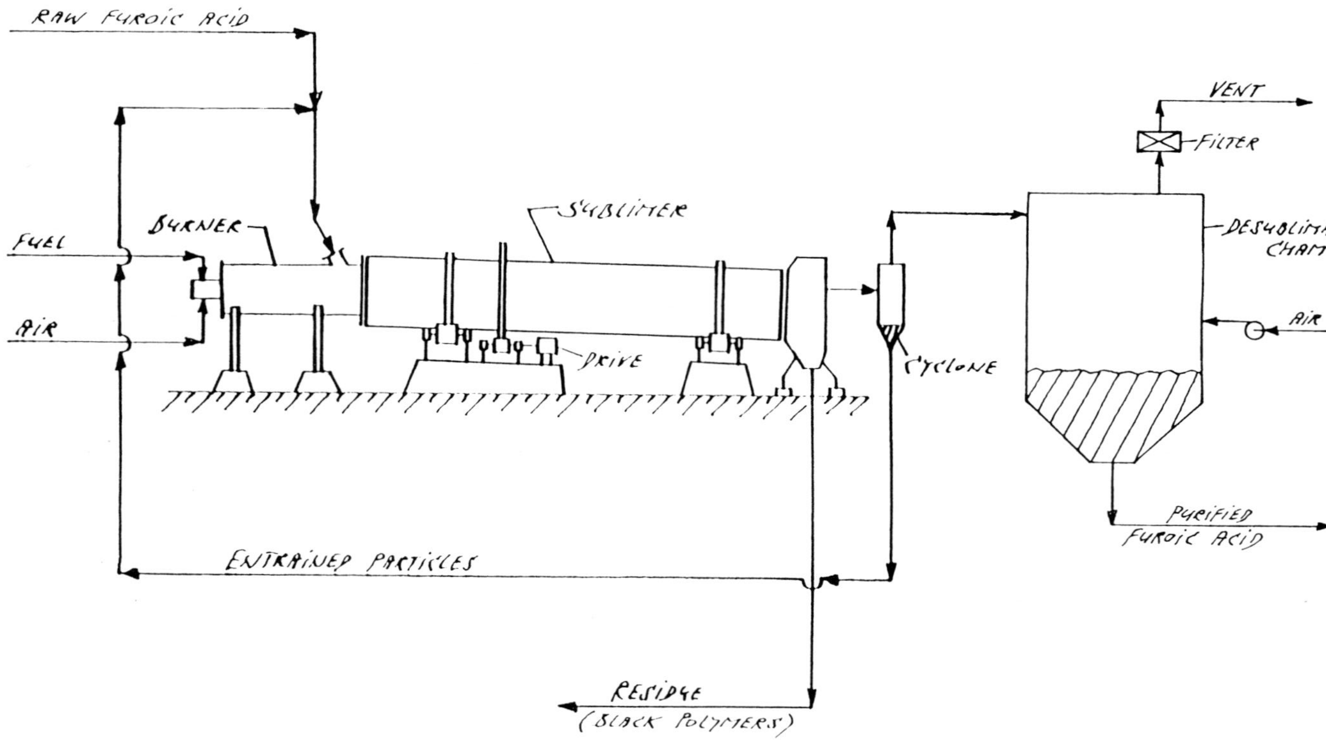

Figure 83. Plant for the Purification of Furoic Acid by Sublimation and Desublimation.

actual yield of furoic acid is taken into account, the sodium hydrogen sulfate production is found to exceed the furoic acid production by a factor between 2.00 and 2.38.

References

[74] W. C. Wilson, Organic Syntheses I (1937) 270-274.

[75] G. Matz, Chemie-Ing.-Techn. 38 (1966) 299-308.

20. Difurfural (5,5'-Diformyl-2,2'-Difuran)

To utilize the outstanding thermal stability of the furan ring for the manufacture of high-temperature polymers, it is necessary to provide it with two "arms" capable of linking with equal or different building block molecules. One way of achieving this is to join two furfural molecules at the 5-position in forming "difurfural" (5,5'-diformyl-2,2'-difuran):

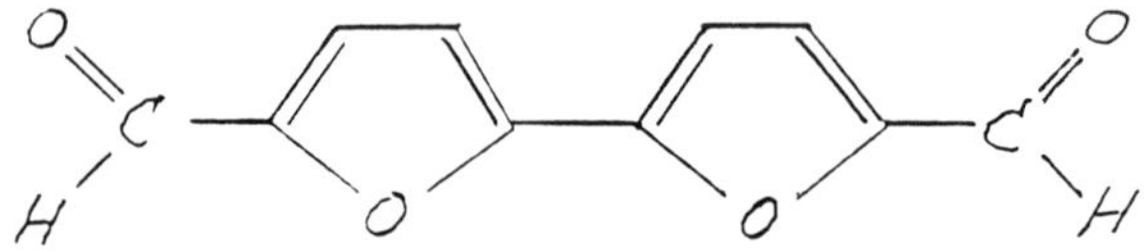

It was known that this can be accomplished by UV irradiation of bromofurfural [76]:

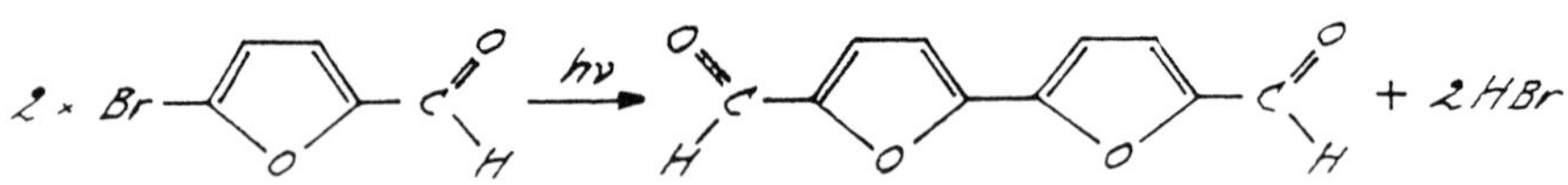

However, as bromofurfural is a very expensive raw material, it was found desirable to improve this process. To this end, Burger [77] invented a rather ingenious procedure which consists in using bromofurfural only as a "trigger substance" for joining arbitrary quantities of furfural. Burger's process consists in a UV irradiation of a mixture of furfural and bromofurfural in the presence of a polyvinylpyridine ion exchange resin (REILLEX 402) and acetonitrile as diluent, the ratios being 416 millimole of furfural : 20.8 millimole of bromofurfural : 0.4 g of REILLEX 402 : 400 ml of acetonitrile. Use is made of a medium pressure mercury arc lamp, and, of course, the irradiation is effected through quartz.

The special feature of Burger's process is the observation that after complete consumption of the bromofurfural, the formation of difurfural continues indefinitely. Thus, after a start with bromofurfural, a continuous addition of furfural results in a continuous precipitation of solid difurfural (melting point 264.5 °C), which can be removed continuously by sedimentation or filtration. By contrast, if the process is started without the bromofurfural, no difurfural is formed.

As in many photochemical processes, the mechanism at work is not firmly established, but application of known principles seems to indicate the following sequence of events:

(1) The bromofurfural is photochemically decomposed to a bromine and a furfural radical:

$$BrFU + h\nu_1 \rightarrow\rightarrow Br\bullet + FU\bullet \qquad (1)$$

(2) A bromine radical abstracts a hydrogen atom from a furfural molecule, thus leading to HBr and another furfural radical:

$$Br\bullet + HFU \rightarrow\rightarrow HBr + FU\bullet \qquad (2)$$

(3) Irradiation of HBr leads to hydrogen and bromine radicals:

$$HBr + h\nu_2 \rightarrow\rightarrow H\bullet + Br\bullet \qquad (3)$$

Reaction (3) is documented in the literature [78] where it is said to take place when HBr is irradiated with 253.7 nm. In Burger's medium pressure mercury arc lamp, this wavelength is strongly represented. Reaction (3) "regenerates" bromine radicals, thus creating a steady state population of this species, so that the formation of furfural radicals by hydrogen abstraction can continue even after all of the bromofurfural has disappeared.

(4) A furfural radical reacts with furfural to form an intermediate radical:

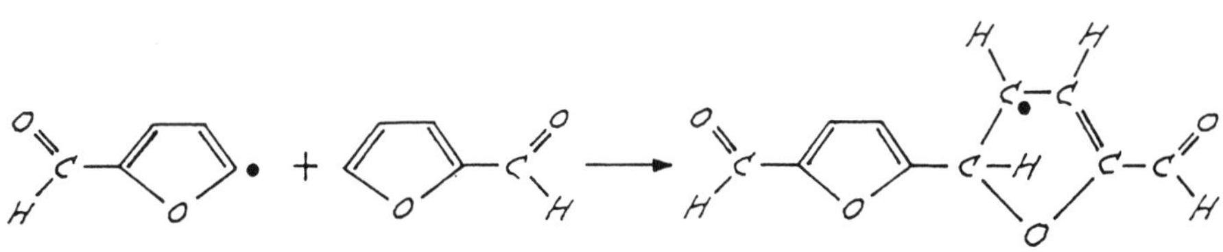

(5) The intermediate radical is oxidized (robbed of a hydrogen atom) by a bromine radical to yield difurfural:

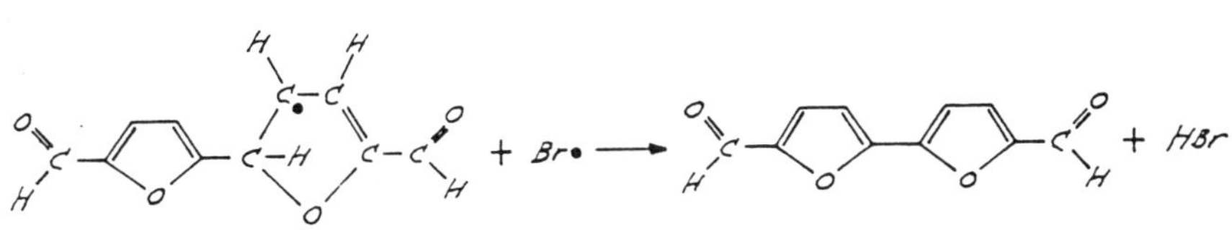

(6) The polyvinylpyridine resin takes out some HBr by forming an ionic salt.

(7) The hydrogen radicals react with ubiquitous oxygen (a diradical) to end up as hydrogen peroxide.

Noteworthily, in this pathway there is no reaction between two furfural radicals as their concentration is not likely to be high enough for such a dimerization to occur.

Thanks are expressed to Professor Dr. J. D. Coyle, formerly at Oxford University, for his kind advice on this pathway.

Figure 84. Pilot Plant for the Production of Difurfural.

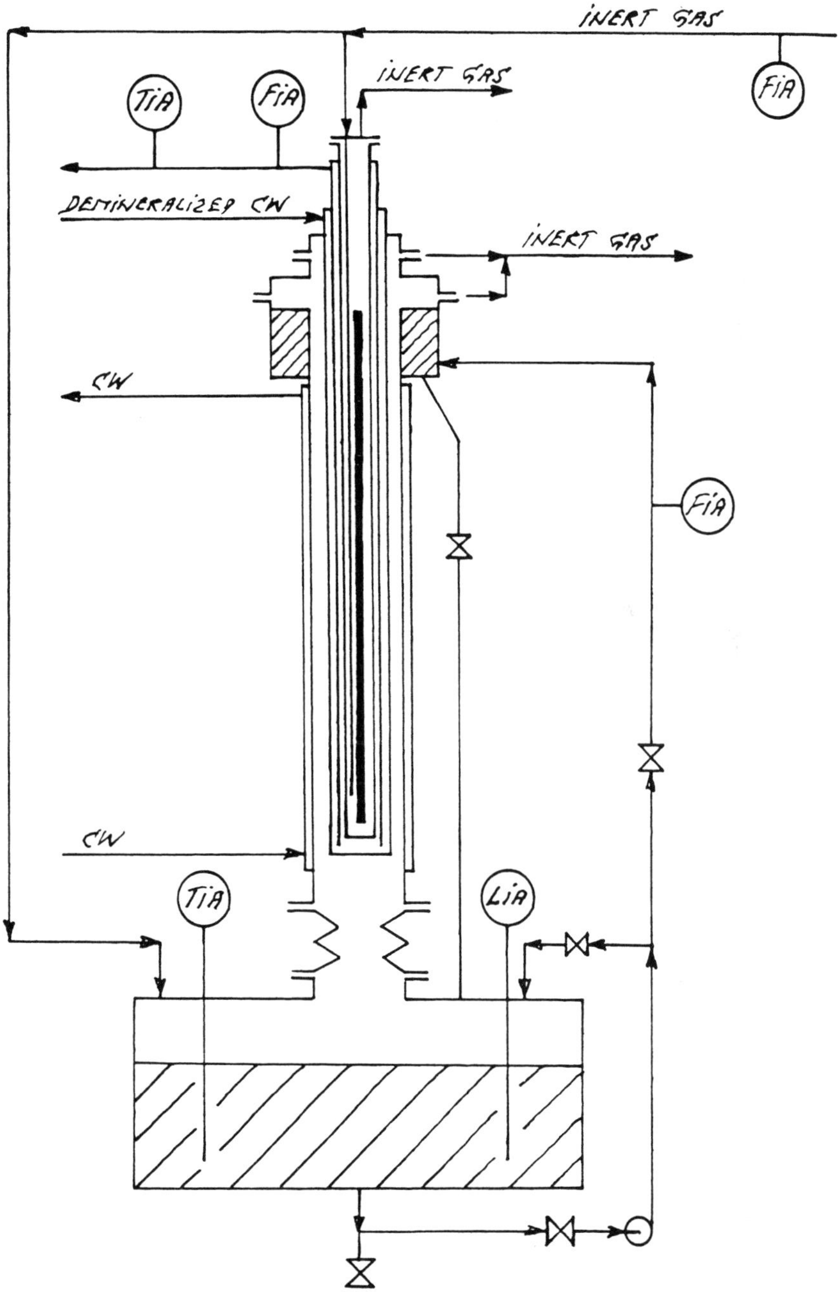

Figure 85. Schematic Cross Section of the Pilot Plant for the Production of Difurfural. Falling Film Design. UV Lamp 4 kW. Circulation Pump 1 m^3/h.

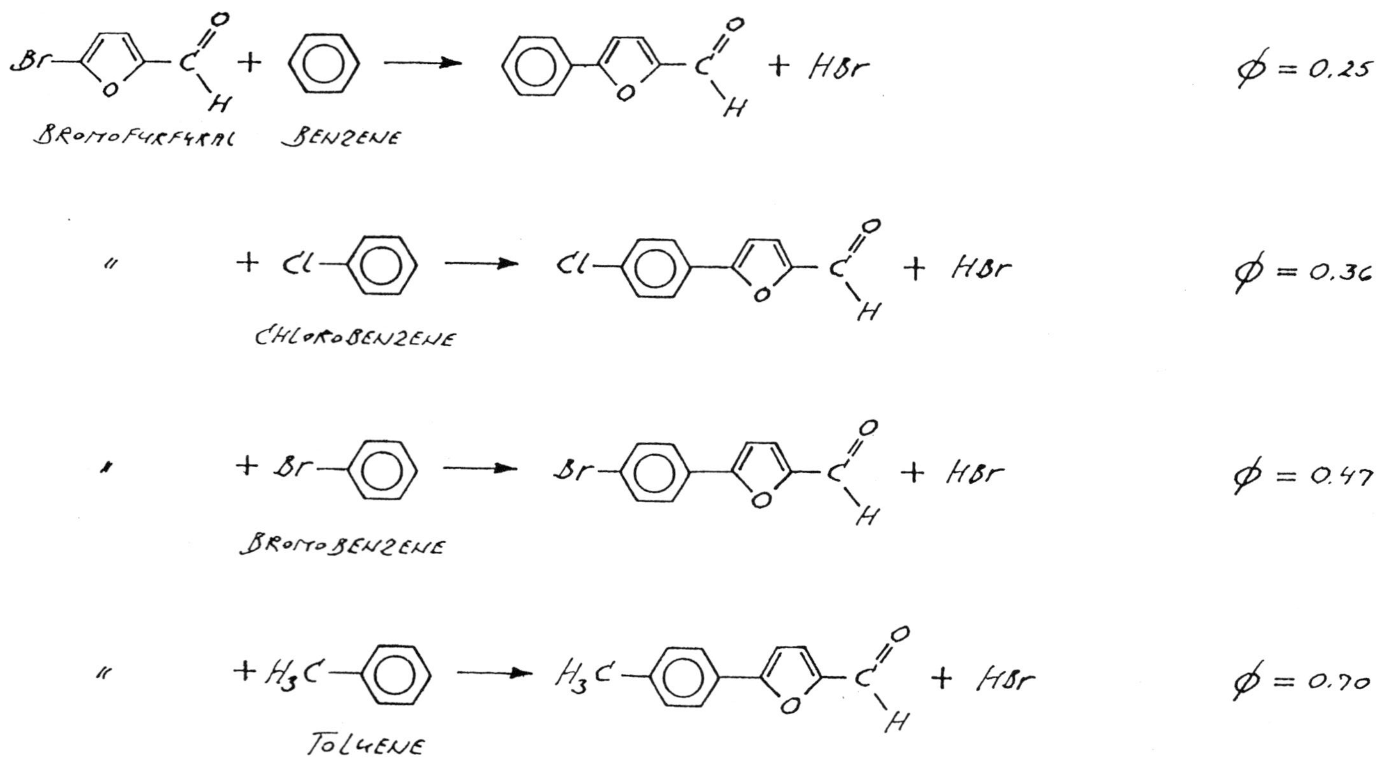

Figure 86. Quantum Yields Φ for the Photochemical Coupling of Bromofurfural with Benzene and various Benzene Derivatives in Acetonitrile [78].

A pilot plant unit used to produce kilogram quantities of difurfural is shown in Figure 84. It was built by RÖDER, JUNG, AND PARTNER of Stuttgart/Germany. A cross section, illustrating an advanced falling film design, is given in Figure 85.

The reaction mixture is charged into a reservoir at the bottom. The polyvinylpyridine resin is contained in a permeable bag suspended in the charge. A circulation pump takes the liquid into an annular chamber at the top from where an overflow forms a falling film on the inside wall of the reactor proper. This falling film is irradiated by a vertical UV lamp arranged along the axis and surrounded by a quadruple wall quartz system cooled by demineralized water. After having passed the irradiated region, the falling film flows back into the reservoir where a filter bag in the suction line of the circulation pump collects the difurfural crystals formed by the process. The unit operates under inert gas, and all electrical parts are explosion-proof. To the extent as difurfural is formed, furfural is replenished continuously.

It is believed that difurfural has the potential of a great future as a building block for high-temperature polymers. Specialty polymer manufacturers are in the process of evaluating difurfural for such applications.

By analogy with the quantum yields for the photochemical coupling of bromofurfural with benzene and various benzene derivatives in acetonitrile as shown in Figure 86 [78], the quantum yield of Burger's difurfural process must be expected to lie far below unity.

References

[76] Ya. R. Tymyanskii, V. M. Feigelman, O. A. Zubkov, V. N. Novikov, M. I. Knyazhanskii, and V. S. Pustovarov, Zh. Org. Khim. 24 (1988) 459-460.

[77] G. J. Burger, Functionalization of Furan Derivatives, Ph.D. Thesis, University of Port Elizabeth, South Africa, 1993.

[78] M. d'Auria and F. d'Onofrio, Gazzetta Chimica Italiana 115 (1985) 595-597.

21. 2-Hydroxyfuranone-(5)

2-Hydroxyfuranone-(5)

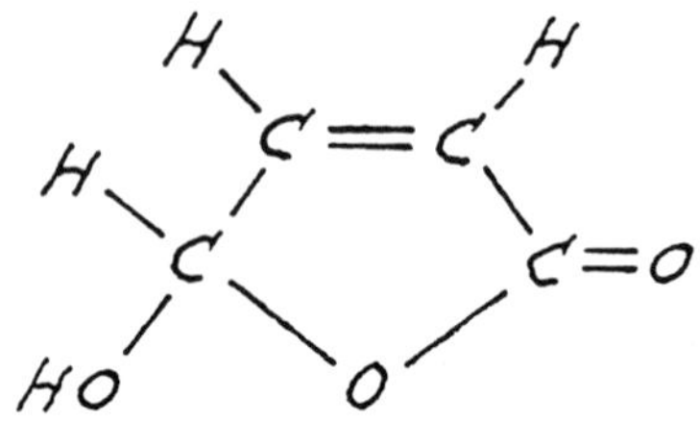

a highly effective insecticide, the active substance in the Mediacea bark, never attacked by insects, can be readily produced by a sensitized photooxidation of furfural, with a very high yield [79]:

$$\text{FURFURAL} + 3/2\ O_2 + h\nu \rightarrow\rightarrow C_4H_4O_3 + CO_2$$

A simple laboratory procedure of this process can be outlined as follows:

330 g (3.43457 mole) of furfural are mixed with 935 g of methanol, and as a photosensitizer for oxygen, 4 g of rose bengal (tetrachlorotetraiodofluorescein) is dissolved in this mixture. A photosensitizer gets excited by picking up light energy, and in the case at hand, it excites ground state oxygen to “singlet oxygen”, commonly written $^1O_2$. This metastable, highly active excited molecular oxygen produces the oxidation reaction desired.

Under stirring, oxygen in fine distribution is passed through this solution, while the latter is irradiated by a 125 watt medium pressure mercury arc lamp for 60 hours. During this time, the temperature is maintained between 15 and 30 °C. Subsequently, the methanol is removed by a vacuum distillation at 20 to 30 mm Hg. The residue is mixed with 250 ml of chloroform and cooled. This results in a crystallization of 2-hydroxyfuranone-(5). Separation of the crystals by filtration and subsequent drying leads to 309 g (3.08778 mole) of 2-hydroxyfuranone-(5). Referred to the furfural input, this corresponds to 89.903 percent of the theoretical yield. A product of high purity is obtained by recrystallization in chloroform containing some activated carbon. The purified product has a melting point of 57.5 °C.

For the solvent, instead of methanol, use can be made of ethanol, propanol, iso-propanol, the various butanols, cyclohexanol, or benzyl alcohol. For the crystallization and recrystallization, instead of chloroform, use can be made of cyclohexane or diethyl ether.

Reference

[79] G. Bolz and W. W. Wiersdorf, German Patent Application No. 2111119 (1971).

22. Acetoin

Acetoin, also known as acetylmethylcarbinol, is the most important derivative of diacetyl. It is a valuable flavor commanding a higher price than diacetyl.

As shown below, acetoin can be readily made from diacetyl, either by catalytic hydrogenation or by electrolytic hydrogenation.

Acetoin is a substance of rather complicated behavior as summarized by the following features:

(1) Liquid acetoin consists of two different molecules in equilibrium: A keto form (A), which is predominant, and an "oxide form" (B):

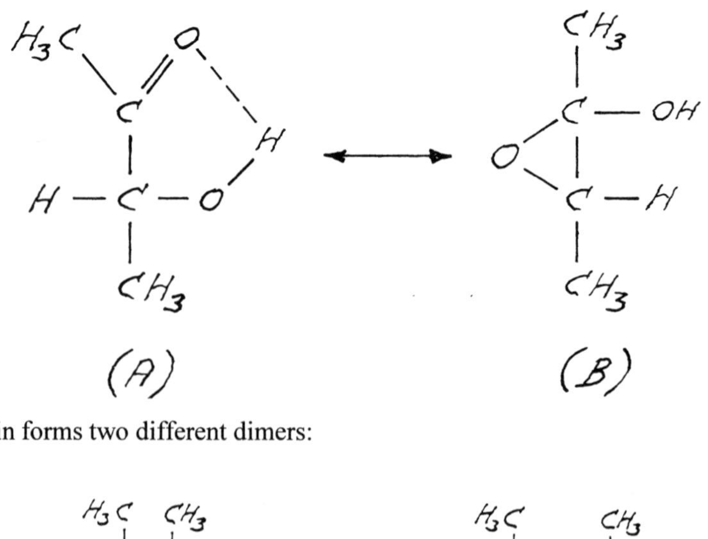

(A) (B)

(2) Acetoin forms two different dimers:

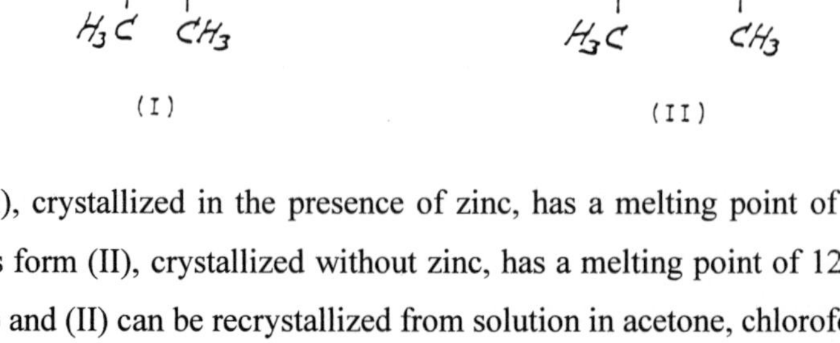

(I) (II)

Form (I), crystallized in the presence of zinc, has a melting point of 90 °C, whereas form (II), crystallized without zinc, has a melting point of 127.5 °C. Both (I) and (II) can be recrystallized from solution in acetone, chloroform, or ethyl acetate, but this does not change their melting point as in these solvents the dimers are only partly depolymerized, so that the remaining dimer molecules in solution act as crystallization nuclei leading to the original form.

By contrast, when the dimer is dissolved in water, diethyl ether, acetic acid, or paraldehyde, it is depolymerized completely, so that there are no crystallization nuclei left in the solution. Hence, it is difficult to recrystallize the dimer from such solutions. When crystallization finally does take place, it leads to form (II) unless a crystal of form (I) is used as a primer.

(3) Liquid acetoin stored for a long period of time has a well-developed ketonic absorption band at 275 nm, indicating the prevalent form (A), whereas liquid acetoin freshly prepared by melting or dissolving the dimer shows no such absorption, which indicates that based on the dimer structure the latter liquid can be expected to consist of the oxide form (B), and that the equilibrium between the keto form and the oxide form is reached very slowly (in a matter of days).

Using the rotatory power of acetoin, due to the asymmetric carbon atoms in the keto form (A) and the oxide form (B), particularly interesting observations on acetoin transformations were made by Dirscherl and Schöllig [80]. Although these authors refrained from interpreting their seemingly strange findings, the knowledge of today permits a complete explanation of their results on the basis of the following facts:

(a) The keto form (A), because of its intramolecular hydrogen bond, is more volatile than the oxide form (B), so that distillation produces a distillate rich in (A) and a sump fraction rich in (B).

(b) Because of their structure, the dimers can form only from the oxide form (B).

(c) The keto form has a higher rotatory power than the oxide form.

Against this background, the seemingly weird findings of Dirscherl and Schöllig, reproduced in Figure 87, can be explained in all details:

Observation 1: When acetoin produced by fermentation and then recovered from the fermentation broth by distillation was left standing for a day, the rotation first decreased. When it reached a minimum at point K, dimer crystals started precipitating, and the rotation started to increase.

Explanation: As the acetoin had been recovered from the fermentation broth by distillation, its

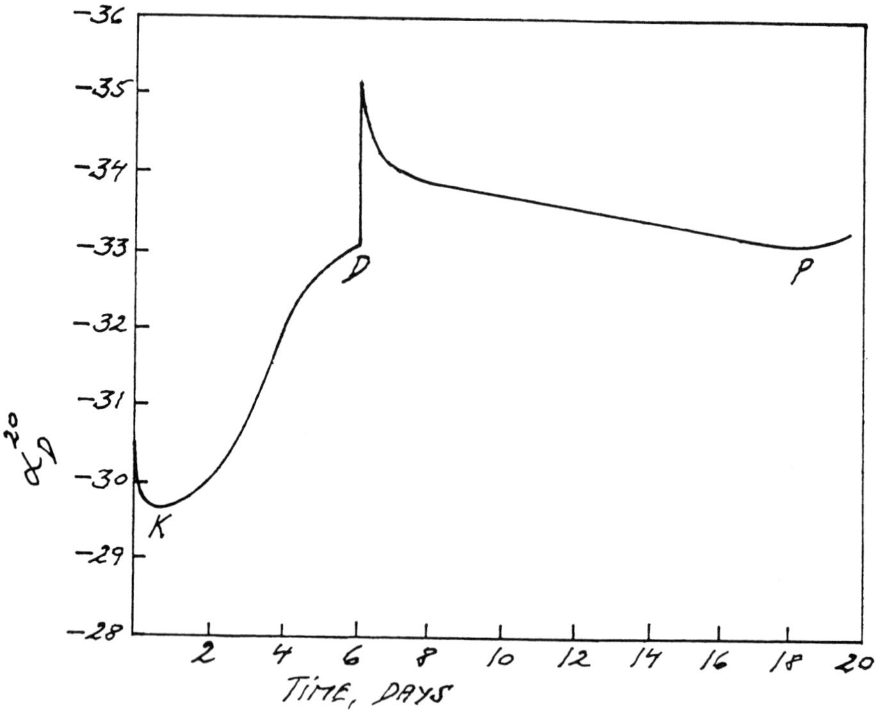

Figure 87. Changes of the Rotatory Power of Acetoin as Observed by Dirscherl and Schöllig [80].
K and P: Start of Dimer Crystal Precipitation.
D: Start of Vacuum Distillation.

content of the keto form was greater than in equilibrium, so that it slowly reestablished equilibrium by converting keto form molecules to oxide form molecules, which lowered the rotation as the keto form has a higher rotation than the oxide form. Then, when at point K a critical concentration of oxide form molecules was reached, the latter started forming dimer molecules. As the consumption of oxide form molecules by crystallization was faster than the replenishment of oxide form molecules from keto form molecules, the rotation increased.

Observation 2: When at point D the acetoin was submitted to a distillation, this abruptly increased the rotation of the distillate, and the sump fraction rapidly solidified to dimer.

Explanation: As the keto form molecules are more volatile than the oxide form molecules, the distillate was overly rich in keto form molecules, thus showing an increased rotation. At the same time, the sump fraction, depleted of keto form molecules, got overly rich in oxide form molecules, and for that reason it readily crystallized to dimer.

Observation 3: When the distillate was left standing for 13 days, its rotation diminished, at first sharply and then more gradually until at point P it started increasing again.

Explanation: In the distillate, the content of keto form molecules was greater than in equilibrium, so that it reestablished equilibrium by converting keto form molecules to oxide form molecules, which lowered the rotation. Then, when at point P a critical concentration of oxide form molecules was reached, the latter started forming dimer molecules, and as the consumption of oxide form molecules by crystallization was faster than the replenishment of oxide form molecules from keto form molecules, the rotation increased on account of an increasing concentration of keto molecules.

Dirscherl and Schöllig also discovered that the dimer formation of acetoin is catalyzed by glass containing sodium ions, whereas no such catalysis was observed in a quartz vessel. Thus, it is not surprising that acetoin readily dimerizes in normal glass bottles used for shipment.

22.1. Catalytic Hydrogenation of Diacetyl

Acetoin can be readily made by a simple hydrogenation of diacetyl, using palladium on carbon or alumina as the catalyst. The process is based on a publication by Skibina, Ioffe, and Artamonov [81]. A laboratory setup suitable for carrying out this reaction is shown in Figure 88. Hydrogen is passed through a diacetyl evaporator kept at 25 °C by means of a thermostat, and the resulting gaseous mixture is fed through the catalyst bed heated by a tube furnace to 125 °C. The conversion to acetoin is in the order of 95 percent, the rest being unconverted diacetyl and butanediol as a minor by-product. An industrial version of the process is shown in Figure 89. The final distillation must be carried out under nitrogen as

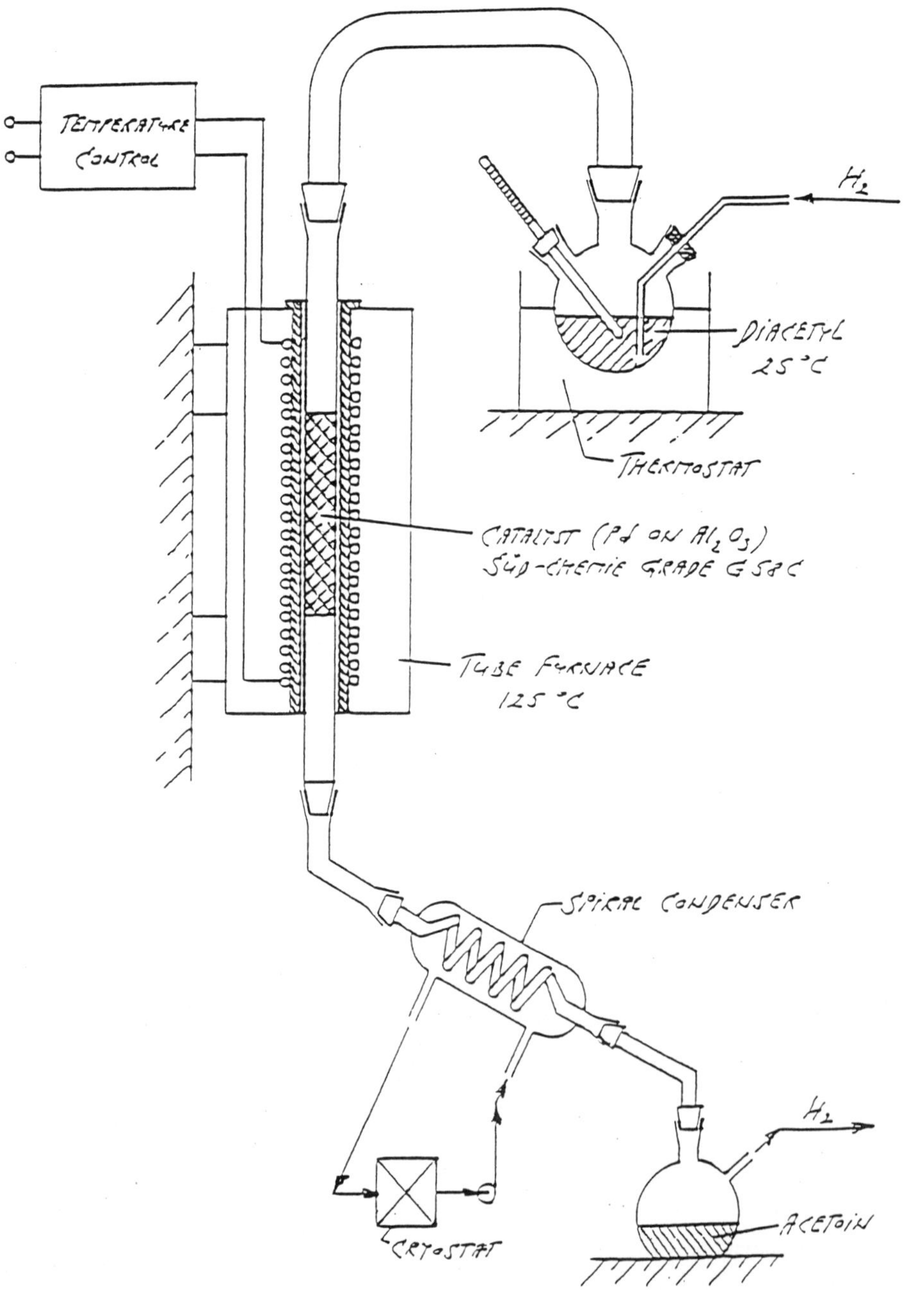

Figure 88. Laboratory Setup for the Catalytic Hydrogenation of Diacetyl to Acetoin.

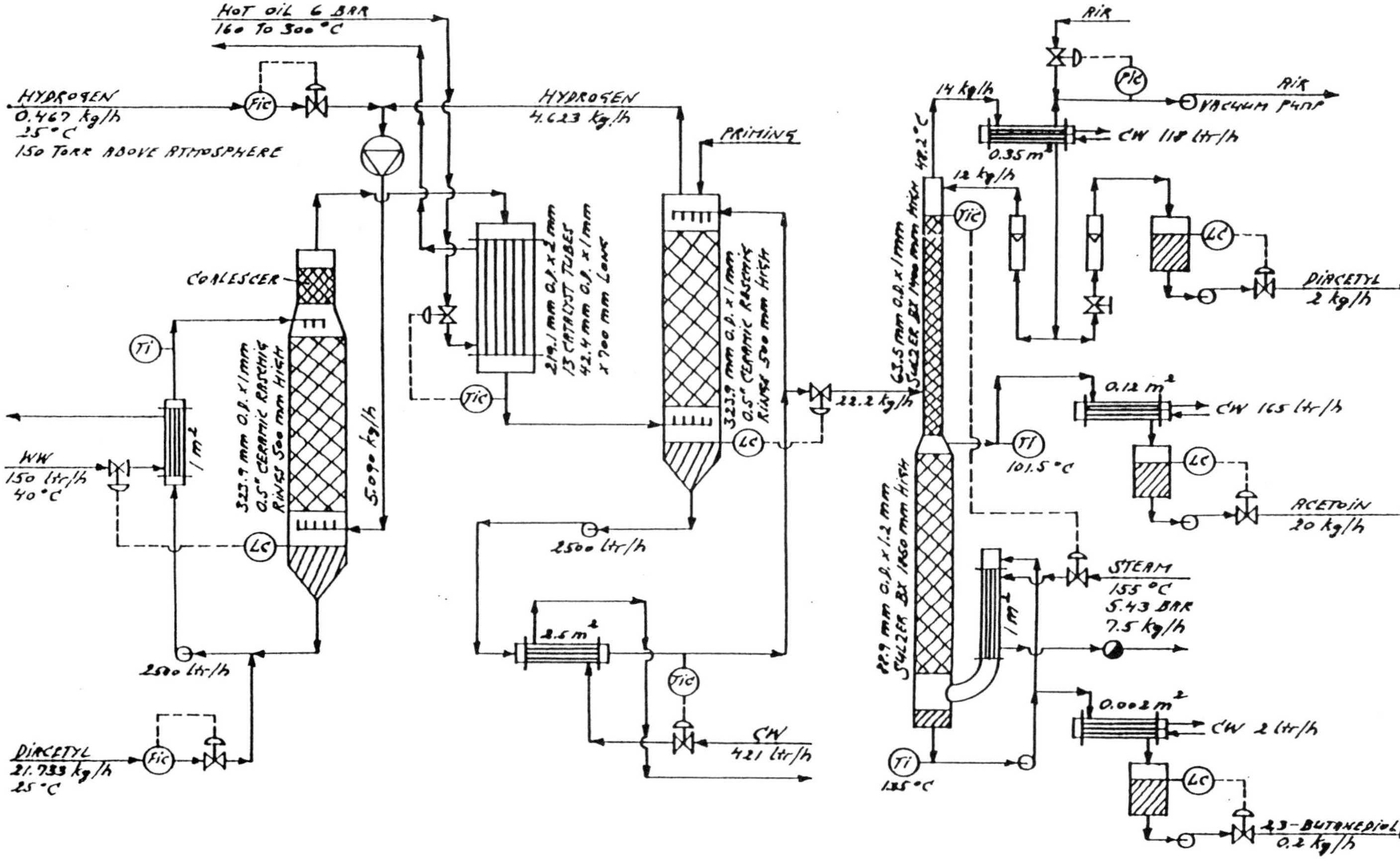

Figure 89. Plant for the Catalytic Hydrogenation of Diacetyl to Acetoin.

distilling acetoin in the presence of air converts it to diacetyl [82].

22.2. Electrolytic Hydrogenation of Diacetyl

A selective hydrogenation of diacetyl to acetoin can also be carried out in an electrolytic cell using a cathode with a high hydrogen overvoltage [83]. It is recommended to start with a solution consisting of

12.0 % by weight of diacetyl,
83.6 % by weight of water, and
4.4 % by weight of sulfuric acid.

The cell characteristics should be as follows:

Cathode: Nickel with nickel shot. The latter was found to greatly increase the rate of diacetyl conversion.

Aplied potential difference: 2.2 volts per cell.

Current density: 0.0665 amps/cm^2.

From a reservoir, the solution is pumped through the cells and through a cooler. The latter must be designed to keep the temperature in the reservoir at 25 ^{o}C. Complete conversion of the diacetyl is obtained after 24 hours.

Although the electrolytic process is very simple, complications arise by the fact that

(1) the acetoin must be extracted from the reaction mixture,
(2) some acid is extracted as well, and
(3) the extractant must be recovered by distillation.

The most suitable extractant is believed to be dichloromethane. A respective plant is shown in Figure 90. As can be seen, due to the recovery procedures, the production of acetoin by an electrochemical hydrogenation of diacetyl is more complicated than the catalytic process.

22.3. Preferred Commercial Form

On one hand, there is an equilibrium between keto form molecules and oxide form molecules, and on the other hand there is an equilibrium between oxide form molecules and dimer molecules. A superposition of these two equilibria means that from a liquid monomer mixture of keto form and oxide form molecules obeying the first equilibrium, some solid dimer crystals must form to satisfy the second equilibrium. However, this dimer formation

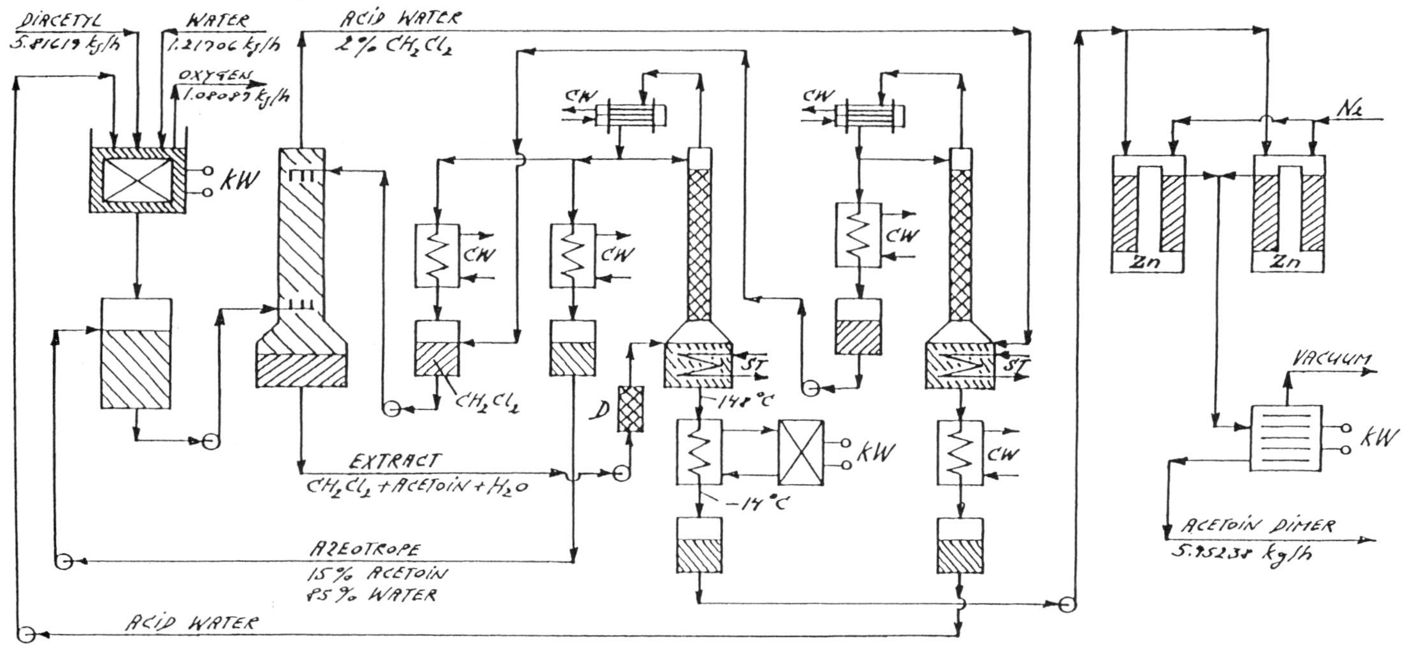

Figure 90. Plant for the Electrolytic Hydrogenation of Diacetyl to Acetoin.

consumes only oxide form molecules, thereby disturbing the first equilibrium, so that new oxide form molecules are generated from keto form molecules. This in turn permits the formation of more dimer molecules consuming oxide form molecules, and so forth, until very slowly, in long storage or shipment, an "equilibrium slurry" of dimer crystals in a liquid equilibrium mixture of keto form and oxide form molecules is reached. The slurry leads to a pasty sediment giving the product an ugly appearance. Although this sediment, being the dimer, is not an impurity, it is frequently misunderstood as such, and for this reason it has become customary to sell the dimer rather than a liquid monomer mixture inevitably converting into an unsightly sludge. To this end, the liquid monomer mixture originally obtained in the process is crystallized to the dimer in the presence of zinc, and these crystals are dissolved in acetone and recrystallized. This yields a beautiful white powder which is stable indefinitely. It can be readily converted to a liquid by mere heating above the melting point.

The liquid obtained by melting the dimer consists at first of oxide form molecules which require days until the equilibrium mixture of keto form and oxide form molecules is reached. During this conversion, the smell changes as on account of the unlike structure the smell of oxide form molecules differs from that of keto form molecules. This must be considered when the liquid is submitted to olfactory tests.

References

[80] W. Dirscherl and A. Schöllig, Berichte der Deutschen Chemischen Gesellschaft 71 (1938) 418-423.

[81] E. M. Skibina, Yu. M. Levin, I. I. Ioffe, and P. A. Artamonov, Zhur. Prikladnoi Khimii 19 (1976) 1554-1558.

[82] H. von Pechmann and F. Dahl, Berichte der Deutschen Chemischen Gesellschaft 23 (1890) 2421-2427.

[83] B. Müller, H. J. Dietz, H. Matschiner, S. Engelmann, and H. Herzberg, Patent of the German Democratic Republic No. DD 237 683 A1 (1986).

23. Pyrazines

When a solution of diacetyl or 2,3-pentanedione in diethyl ether is slowly added to a solution of ethylene diamine in diethyl ether while maintained at 0 °C under an inert gas atmosphere, an extremely exothermic reaction results in 2,3-dimethyl-5,6-dihydropyrazine or 2-ethyl-3-methyl-5,6-dihydropyrazine, henceforth called I and II, as shown in Figure 91.

I and II are substances with a disgusting smell, apparently without any direct application, but when they are dehydrogenated, they yield 2,3-dimethyl pyrazine and 2-ethyl-3-methyl pyrazine, respectively, which are extremely valuable flavor compounds. The dehydrogenation is carried out by bubbling nitrogen through a thermostatically heated bath of I or II, and by passing the resulting gaseous mixture through a copper chromite catalyst held at 300 to 350 °C [84].

Alternatively, when I or II are hydrogenated, they yield 2,3-dimethyl hexahydro-pyrazine or 2-ethyl-3-methyl hexahydropyrazine, also called 2,3-dimethyl piperazine and 2-ethyl-3-methyl piperazine, respectively, which are used as drugs against diseases caused by intestinal worms.

A compilation of all of these substances is given in Figure 92. Against this background, diacetyl and 2,3-pentanedione are seen to be starting materials for a family of extremely high-priced fine chemicals.

Reference

[84] I. Flament and M. Stoll, Helvetica Chim. Acta 50 (1967), Fascilulus 7 No. 180, 1754-1758.

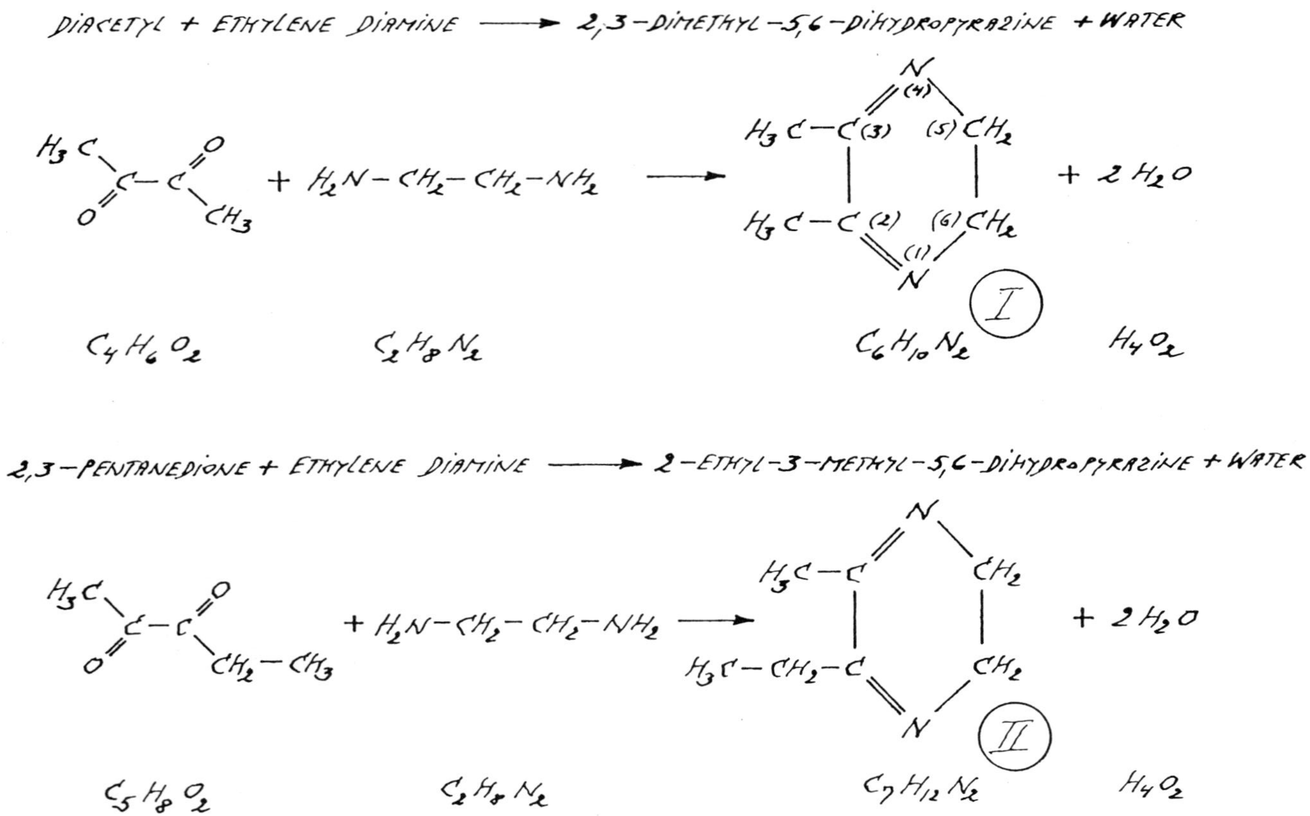

Figure 91. The Formation of Dihydropyrazines.

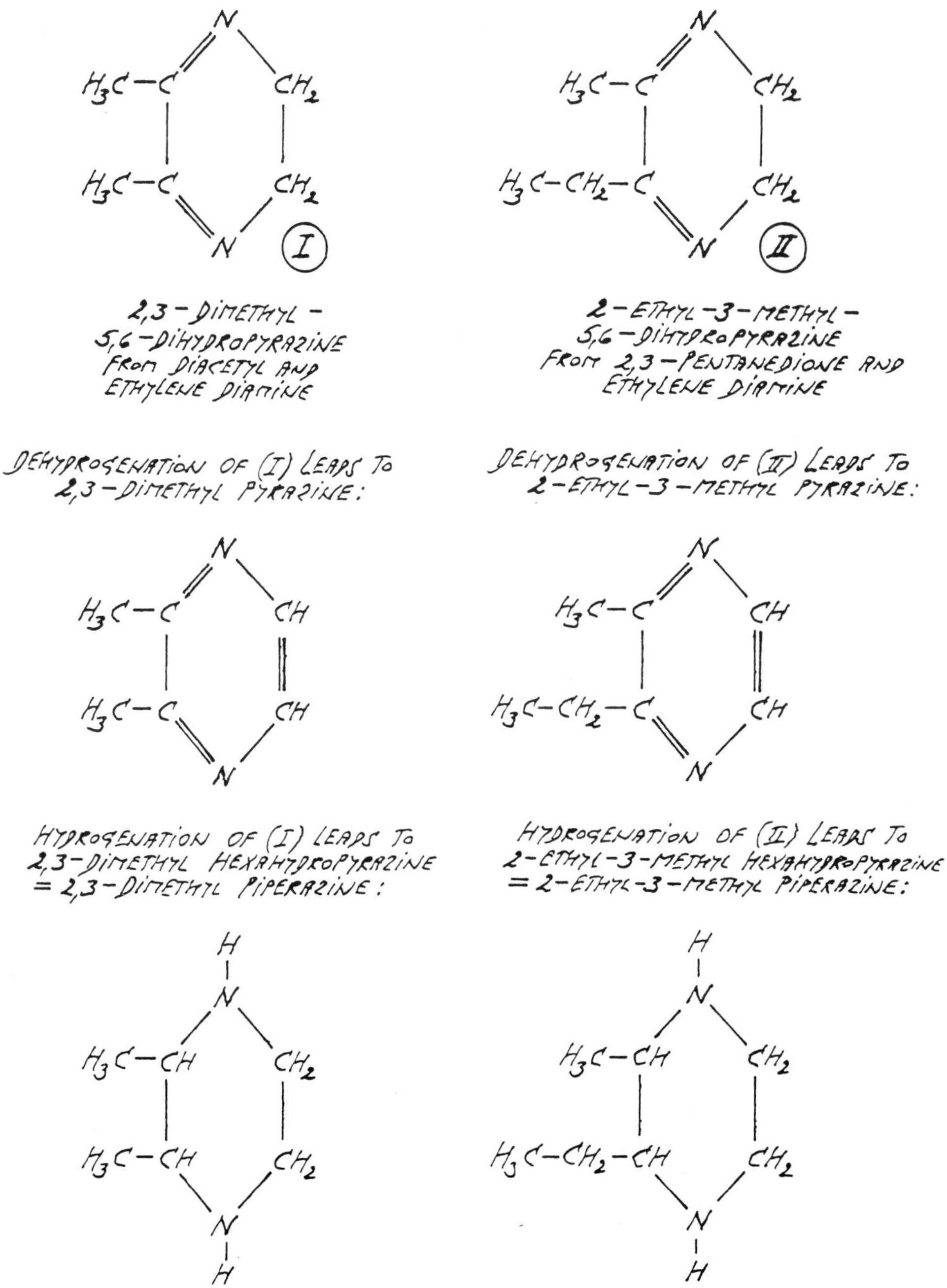

Figure 92. Compilation of the Pyrazines Derived from Diacetyl and 2,3-Pentanedione.

24. Tetrahydrofuran

Tetrahydrofuran (THF) is made by hydrogenation of furan:

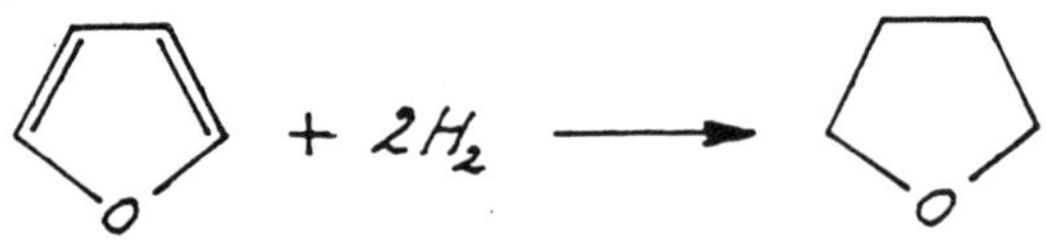

Thus, with furan being made from furfural, THF is a "second generation descendant" of furfural. The process is commonly carried out at 100 °C and 20 bar, using a catalyst consisting of 5 % palladium on microporous carbon [85].

Interestingly, the same catalyst is used for the conversion of furfural to furan, so that the question may be asked why THF is not made directly from furfural. The explanation lies in the fact that the conversion of furfural to furan liberates carbon monoxide which has such a high heat of adsorption on palladium that hydrogen cannot be adsorbed simultaneously. The heat of adsorption of carbon monoxide on palladium exceeds the heat of adsorption of hydrogen on palladium by 14 kcal/mole (58.576 kJ/mole).

The hydrogenation of furan to THF is an energetically very strongly favored reaction. The enthalpy change of the reaction in the liquid phase is 153.8 kcal/mole (643.499 kJ/mole), and in the vapor phase it is 149.4 kcal/mole (625.090 kJ/mole), which is more than twice as much as for the hydrogenation of furfural to furfuryl alcohol.

The guideline design value for the process is 150 liter/h of THF per kg of palladium.

Noteworthily, the support of the palladium by microporous carbon was found to be of great importance. With palladium on Al_2O_3, the reaction rate amounts to only 25 percent of the value for palladium on microporous carbon, and with palladium on polystyrenedivinylbenzene, no THF is obtained at all. This seems to indicate that the electrical conductivity of the support plays a decisive role in the process.

With microporous carbon as support, the reaction is perfectly selective, whereas with other supports some by-products are formed.

The only negative aspect of the process is the fact that the catalyst starts losing its activity after approximately 10 hours of operation, thus requiring rather frequent regenerations. The reason for the decreasing activity is believed to be a gradual growth of the palladium crystallites.

Reference

[85] C. Godawa, A. Gaset, P. Kalck, and Y. Maire, Journal of Molecular Catalysis 34 (1986) 199-212.

25. Polytetrahydrofuran

Tetrahydrofuran, just as other cyclic ethers, can be submitted to a cationic polymerization. It leads to polytetrahydrofuran (PTHF). Schematically, this can be expressed as follows:

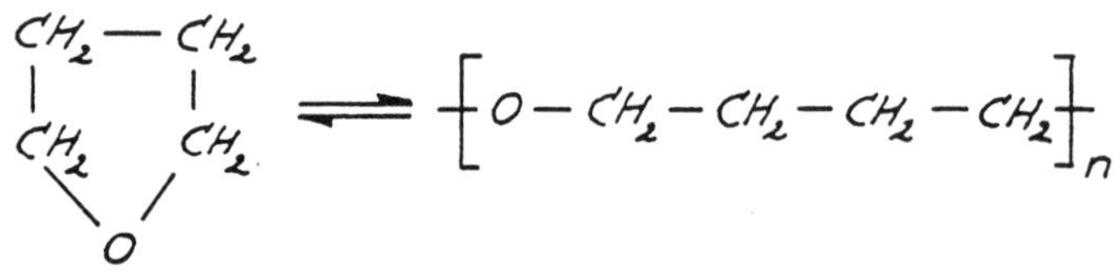

The mechanism of the reaction is determined by the characteristic of the initiator. Of the numerous known initiator systems, so far only four are being used for the industrial production of PTHF:

(1) Fluosulfonic acid (FSO_3H).

(2) Antimony pentachloride ($SbCl_5$).

(3) Siliceous earth plus acetic anhydride.

(4) Perfluosulfonic acid resin plus acetic anhydride.

The first two systems initiate single phase processes. Their advantage is the fact that chain termination with water leads directly to PTHF with OH end groups whereas the systems (3) and (4) give the PTHF in the form of its diacetate which must be converted to the diol in an additional step. On the other hand, the systems (1) and (2) have the disadvantage of yielding colored products. In addition, system (1) leads to huge quantities of calcium sulfate mixed with calcium fluoride, and in the case of sytem (2) the waste product is a mixture of antimony oxide and calcium oxide, which requires a legally acceptable disposal.

The process (3), developed by the BASF, yields an almost colorless polyether, and the initiator (siliceous earth), after being separated in a centrifuge, can be reused.

The process (4), developed by DU PONT, is associated with extensive experience in fluorine chemistry and requires a very special know-how for the production of the initiator resin.

The sytem (3) is viewed as the most elegant process, and for this reason it is discussed in detail. The systems (1) and (2) are treated briefly, whereas the system (4) is not treated at all.

25.1. Ring Opening and Addition of Opened Rings

The polymerization of THF involves an opening of the ring and a putting together of such opened rings. To initiate this process, it is first necessary that in one of the many available THF molecules the two carbon atoms bonded to the ring oxygen are being partially depleted of electrons. To this end, an attack agent Δ (an initiator) is added which is capable of forcing one of the two lone electron pairs of the ring oxygen to form a bond. In this fashion, the ring oxygen is partially depleted of electrons, which means that the ring oxygen assumes a positive charge, and inasmuch as oxygen is more electronegative than carbon, it counteracts this electron depletion by withdrawing electrons from the two carbon atoms it is bonded to. Consequently, in the THF ring attacked by Δ, the two carbon atoms next to the oxygen atom assume a positive charge:

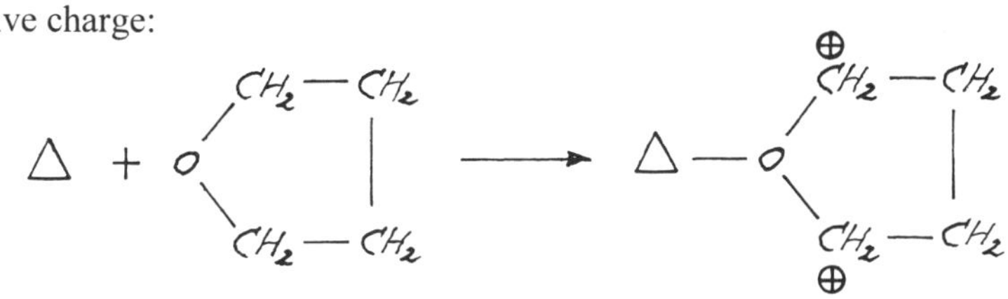

(1)

Once reaction (1) has taken place, collision of the oxygen of an ambient normal THF molecule with one of the positively charged carbon atoms leads to a reaction characterized by the fact that a new bond is formed with the positively charged carbon atom while the latter abandons one of its original bonds. In the case considered, the bond abandoned turns out to be the bond to the original trivalent oxygen, since this is the weakest of all the bonds around. The C-H bonds are very much stronger than C-C and C-O bonds, and the C-O bond partially depleted of electrons is weaker than the C-C bond.

In the brief transition state, when the new bond is being established while the old bonds still persist, the carbon atom attacked has five bonds. This means that the normally sp^3 hybridized carbon atom (tetrahedral bond arrangement as in methane) assumes a bond configuration where the two hydrogen atoms and the second ring carbon are located in a plane (sp^2 hybridization) while the oxygen of the attacking THF molecule and the oxygen of the attacked ring are at right angles to this plane and bonded by the two lobes of the unhybridized p orbital that is thereby made available. This puts considerable strain on the attacked ring and facilitates the rupture of the bond to the original trivalent oxygen.

After establishment of the new bond, a positive charge is now on the oxygen atom in the added ring, as this oxygen has become trivalent, and this positive charge is transferred on the three carbon atoms bonded to this oxygen:

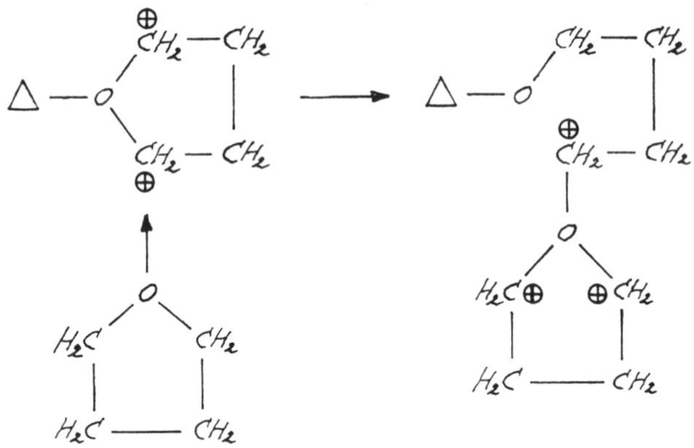

(2)

As can be seen, the first ring has been opened, a second ring has been added, and this second ring is destabilized by positive charges. If in this situation one of the two positively charged ring carbon atoms collides with the oxygen of a normal THF molecule, the process outlined by reaction (2) is repeated, thus leading to a further ring opening coupled with the attachment of another ring. If, on the other hand, the positively charged carbon atom in the newly formed chain collides with the oxygen of a normal THF molecule, the latter is being attached while the originally added ring is freed to return into the ambient system as a monomer molecule. The net result of this latter transformation is obviously zero, so that this type of process need not be discussed any further. Hence, the chain creation and propagation due to ring opening and ring addition is the only observable change.

The driving force (decrease in free energy) for this change is due to the increase in entropy as caused by the increase in disorder when the orderly ring structure is converted to the disorderly chain structure, with more degrees of freedom leading to a higher specific heat and thus to a higher entropy.

25.2. Effect of Acetic Anhydride

Due to its oxygen atoms, acetic anhydride is an electron pair donor in very much the same fashion as THF. The "donor number" DN (defined as the negative enthalpy of reaction for a 1 : 1 adduct formation between antimony pentachloride and the electron pair donor) is 10.5 kcal/mole (43.932 kJ/mole) for acetic anhydride and 20.0 kcal/mole (83.680 kJ/mole) for THF [86]. Hence, acetic anhydride is a serious competitor for reacting with the positively charged carbon atoms in reaction (2) of the preceding chapter, especially when due to progressing polymerization the concentration of THF in the reactor solution decreases.

The reaction of acetic anhydride with a growing PTHF chain leads to the attachment of an acetate end group on the existing chain, and to a new chain embryo consisting of an acetate starter group bonded to a THF molecule by trivalent oxygen. This reaction is shown below, where for the sake of clarity the hydrogen atoms bonded to the carbon atoms are now omitted.

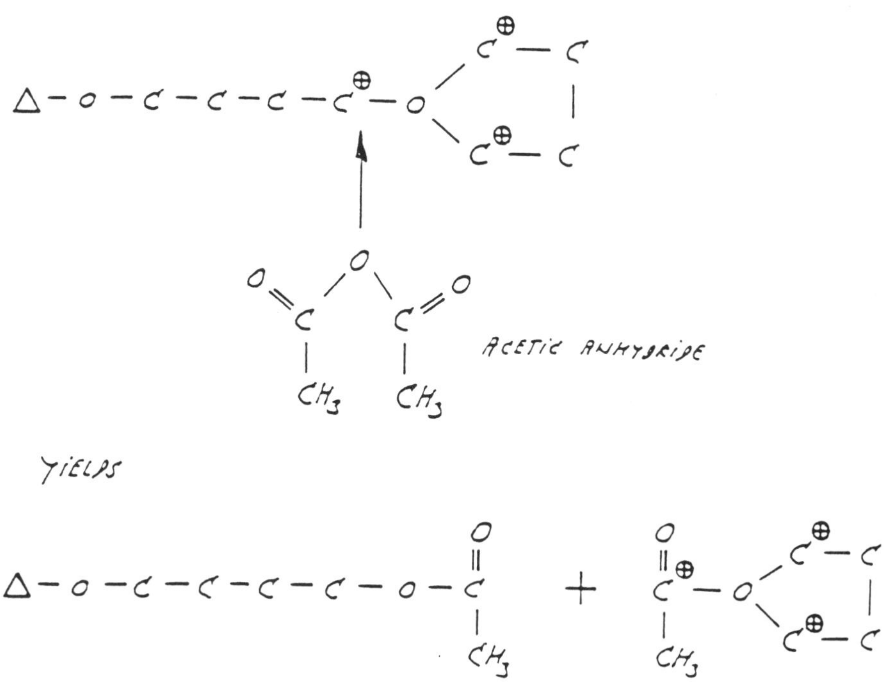

As can be seen, with the acetic anhydride being split, the "old" chain is terminated while a new chain starts growing. Such a process is called "chain transfer", and the agent causing it, in this case the acetic anhydride, is known as the "polymerization modifier".

From the mechanism described, it is readily realized that the average molecular weight of the end product must decrease with increasing modifier concentration. However, it must be noted that the average molecular weight also depends on the yield. At a low yield, the final ratio of monomer concentration to modifier concentration is greater than in the case of a high yield, so that the chain transfer effect in the final stages of the process will be the smaller the smaller the yield.

When siliceous earth is used as the initiator, with acid sites Δ in its surface being the active points, then without acetic anhydride the average molecular weight would become too high as there are simply not enough acid initiator sites to start many chains. Many acid sites would lead to many chains, while few acid sites would lead to few chains, so that with few acid sites a given number of monomer molecules leads to few long chains and thus to a high molecular weight.

In addition to its effect on the molecular weight, acetic anhydride also plays an important role in freeing a PTHF chain from the solid siliceous earth initiator. In view of acetic anhydride having a sizable donor number, as given above, one of the lone electron pairs of its bridge oxygen is capable of forming a bond with an acid site of the siliceous earth in very much the same fashion as THF, so that this oxygen becomes trivalent and positively charged. On account of the high electronegativity of oxygen, the positive charge is transmitted to the carbon atoms bonded to this oxygen. Consequently, each of these positive carbon atoms becomes susceptible to being attacked by a lone pair of electrons from the oxygen linking a neighboring PTHF chain to an acid site:

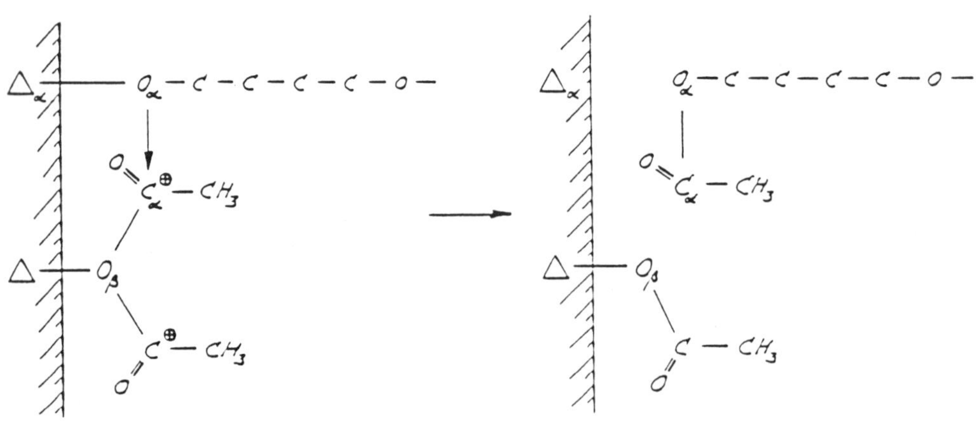

In the resulting reaction shown, the oxygen O_α forms a bond with the carbon atom C_α and abandons its previous bond with the acid site Δ_α, while C_α abandons its previous bond with O_β. The outcome is a freed PTHF chain with an acetyl end group, and a CH_3COO group attached to an acid site. The mechanism is analogous to a normal acetylation reaction such as

$$\text{ethanol} + \text{acetic anhydride} \rightarrow\rightarrow \text{ethyl acetate} + \text{acetic acid} \qquad (A)$$

Considering that the free energies of formation of ethanol, acetic anhydride, ethyl acetate, and acetic acid at standard conditions (1 ATM, 25 °C) are –40.1, -114.0, -90.0, and –78.1 kcal/mole (-167.778, -476.976, -376.560, and –326.770 kJ/mole), respectively, the driving force of reaction (A), i.e. the decrease in free energy, is 14.0 kcal/mole (58.576 kJ/mole). Approximately the same value can be expected to apply to the "chain liberation reaction".

25.3. Polymerization with Siliceous Earth

A laboratory procedure for the polymerization of THF with siliceous earth can be outlined as follows:

The experimental setup to be used is shown in Figure 93. The reactor is a 2-liter 3-neck round bottom glass flask 1. Oxygen is to be removed by flushing with pure nitrogen for 12 hours. During this time, the flask is heated to 200 °C by means of an electrical heater to remove traces of water from the glass surfaces. Under increased nitrogen flow, column 2 is removed and replaced by a ground glass stopper. The column 2 is filled with a 3 angstrom molecular sieve previously dried at 140 °C for 3 hours under vacuum. The column, with the tap 4 closed, is clamped on a stand over an open beaker and slowly filled with absolutely dry cyclohexane. The latter is drained, and the column is then filled with THF of maximum available purity. After 5 minutes, this THF is drained, and new THF is filled in and drained twice in the same fashion. Then the column is replaced onto the reactor flask 1. Now the measuring and feeding cylinder 3 is filled with 250 ml of THF, a nitrogen input line is attached, and the cylinder is placed on the column 2. By opening tap 4, the column 2 is flushed with nitrogen. After 15 minutes, tap 5 is opened so that 250 ml of THF flow into the flask. Under increased nitrogen flow, 8 ml of acetic anhydride are added through a funnel at point 6. Then the funnel is replaced by a powder funnel to add 25 g of siliceous earth previously dried at 130 °C for 12 hours under vacuum, preferably on an aluminum foil shaped to form a chute. The siliceous earth should be a product named TONSIL manufactured by SÜD-CHEMIE of Munich/Germany. Before being added to the flask 1, the TONSIL powder must

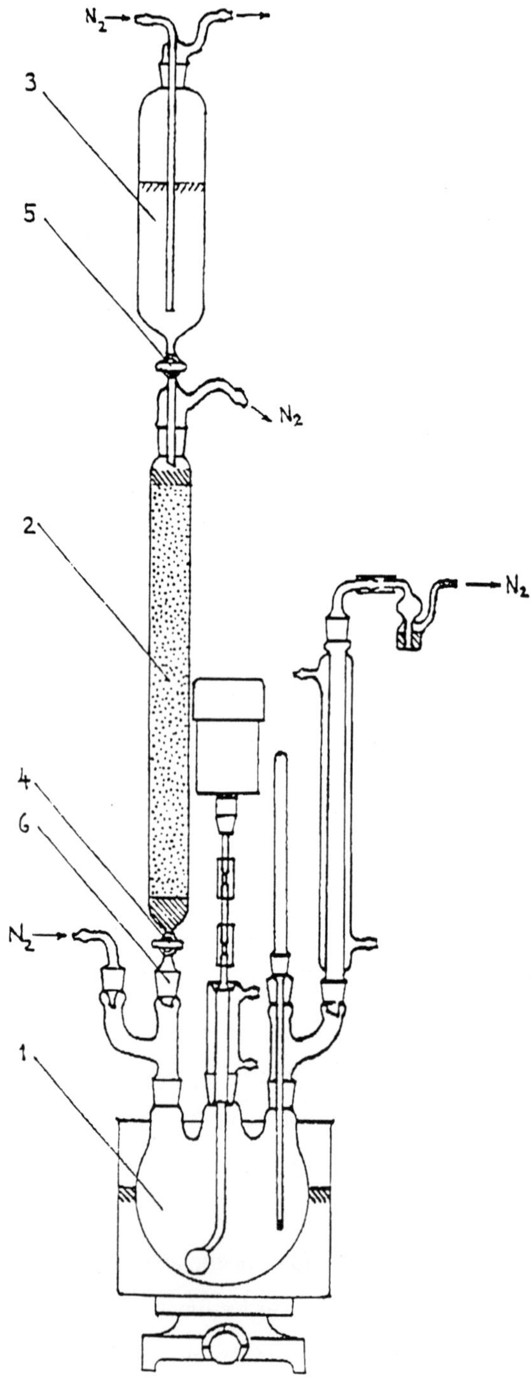

Figure 93. Laboratory Setup for the Polymerization of Tetrahydrofuran.

have been allowed to cool to a temperature between 20 and 40 °C. By means of a water bath, the reaction flask is then heated to 40 °C and held at this temperature for a period of 3 hours. The start of the polymerization can be noted by a small rise of the internal temperature, which must be compensated by lowering the temperature of the water bath. The viscosity will increase noticeably after 30 minutes.

After 3 hours, the degree of conversion will be 50 to 60 percent. At this point, the content of the reactor is removed, and the TONSIL is separated from the polymer solution by centrifugation. After being dried and stored in the absence of oxygen, it can be reused.

With the separation of the TONSIL, the polymerization is terminated. The remaining monomer and the acetic anhydride are distilled off in a rotary evaporator, at the end under vacuum. The remaining viscous liquid is pure PTHF in the form of its diacetate.

In general, the PTHF is needed in the diol form (with OH end groups). This form is called PTMEG which stands for polytetramethylene ether glycol. It is also known as POLYMEG (a trade name of QUAKER OATS). Conversion of the diacetate form to the diol form can be carried out as follows:

The polymer is dissolved in 100 ml of methanol and mixed with a solution of 15 g of NaOH in 100 ml of methanol. In a round bottom flask equipped with a feed pipe and a reflux condenser, the mixture is boiled under nitrogen for a period of 2 hours. Then the methanol is distilled off as far as foam formation will permit. After addition of 200 ml of toluene, the mixture is distilled under a weak vacuum until the distillation temperature becomes constant. After cooling and addition of another 200 ml of toluene, the mixture is extracted with 100 ml quantities of water in a separating funnel until the aqueous phase is neutral. The toluene phase is dried with anhydrous sodium sulfate and clarified before the toluene is removed in a rotary evaporator. The remaining viscous liquid is an almost colorless PTMEG having a mean molecular weight between 800 and 1000.

The given conversion from the diacetate form to the diol form is shown below. It leads to sodium acetate.

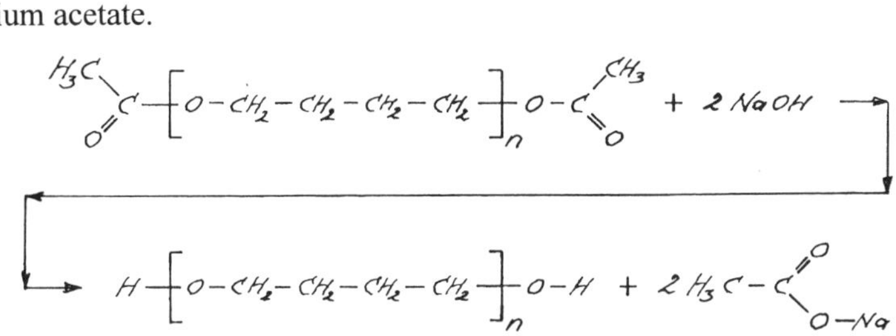

When large quantities are to be processed, the last step (diacetate to diol) is more conveniently carried out as a transesterification, with an excess of methanol and CaO as a catalyst [87]. This process is shown below. It leads to methyl acetate.

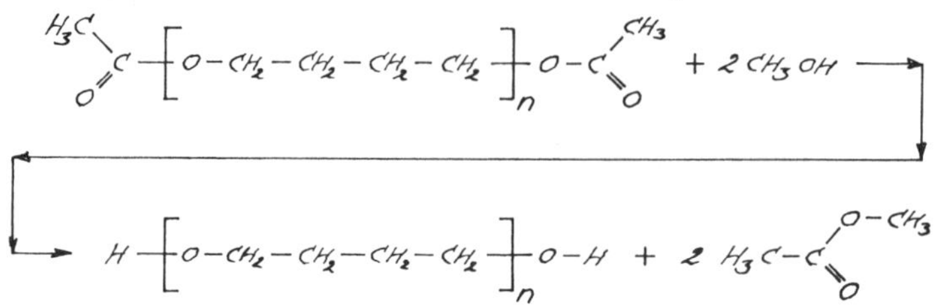

A respective production plant is outlined in Figure 94. The processing sequence consists of four stages requiring an equal time of approximately 6 hours. If reactor R_1 has a nominal capacity of 6 m^3 and a usable volume of 4 m^3, and if the units R_D, R_2, and R_3 are sized accordingly, then one such processing street can deliver approximately 8 metric tons of PTMEG per day of 24 hours if the degree of conversion is 50 percent. Hence, a 10 000 tons/a facility would require four of these production streets. In this design, enough leeway is incorporated to permit producing several different grades (different mean molecular weights).

Of course, if a supply of suitable THF can be guaranteed, reactor R_D (drying) can be omitted.

After an initial thorough flush with high-purity nitrogen, the entire plant must be kept under a slight positive nitrogen pressure. The initial flush is effected by repeated evacuation and refilling with nitrogen. When vacuum units as the evaporators are brought to atmospheric pressure to permit discharging, this is always done with nitrogen, never with air.

25.4. Polymerization with Fluosulfonic Acid

Fluosulfonic acid

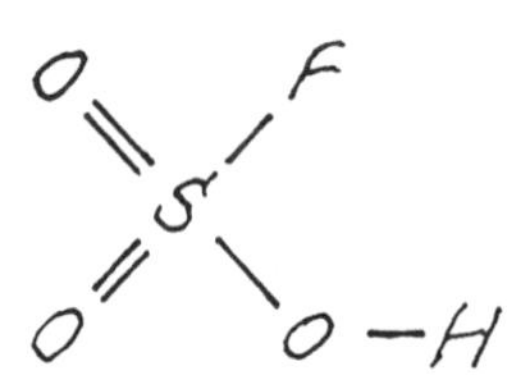

is a colorless liquid having a freezing point of –89.0 °C and a boiling point of +162.7 °C. Due to the high electronegativities of fluorine and oxygen (4.0 and 3.5, respectively), the hydrogen atom is very strongly positive, in complete analogy to the "acid sites" of siliceous earth. Thus, in the ring opening and ring addition mechanisms described in subchapter 25.1, with

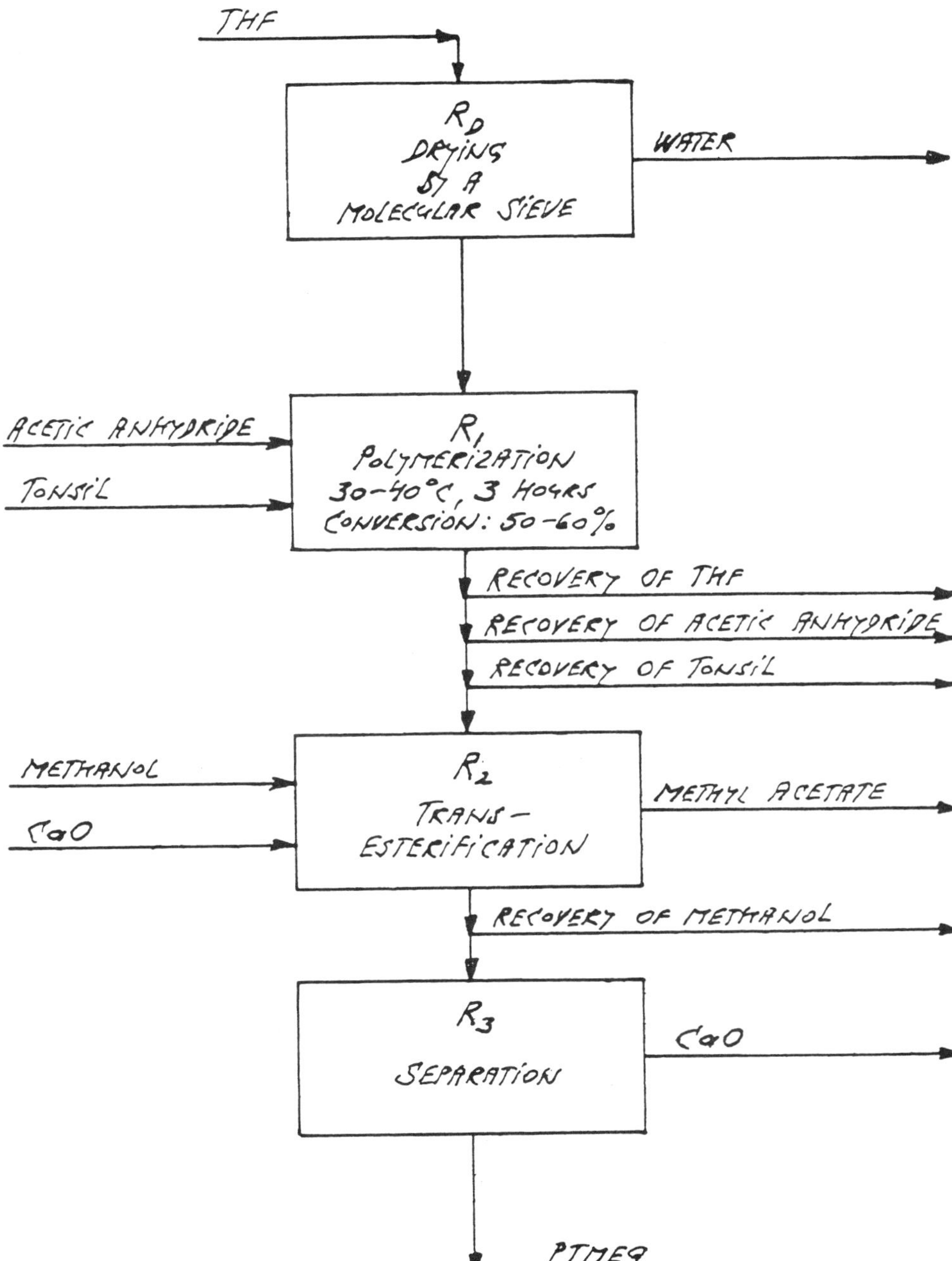

Figure 94. Outline of a PTMEG Production Plant.

fluosulfonic acid as the polymerization catalyst, the initiator Δ is the strongly positive hydrogen atom. Of course, contrary to an aqueous system, in THF there is no dissociation of the fluosulfonic acid, so that the latter acts in molecular form.

Unlike the rather ill-defined "acid sites" of siliceous earth, with fluosulfonic acid the intensity of polymerization catalysis (i.e. the number of chain initiations) is clearly dictated by the quantity of fluosulfonic acid added to the THF. The more fluosulfonic acid is added, the more chains are initiated, which means that the more fluosulfonic acid is added, the lower will be the molecular weight of the polymer product.

Against this background, the process for polymerizing THF with fluosulfonic acid can be outlined as follows:

The laboratory setup to be used is the same as shown in Figure 93 (page 192), and the procedure for charging the reactor with 250 ml of THF is the same as described for the polymerization with siliceous earth. Then, the THF in the reactor flask must be cooled to –10 °C. Subsequently, an absolutely dry piston pipette is used to withdraw 30 ml of fluosulfonic acid from the storage bottle, and to slowly inject this initiator into the reaction flask while cooling is maintained. Depending on the purity and the pretreatment of the THF employed, the fluosulfonic acid will cause a more or less pronounced discoloration. When the addition procedure is completed, a water bath is used to take the temperature in the reaction flask to 25 °C and to maintain this level for a period of 6 hours. During this time, the viscosity will be seen to increase significantly.

To terminate the polymerization, 50 ml of water are added while the flask is being cooled and stirring is accelerated. The "enforced termination reaction" between the growing macrocation and water is shown below:

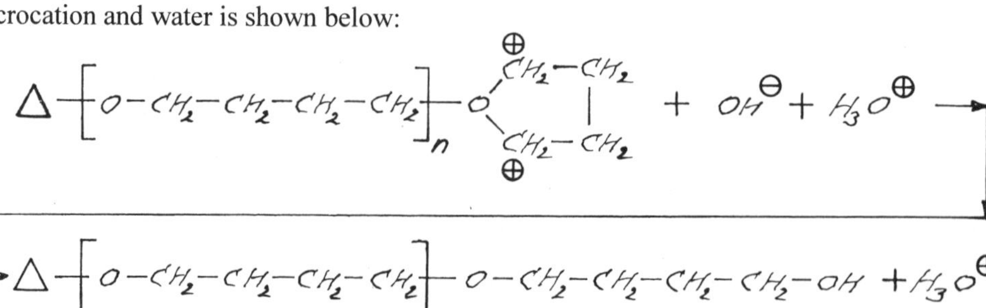

The macrocation is neutralized by a hydroxyl ion, thus leading to an OH end group, so that the final product is PTMEG. This reaction is the reason why, in the procedures described,water had to be scrupulously eliminated from the system as otherwise polymerization would not take place.

After addition of the water, the mixture in the reactor is heated to 80 °C for a period of 30 minutes. Due to this treatment, the major portion of the nonpolymerized THF (boiling point 65 °C) will be distilled off. After cooling, 500 ml of toluene are added, and the mixture is submitted to an extraction with 100 ml quantities of water in a separating funnel until the aqueous phase is neutral. The toluene phase is then mixed with 10 g of bleaching earth and submitted to filtration or centrifugation before being freed of the toluene in a rotary vacuum evaporator.

25.5. Polymerization with Antimony Pentachloride

Antimony pentachloride, $SbCl_5$, is a liquid having a freezing point of 2.8 °C and a boiling point of 140 °C. When absolutely pure, the liquid is colorless, but commercial $SbCl_5$ is usually yellowish. $SbCl_5$ has a high density (2.35 g/cc), and in humid air its vapor reacts with water to form visible fumes of hydrochloric acid mist and Sb_2O_5 particles:

$$2\ SbCl_5 + 5\ H_2O \rightarrow\rightarrow 10\ HCl + Sb_2O_5$$

As catalyst for the polymerization of THF, $SbCl_5$ works by self-ionization [88] according to the equilibrium

$$2\ SbCl_5 \longleftrightarrow SbCl_4^{\oplus} + SbCl_6^{\ominus}$$

where the cation $SbCl_4^+$ takes the role of the initiator Δ. The given equilibrium is proven by the fact that $SbCl_5$ is an electrical conductor.

The polymerization of THF with $SbCl_5$ starts in the same fashion as with fluosulfonic acid except that 250 ml of THF require only 15 ml of $SbCl_5$ (instead of 30 ml of fluosulfonic acid). The addition of the $SbCl_5$ causes the THF to assume a yellow to brown color. Some toluene is added to prevent any sudden rise in viscosity. A water bath is employed to take the mixture of THF, $SbCl_5$, and toluene to 40 °C for a period of 3 hours. During this time, the viscosity will be seen to increase significantly.

To terminate the rection, 40 ml of aqueous $Ca(OH)_2$ solution is added. Then the $SbCl_5$ is decomposed by boiling the mixture under reflux for a period of 30 minutes. This leads to a formation of Sb_2O_3 and to a venting of chlorine. Subsequently, the mixture is concentrated in a rotary evaporator until all of the water is distilled off. The remaining mixture, which is turbid because of its $Ca(OH)_2$ and Sb_2O_3 particles, is diluted with toluene and mixed with 5 to 10 g of bleaching earth before being submitted to centrifugation. Finally, the clarified toluene solution thus obtained is concentrated in a rotary evaporator, at the end under vacuum. As in the case of fluosulfonic acid as initiator, the chains have OH end groups, so that the product is PTMEG.

25.6. Discussion of the Initiators

In subchapter 25.1, it was shown that the opening of the THF ring requires an initiator Δ capable of forcing one of the two lone electron pairs of the ring oxygen to form a bond. In other words, Δ must be an agent featuring a positive charge, but it must satisfy two additional conditions:

(1) It must be free of water as the latter prevents polymerization in neutralizing any cation by hydroxyl ions.

(2) It must permit being removed from the system with ease as it cannot be left in the polymer.

Fluosulfonic acid provides the positive charge by its hydrogen atom highly depleted of electrons, and it can be removed, though not conveniently, by extraction. Antimony pentachloride provides the positive charge by its self-dissociation into $SbCl_4^+$ and $SbCl_6^-$, and can be removed by decomposition.

In the case of siliceous earth, being a solid, the removal by centrifugation is particularly easy, but the nature of its positive charge is not obvious. The TONSIL of SÜD-CHEMIE is produced from a mined montmorillonite clay having the composition $Al_4Si_8O_{20}(OH)_4$. As mined, this clay has no catalytic activity. The latter is imparted to the mineral by subjecting it to an acid treatment. The raw material is reduced to a fine powder and then boiled in a strong aqueous solution of hydrochloric or sulfuric acid. After this treatment, which lasts 3 to 4 hours, the material is washed, drained, and finally dried at a temperature in the order of 200 °C. In the acid processing, roughly half of the aluminum is leached out while the silicon remains untouched. This results in a negative lattice charge, so that hydrogen ions from the acid become incorporated in the system to restore neutrality. Thus, the TONSIL is an "acid-activated clay" and represents a "solid acid". Against this background, the question arises as to how the proven catalytic activity of an acid-activated clay can be characterized by a meaningful measurement. For a liquid, there is only one acidity, usually defined as moles of

H_3O^+ per liter, but for a “solid acid” such as acid-activated clay a sharp distinction must be made between “soluble acidity” and “local acid strength”. The “soluble acidity” can be readily measured by convential techniques such as titration or gas volumeter analysis. As to titration, the clay can be dispersed in water, and any acidity thus liberated can be neutralized. On this basis, Thomas, Hickey, and Stecker [89] found that raw montmorillonite yielded 0.41 milliequivalents of acid per gram of dry clay, while after acid treatment (removal of half of the aluminum) this value rose to only 1 milliequivalent per gram. If the clay were a liquid with the density of water, these results would mean hydrogen ion concentrations of 0.41×10^{-6} and 1×10^{-6} mole per liter, which corresponds to pH values of 6.39 and 6.00, respectively. Thus, even for the acid-activated clay the “soluble acidity” is extremely small, and cannot possibly explain the proven catalytic effect of this material. It does, however, explain the fact that TONSIL can be swallowed without harm.

In other words, as far as characterizing the catalytic activity is concerned, the “soluble acidity” as determined by conventional techniques is absolutely meaningless. The reason for this can be seen in two plausible aspects:

(1) Most of the hydrogen ions do not go into solution when the clay is dispersed in water. They rather remain fixed in the lattice, so that conventional techniques fail to count them.

(2) The hydrogen ion activity (or positive charge density) is not uniform over the surface but must be seen as a rugged relief with pronounced peaks:

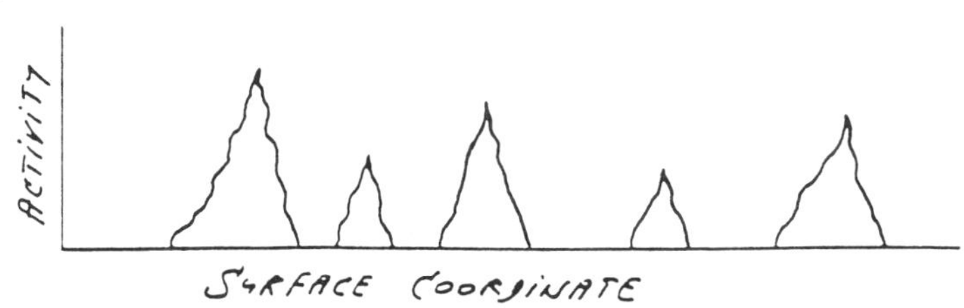

Against this background, any meaningful characterization of the catalytic activity of acid-activated clay imposes the following conditions:

(a) The measurement must be capable of determining the activity in situ.

(b) The measurement must establish the “height” of the activity peaks and their relative area.

The first step in this direction was made by Walling [90]. He measured the acid strength of a solid by determining its ability to change the color of various indicator dyes adsorbed on the surface. He introduced the fundamental idea that activities and equilibria can be formulated not only for mobile species but also for systems in which some of the species are fixed. As measuring agents, he used certain basic indicator dyes that gave deep colors when exposed to strong acids. Dye molecules dissolved in an appropriate liquid will be adsorbed on the entire surface of the solid, but the color change will be experienced only by those molecules which are in contact with an area of high acidity (high positive charge density). If there are such areas, the color will be seen in very much the same fashion as in a photograph which implements the color sensation by a multitude of individual dots. The process can be compared to the printing of a relief map. If all areas higher than 1000 m are printed blue, and if only these areas are printed, the paper will show a few blue dots if there are a few peaks that high, and no blue at all if in the region represented by the map there are no peaks of this altitude. Thus, any blue is proof for the existence of peaks higher than 1000 m, and with other colors the argument can be extended to other altitudes. By the same token, when the color of a basic dye is known to change when the hydrogen ion activity exceeds 0.1 mole/liter (pH = 1), and if this dye, when adsorbed on the surface of a solid, does experience this color change, then it is proven that the surface has acid sites stronger than the hydrogen ion activity in a solution of pH = 1.

Using a series of indicators with pK_a values in the range between +3.3 and +0.42, Walling was able to show that acid-activated clays have local acid activities greater than what is found in a solution with a pH = 0.1.

A few years later, Benesi [91] extended Walling's technique by using indicators covering a pK_a range from +6.8 to –8.2. The pK_a value of –8.2 refers to anthraquinone (basic form colorless, acid form yellow). Benesi found that acid-activated clays did give anthraquinone a yellow color, so that their surface contains acid sites which are more strongly acid than a 92 % aqueous sulfuric acid solution. How much stronger they are he could not say as he did not have an indicator beyond this range. His findings explain, however, why a perfectly harmless powder such as TONSIL catalyzes the polymerization of THF: It has acid sites with an acidity (positive charge density) stronger than that of concentrated sulfuric acid.

In the meantime, indicators covering the range from pK_a = +6.8 to –16.04 were found [92].

The acid strength determination is made by placing approximately 0.2 ml of the acid-activated clay into a test tube, adding 2 ml of a nonpolar solvent containing 0.2 ml of the indicator, and shaking briefly. Adsorption proceeds rapidly, and the change in color between the basic and the acid form of the indicator, if it occurs, is most striking. The nonpolar indicator solvents commonly used are benzene, cyclohexane, and isooctane.

Even the number of acid sites can be measured. The procedure consists in mixing a weighed clay sample with an indicator solution, and in submitting this slurry to a titration with n-butylamine [93]. Before the titration, indicator molecules adsorbed on sites having acidities beyond the conversion point of the indicator will have changed to the acid form. Since the basicity of n-butylamine is very much greater than the basicity of the indicators, it follows that n-butylamine molecules will replace the indicator molecules on the acid sites mentioned, so that the quantity of titer liquid required to restore the basic color of the indicator represents a measure of the number of acid sites having a strength in excess of the value associated with the indicator. Hence, by employing such a procedure with various indicators, it is possible to determine the number of acid sites of various strength.

A typical reaction taking place in this procedure can be written schematically as follows:

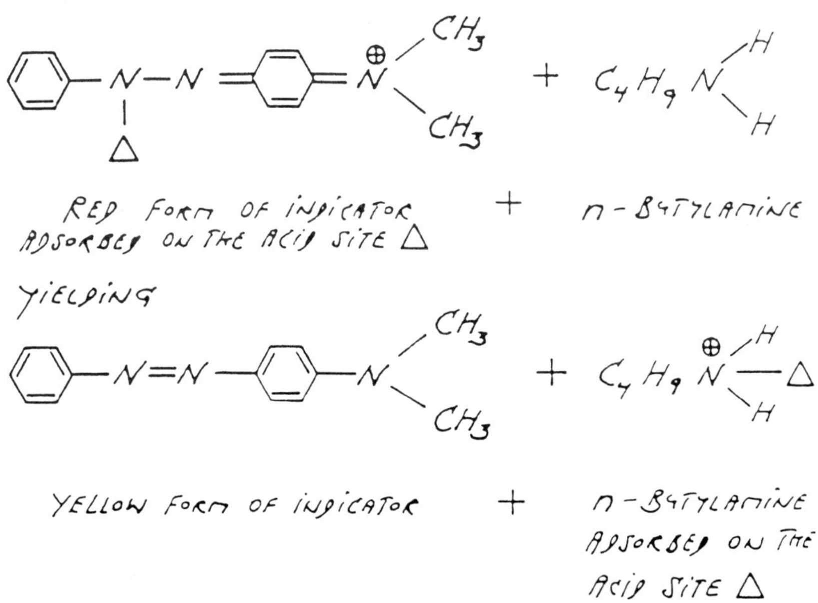

The indicator shown is p-dimethylaminoazobenzene (also known as “dimethyl yellow”). Its basic color is yellow, and its acid color is red. Its pK_a is +3.3. The red color of the acid form is due to the fact that the acid site Δ withdraws electrons from the indicator molecule, so that the latter becomes electron-deficient with respect to the basic form. On the other hand, when n-butylamine reacts with an acid site Δ, it too is being depleted of electrons, but no color sensation is noted since n-butylamine does not absorb light in the visible spectrum. Hence, n-butylamine is added until the normal yellow form of the indicator is restored, and the quantity of n-butylamine required to achieve this result is proportional to the number of acid sites having produced the red color.

Inasmuch as the number of acid sites and their strength determine the number of chains produced in the polymerization of THF with siliceous earth, analyses of the given type are a prerequisite for correlating the catalytic activity of the clay with the molecular weight of the product. The other initiators, not being subject to inherent variations imposed by a mining field, obviously do not require such an elaborate "initiator quality control", but the harmlessness of the clay as opposed to the hazardous nature of the other initiators is such a huge advantage that only the polymerization with siliceous earth can be recommended. After all, confirming this argument, the BASF, one of the largest PTHF producers in the world, is working by this process.

The original patent on the polymerization of THF with siliceous earth, granted to C. Dörfelt in 1967 [94], with FARBWERKE HOECHST AG as assignee, has long expired.

25.7. Quality of the THF Input

Because of the ease with which THF forms peroxides, it is normally shipped and stored in the presence of a stabilizer such as 2,6-di-t-butyl-4-methylphenol (butylated hydroxytoluene, BHT). In addition, THF readily absorbs water and oxygen. Inasmuch as these impurities are detrimental to polymerization, it is imperative to purify the THF before it can be used as a polymerization monomer.

It is strongly recommended that the THF first be checked for the amount of peroxide or hydroperoxide before any further purification is attempted. If more than 0.1 % is found, the peroxide must be decomposed. This can be done by agitating with flake caustic soda. It has also been suggested to effect the deperoxidation by means of self-indicating molecular sieves [95], but inasmuch as this must result in a peroxide enrichment in the molecular sieve, this procedure is viewed as a source of danger which has not been sufficiently investigated so far.

Water in THF is the second most important problem as it makes polymerization impossible. The vapor/liquid equilibrium diagram of the THF/water system at atmospheric pressure is shown in Figure 95.

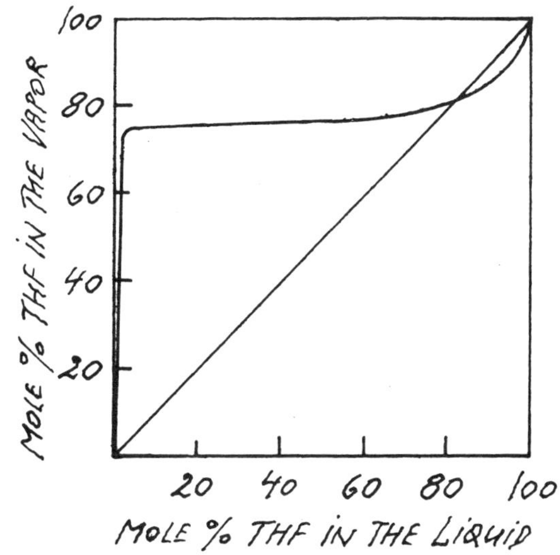

Figure 95. The Vapor/Liquid Equilibrium Diagram of the THF/Water System at 1 ATM.

There is an azeotrope at 18.3 % water by mole (5.3 % water by weight). Thus, if the water content is below 5.3 % by weight, which is usually the case, most of the water can be removed by azeotropic distillation, as in the case of furfural. To remove the last traces of water, the best procedures are the following:

(1) Passage through a molecular sieve.

(2) Distillation, under nitrogen, from sodium or potassium metal.

(3) Distillation, under nitrogen, with toluene as entrainer. This leads to a low-boiling azeotrope, the effect being similar to the well-known dehydration of ethanol with benzene or cyclohexane.

25.8. Applications

The applications of PTHF, in the form of PTMEG (i.e. with OH end groups), lies in the manufacture of elastomers. On an atomic scale, the PTHF chain is not straight but consists of straight sections joined by “angled links” (ether bridges) in between as shown below:

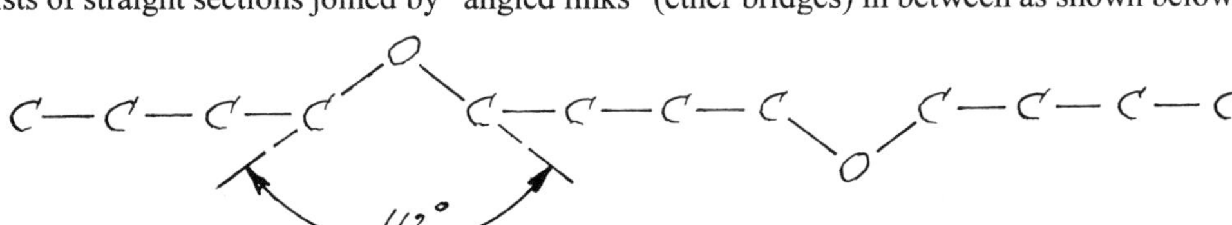

The normal C – O – C bond angle is 112 ° [96], so that under tension the oxygen bridges are "flexed" to larger angles, thereby permitting an elastic (reversible) lengthening of the chain in longitudinal direction. To give stability in directions perpendicular to the chain axis, the PTHF chains are reacted with diisocyanates as this leads to a structure conducive to interchain hydrogen bonding:

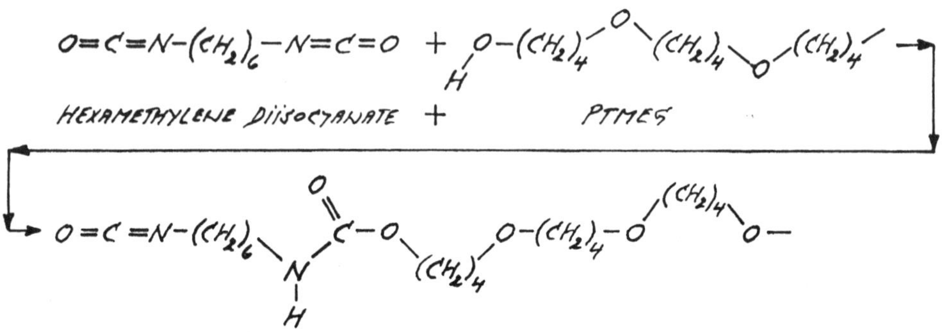

The C – N – C bond angle of 109 ° [96] is seen to add to the elasticity of the chain. Linkage of such structures with hydrazine leads to fibers as they are used for "stretch pants".

References

[86] C. Reichardt, Solvents and Solvent Effects in Organic Chemistry, VCH Verlagsgesellschaft, Weinheim, 1988.

[87] G. Pruckmayr, Alcoholysis Process for Preparing Poly-(Tetramethylene Ether) Glycol, US Patent No.4,230,892 (1980).

[88] P. Rempp and E. W. Merrill, Polymer Synthesis, Huethig & Wepf Verlag, Basel, 1986.

[89] C. L. Thomas, J. Hickey, and G. Stecker, Ind. Eng. Chem. 42 (1950) 866-871.

[90] C. Walling, J. Amer. Chem. Soc. 72 (1950) 1164-1168.

[91] H. A. Benesi, J. Amer. Chem. Soc. 78 (1956) 5490-5494.

[92] K. Tanabe, M. Misano, Y. Ono, and H. Hattori, New Solid Acids and Bases, Elsevier Science Publishers, Amsterdam, 1989.

[93] O. Johnson, J. Phys. Chem. 59 (1955) 827-831.

[94] C. Dörfelt, Verfahren zur Herstellung von Butylenglykoldiacetaten, German Patent No. 1 226 560 (1967).

[95] D. R. Burfield, J. Org. Chem. 47 (1982) 3821-3824.

[96] J. March, Advanced Organic Chemistry, John Wiley & Sons, New York, 1992.

26. Xylose

Xylose is not a by-product of furfural but its precursor. On account of this, its production is governed by the very kinetics of furfural formation, but with the aim of avoiding the latter as best as possible. However, the technically most important difference between xylose production and furfural production is the fact that furfural, because of its low-boiling azeotrope with water, is readily recovered as a vapor, whereas xylose, being nonvolatile, ends up dissolved in the liquid reaction medium, together with many other unwanted by-products, from where a recovery in a sufficiently pure form is not as easy as in the case of separating a product from a vapor mixture. Consequently, xylose plants are far more complicated, and therefore more costly, than furfural plants.

26.1. Xylose from Agricultural Raw Materials

Conventional xylose plants as operated in China use simple batch techniques employing a 2-step treatment of corn cobs with sulfuric acid, step 1 being considered a wash to remove "gums" detrimental for the xylose crystallization [97], and step 2 representing the hydrolysis proper.

Step 1 starts with a mixture of 3 parts by weight of solids in 10 parts by weight of 0.1 % aqueous sulfuric acid, and involves cooking by steam injection at 100 °C for 1.5 hours. The resulting hydrolyzate, containing up to 25 percent of the available pentose, is drained and discarded.

Step 2 starts with a mixture of 3 parts by weight of 1.7 % aqueous sulfuric acid, and involves cooking by steam injection at 120 °C for 3 to 3.5 hours.

According to Dunning and Lathrop [98], the "gums" can also be removed by an extraction with butyl or isoamyl alcohol, but inasmuch as this would greatly complicate the plant by requiring a solvent recovery, this alternative is not being used on an industrial scale, although it would significantly reduce the pentose losses incurred by the conventional step 1.

The given solid/liquid weight ratio of 3 : 10 in the reactors corresponds to 23 percent solids. This means that the reactor content is not a slurry but a packing of solids wetted to some extent, so that separation of the hydrolyzate from the residue must be effected by flushing with steam. With this mode of processing, the final hydrolyzate of step 2 has a xylose concentration in the order of 5 percent. To permit crystallization, this solution must be concentrated in evaporators, which explains why in the hydrolysis the relative quantity of water employed is kept as low as possible.

Depending on the requirements for the final product, a recrystallization can be used to improve the purity.

A flow diagram of a Chinese xylose plant is shown in Figure 95. The raw material is corn cobs. The hydrolyzate of step 2 is neutralized with alkali and then submitted to a discoloration with activated carbon at 80 °C. After separation of the spent carbon in a filter press, the hydrolyzate passes two ion exchange columns before it enters a 3-stage vacuum evaporator, where the xylose concentration is taken from 5 % to 26 % in the first stage, from 26 % to 43 % in the second stage, and from 43 % to 80 % in the third stage, with another ion exchange treatment between the first and the second stage. The concentrate enters a cooled crystallizer. The resulting slurry containing 50 percent by weight of crystals is fed into a filter centrifuge producing a drained crystal cake having a xylose content of 97 percent. The residual moisture is removed in a vibratory drier.

26.2. Xylose from Sulfite Waste Liquor

The process described in the preceding subchapter may be considered obsolete as since the advent of industrial chromatographic separation, xylose can be obtained much more economically from sulfite waste liquor, which is an ideal starting material for the following reasons:

(1) It is available at essentially no cost, and by being freed of the xylose, the sulfite waste liquor becomes more suitable for making "lignosulfonate" (waste liquor concentrate) since in the latter the presence of pentose is detrimental for most applications.

(2) The xylose is already in solution, so that there is no need for handling bulky raw materials and for treating them in voluminous reactors. The cost for hydrolyzing is already absorbed in making pulp.

A Finnish company named CULTOR has developed an industrial chromatographic separation of xylose from sulfite waste liquor, and based on this accomplishment, they now dominate the xylose market. As their process is proprietary, no details can be given, but the general scheme

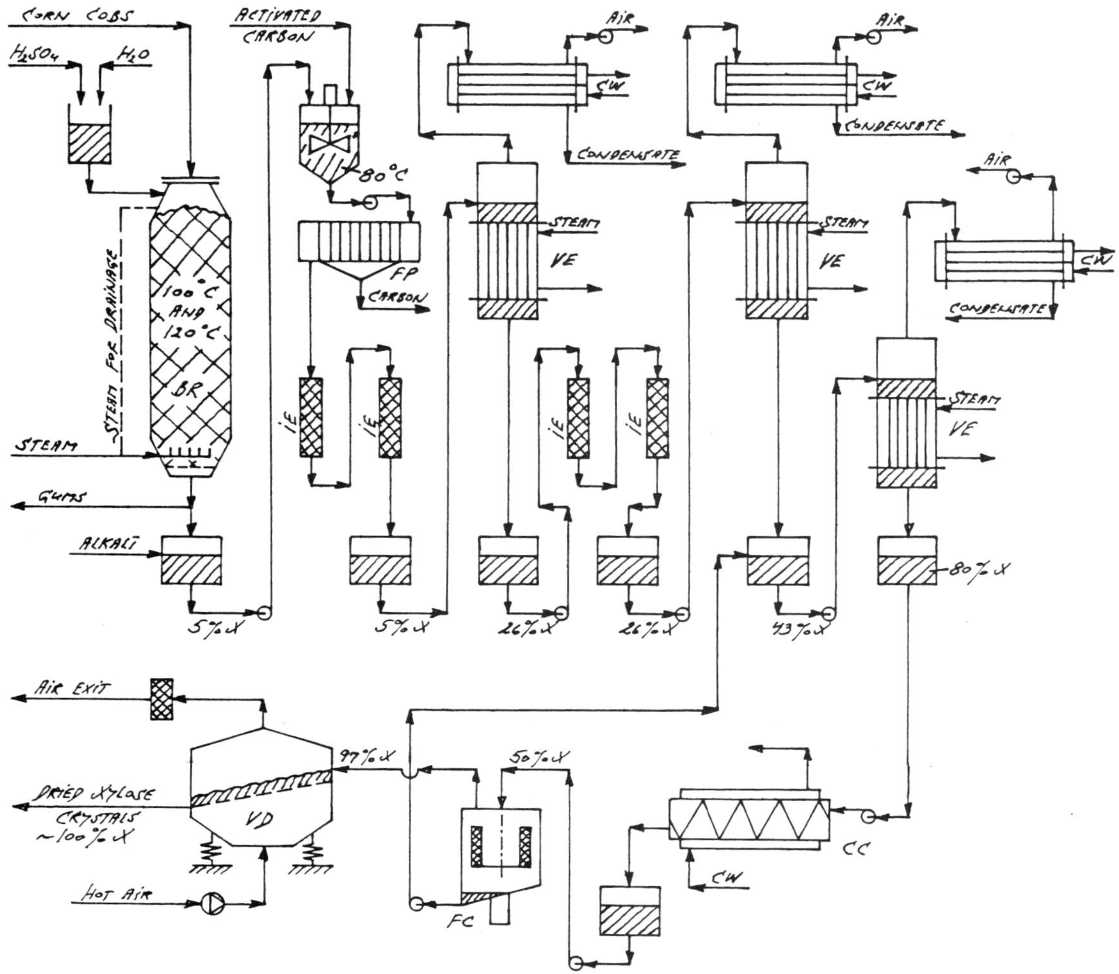

Figure 95. Schematic of a Xylose Plant at Jiaozuo (China).

Capacity: 1800 tons of xylose per annum.

10 reactors 2.45 m in diameter by 7 m high.

BR = Batch reactor

FP = Filter press

IE = Ion exchange

VE = Vacuum evaporator

CC = Cooled crystallizer

FC = Filter centrifuge

VD = Vibratory drier

CW = Cooling water

X = Xylose

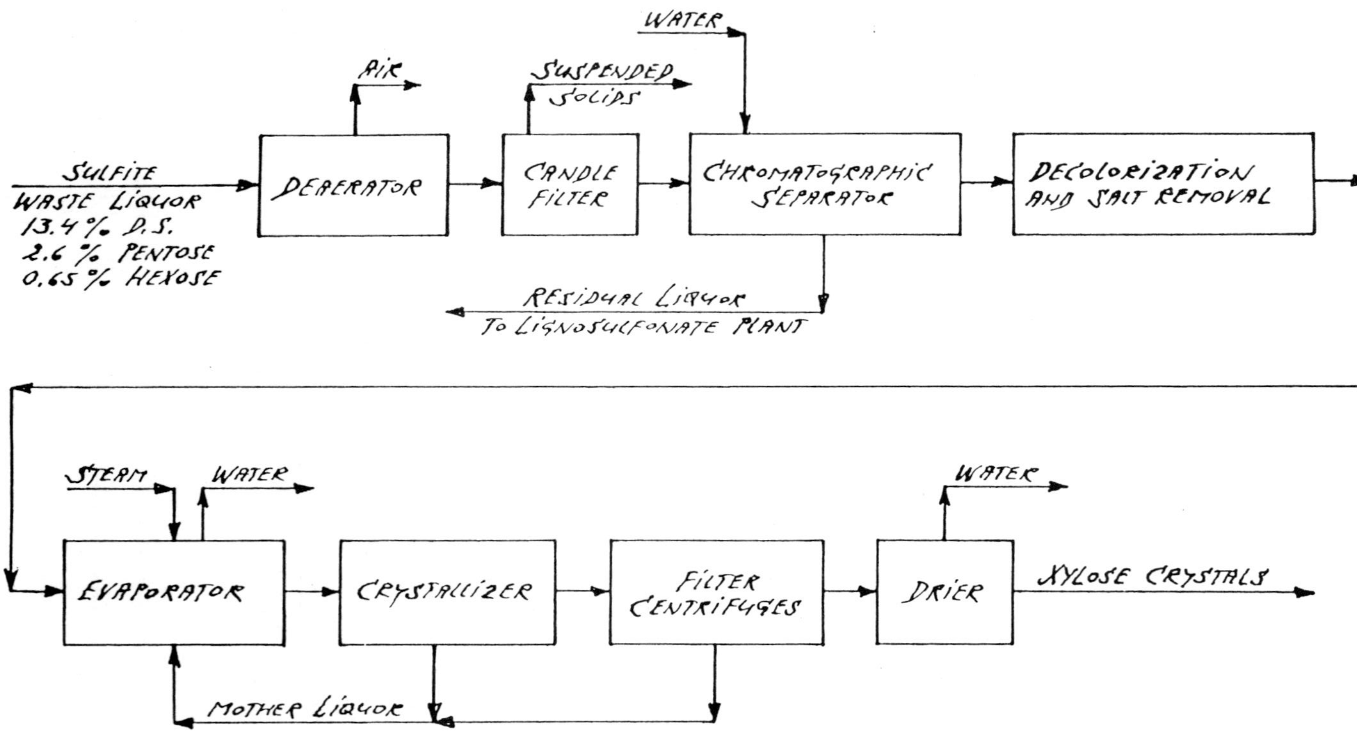

Figure 96. Outline for the Production of Xylose from Sulfite Waste Liquor.

can be expected to look somewhat as outlined in Figure 96. A similar chromatographic separation is commonly used for the production of fructose from an invert sugar solution (a mixture of glucose and fructose).

References

[97] F. B. LaForge and C. S. Hudson, Ind, Eng. Chem. 10 (1918) 925-927.
[98] J. W. Dunning and E. C. Lathrop, Ind. Eng. Chem. 37 (1945) 24-29.

27. Furan Dialdehyde

Furan dialdehyde may be the closest relative of furfural, but strangely enough it is not made from furfural but from fructose. Furan dialdehyde is an obvious competitor of difurfural as a building block for high-temperature polymers based on the thermal stability of the furan ring.

A chemically elegant process for the production of furan dialdehyde was invented by Marval [99]. The process is illustrated in Figure 97. Fructose is dissolved in dimethylsulfoxide and methylisobutyl ketone, and this solution is fed into a heated reactor charged with an acid ion exchange resin as the catalyst. This leads to a dehydration of the fructose to hydroxymethylfurfural. From the resulting solution, the reaction water is removed by distillation as the water/methylisobutyl ketone azeotrope. The sump fraction consisting of hydroxymethylfurfural, dimethylsulfoxide, and methylisobutyl ketone is fed into a second reactor where it is mixed with dicyclohexyl carbodiimide and a solution of phosphoric acid in dimethylsulfoxide. The dimethylsulfoxide acts as the oxidizer, and the dicyclohexyl carbodiimide acts as a water trap to yield furan dialdehyde, dimethyl sulfide, and N,N'-dicyclohexyl urea. The latter precipitates, is separated by filtration, and is then regenerated to dicyclohexyl carbodiimide while water is added to the filtrate to get two liquid phases: A light phase consisting of furan dialdehyde and methylisobutyl ketone, and a heavy phase consisting of dimethyl sulfide, water, and phosphoric acid. The light phase is separated by distillation, yielding methylisobutyl ketone in the head and furan dialdehyde in the sump. The reaction sequence is summarized in Figure 98.

The furan dialdehyde is a stable colorless compound having a melting point of 110 °C. Contrary to furfural, even in the liquid state the furan dialdehyde does not undergo discoloration as it does not have the highly reactive hydrogen by which furfural polymerizes. At room temperature, furan dialdehyde is slightly soluble in water (10 g per liter at 17 °C) as well as in cyclohexane, hexane, ligroin, carbon tetrachloride, diethyl ether, and benzene. It is well soluble in ethanol, ethyl acetate, acetone, dimethylsulfoxide, and methylisobutyl ketone, as well as in hot water.

To the knowledge of the writer, so far furan dialdehyde is not yet produced on a

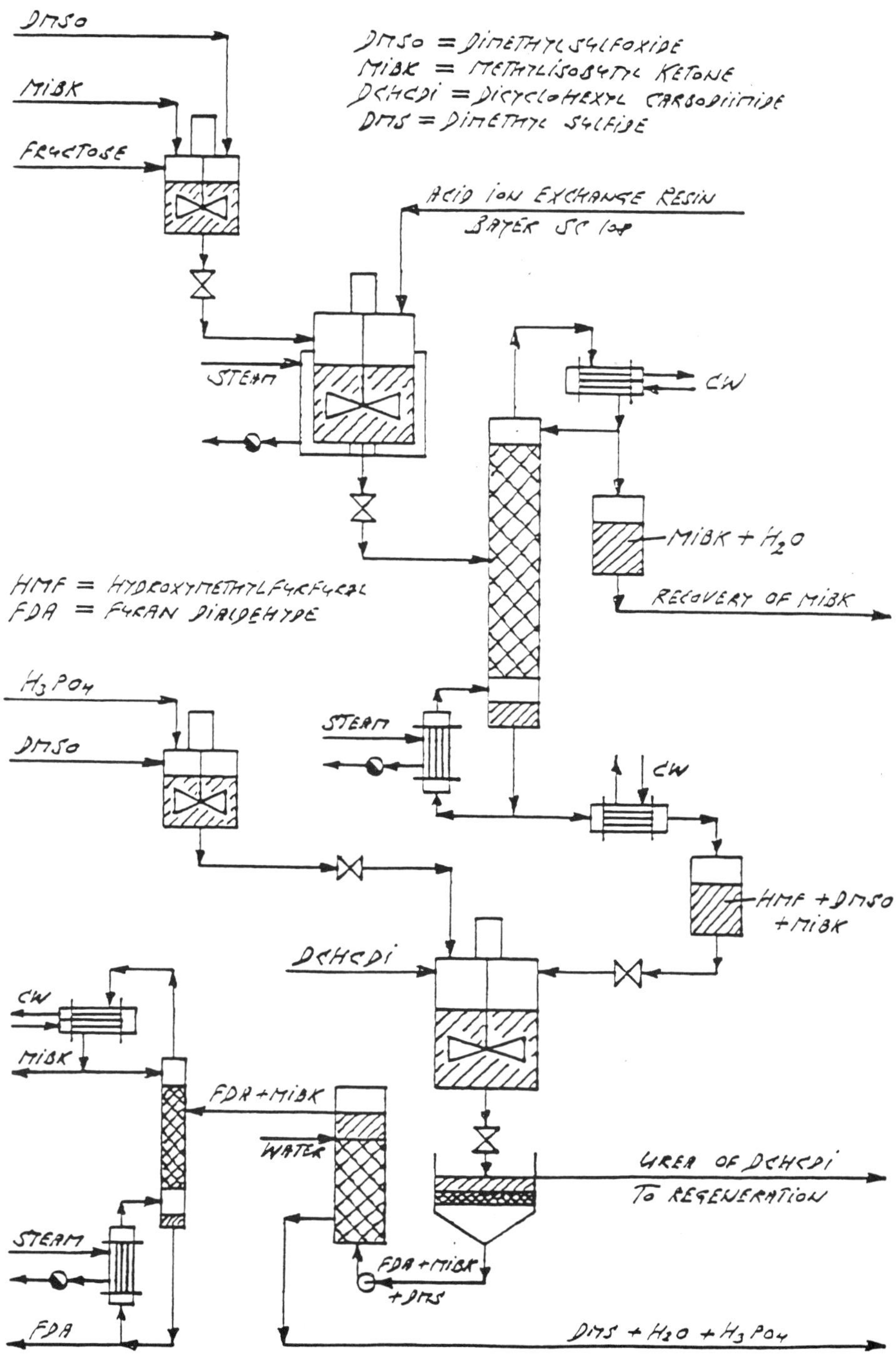

Figure 97. Plant for the Production of Furan Dialdehyde.

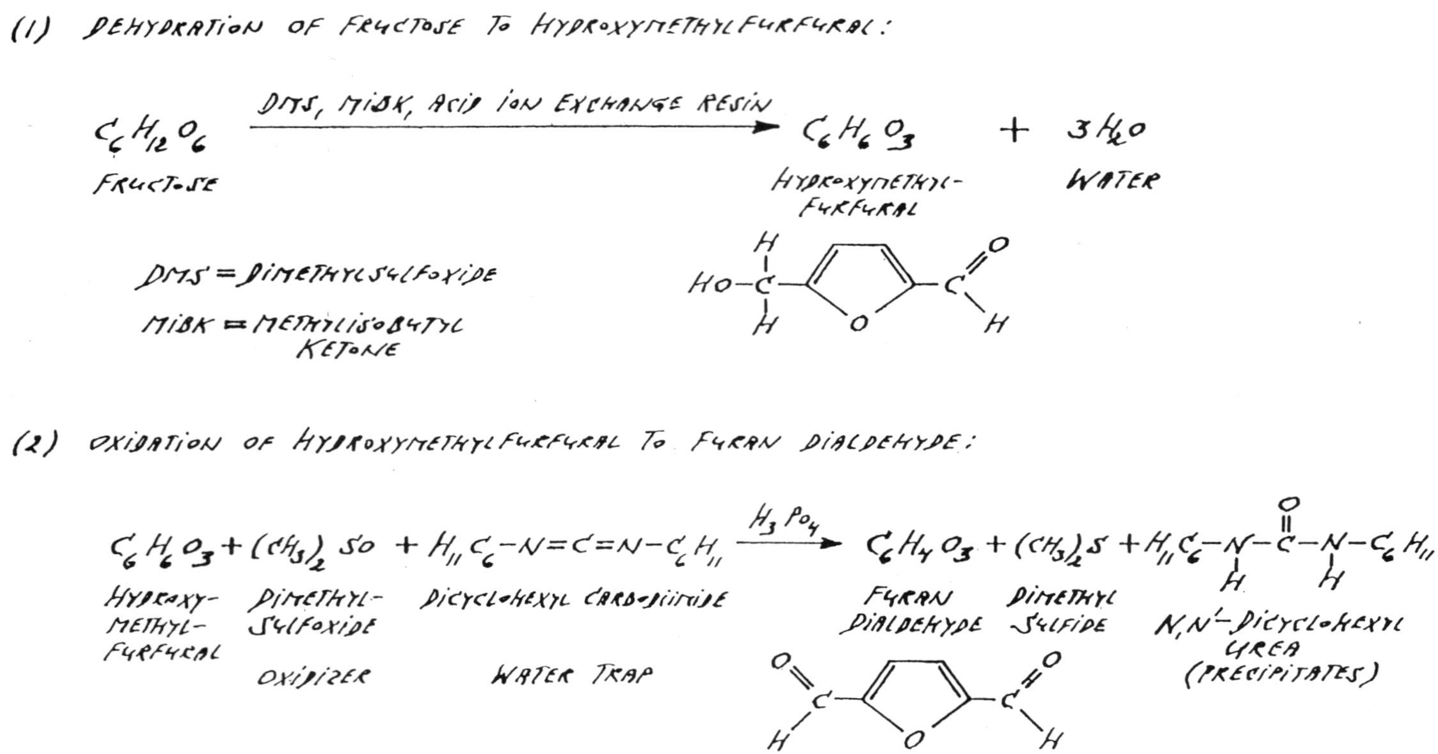

Figure 98. The Reactions Leading to Furan Dialdehyde.

commercial scale.

Reference

[99] M. Marval, Procédé de Synthèse du Furannedicarboxaldéhyde-2,5 à partir du D-Fructose, Ph.D. Thesis, Institut National Polytechnique, Toulouse, 1985.

28. Furan Resins

Furan resins can be obtained from both furfural and furfuryl alcohol.

28.1. Furan Resins from Furfural

In perfect analogy to the formation of phenol/formaldehyde resins, phenol reacts with furfural to form phenol/furfural resins. In phenol, due to its hydroxyl group, the hydrogen atoms in ortho and para positions are highly reactive. Consequently, phenol reacts with furfural to form phenol furfurylol:

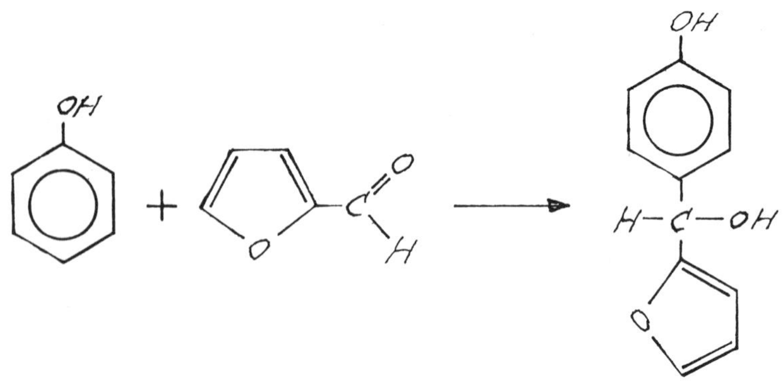

Furfurylol is the group

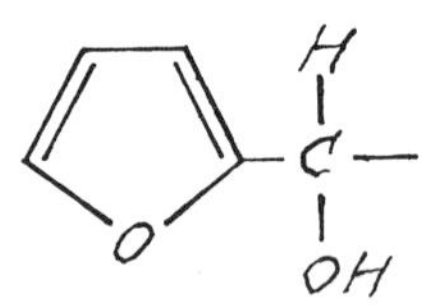

At temperatures above 100 °C, and with appropriate catalysts at much lower temperatures, two molecules of phenol furfurylol react with each other to form an oxygen bridge (ether bridge) and to liberate water as shown in the upper conversion of Figure 99. In addition, also driven by a liberation of water, phenol itself can react with phenol furfurylol as shown in the lower conversion of Figure 99.

These processes can continue as on account of its three highly reactive hydrogen atoms, the phenol can form not only phenol monofurfurylol but also phenol difurfurylol, and phenol trifurfurylol, the latter being shown below, thus permitting the formation of arbitrarily

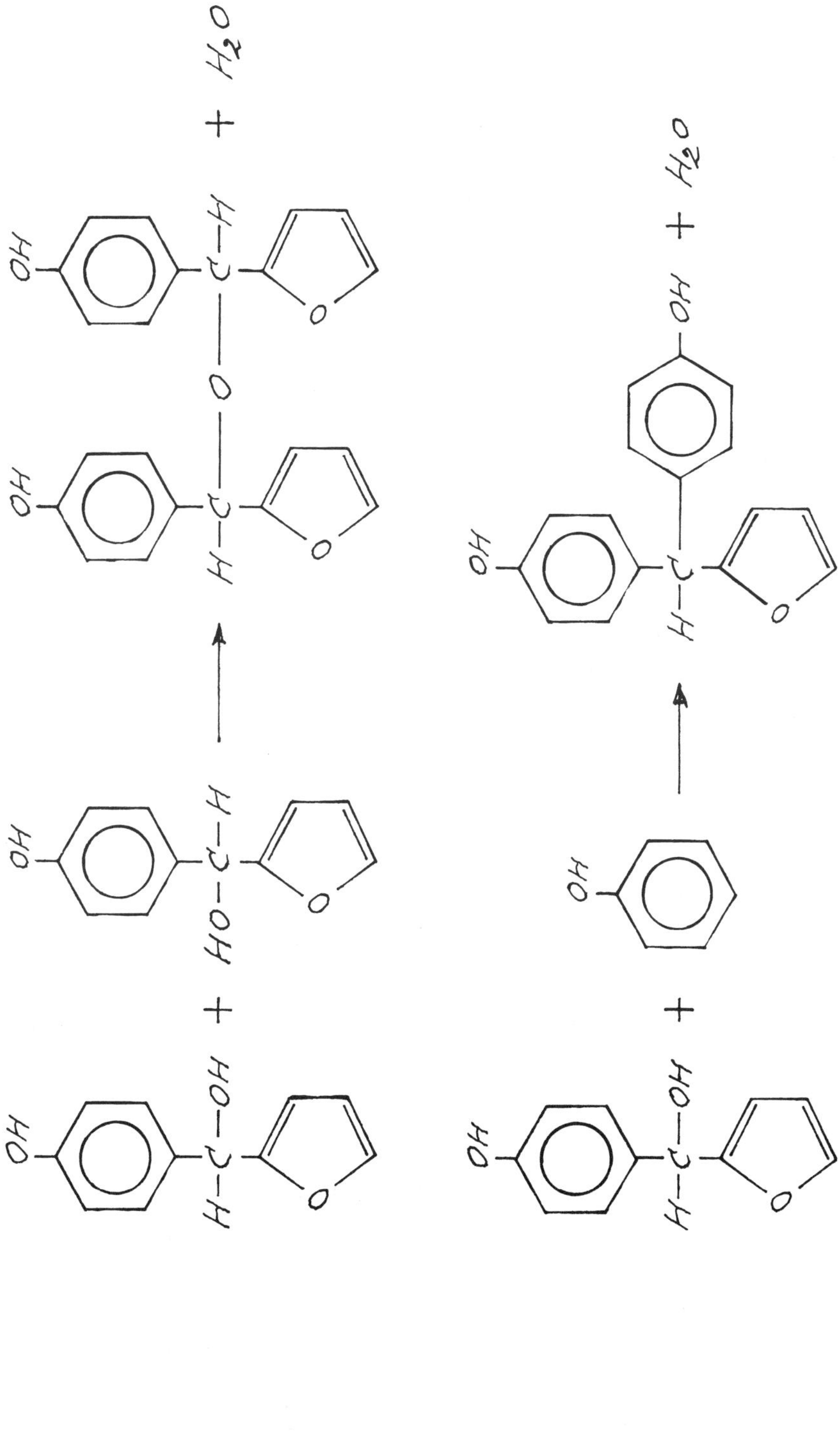

Figure 99. Fundamental Reactions in the Formation of Resins from Phenol and Furfural.

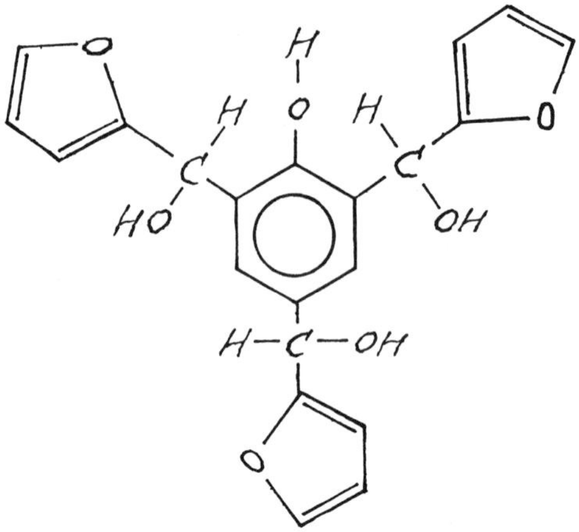

large randomly interlinked systems as shown below, the thermodynamic driving force being the liberation of water. Noteworthily, the hydroxyl group of the phenol does not take part in any of these reactions, its only function being the activation of three hydrogen atoms on the benzene ring.

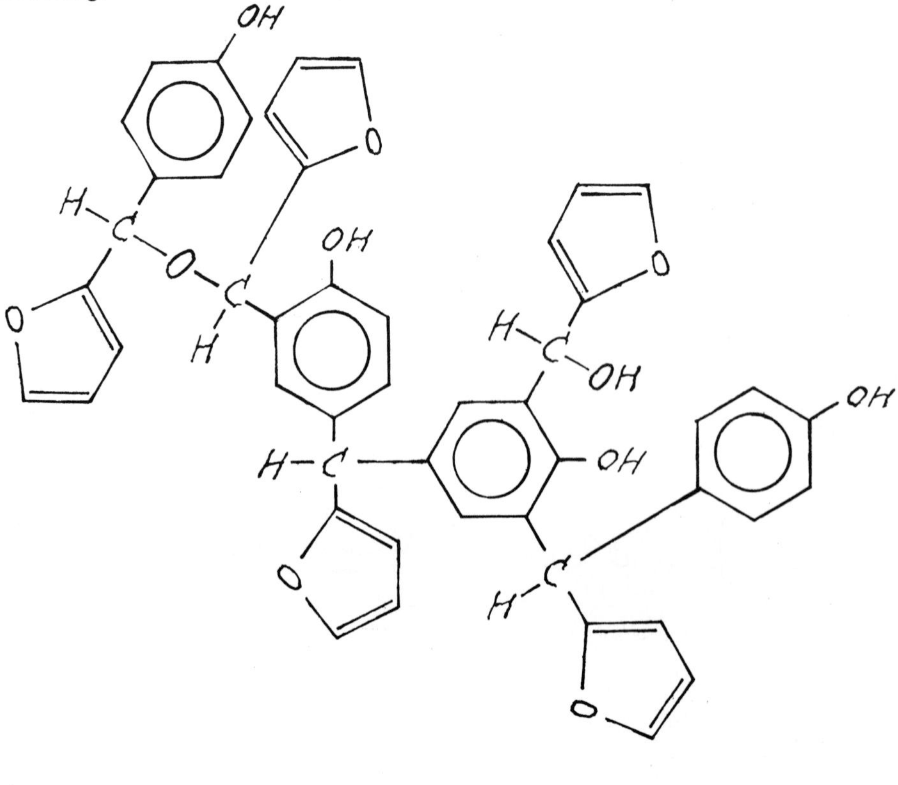

Instead of phenol, one of the cresols can be used as well [100]. In the cases of o-cresol and p-cresol, with one of the reactive ring positions taken up by a methyl group, it is readily understood why the polymerization rate is slowed down. By contrast, with m-cresol the reaction rate is unchanged.

The principal advantages of phenol/furfural resins as compared to phenol/formaldehyde resins can be summarized as follows:

(1) The curing time (setting time) increases with increasing molecular weight of the aldehyde. Thus, under equivalent conditions, phenol/furfural resins cure more slowly than phenol/formaldehyde resins. Consequently, in molding deep draw pieces, the longer flow period of the phenol/furfural resins lessens the tendency of precuring during the molding process. In addition, the longer flow period permits a longer heat transfer, which results in a better appearance and in a better strength of the article. The trade speaks of an improved "flow/cure characteristic".

(2) The phenol/furfural resins have a greater hydrocarbon solvent tolerance than phenol/formaldehyde resins. This is an advantage in the preparation of resin solutions used for the impregnation of paper or cloth for laminating.

(3) The phenol/furfural resins yield products of greatly reduced brittleness.

The initiation of the reactions between phenol and furfural to form the resin can be catalyzed by both acid or alkali. As these reactions are exothermic, heat must be withdrawn at this stage. The final curing of the resin is universally triggered by the application of heat.

Another important furan resin from furfural is obtained by reacting furfural with urea. In analogy to the formation of urea/formaldehyde resins, the reaction proceeds as shown in Figure 100. For a 1 : 1 molar ratio of the input materials, the first step is a formation of monofurfurylol urea. In liberating water, two such molecules combine in forming an oxygen bridge. Inasmuch, however, as the resulting molecule still has active hydrogen atoms on the nitrogen atoms, addition of further furfural to yield furfurylol groups, and further liberation of water accompanied by the formation of oxygen bridges result in an extension of the system in all directions.

When starting with two moles of furfural and one mole of urea, difurfurylol urea is formed instead of monofurfurylol urea, but there, too, active hydrogen atoms on nitrogen atoms remain to permit further additions as outlined.

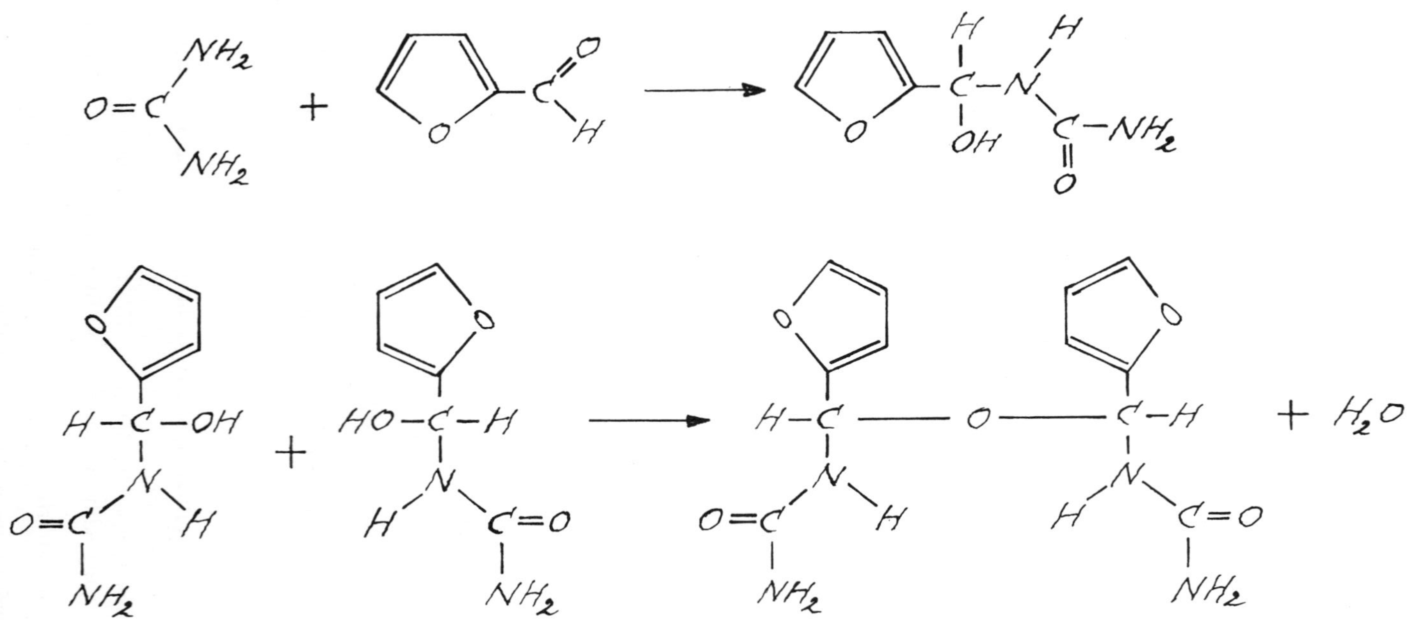

Figure 100. Fundamental Reactions in the Formation of Resins from Urea and Furfural.

28.2. Furan Resins from Furfuryl Alcohol

When furfuryl alcohol is mixed with a small quantity of an aqueous acid solution, the alcohol group of one molecule of the furfuryl alcohol reacts with the active hydrogen from the ring of another molecule of furfuryl alcohol to yield water and a furan ring linked to a furfuryl alcohol molecule by a methylene bridge:

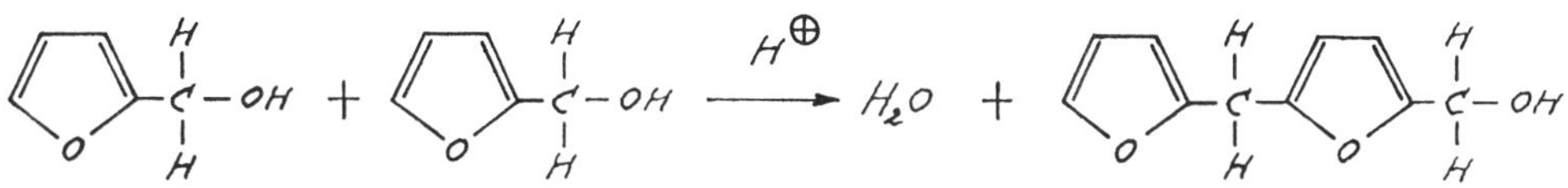

In the same fashion, further furfuryl alcohol molecules can be added indefinitely.

This highly exothermic reaction, potentially explosive when catalyzed by sulfuric, hydrochloric, or nitric acid [101], can be controlled by intensive cooling as well as by an appropriate choice of the pH value, and it can be stopped by neutralization at any desired viscosity. In this way, liquid resins are obtained which are capable of reacting further when submitted to elevated temperatures. In this final "setting reaction", called "heat hardening", the double bonds of a furan ring in one chain react with the double bonds of a furan ring in another chain to yield a tightly cross-linked solid. This is the basis for a multitude of products ranging from cements to composites consisting of such resins mixed with various fillers and then "heat-hardened" to give chemically resistant equipment.

Alternatively, the entire reaction can be carried out "in situ", for instance by impregnating a porous material such as graphite or brick with properly acid-catalyzed furfuryl alcohol, and by then submitting this material to a "temperature program" devised so as to permit the water of the catalyst as well as the reaction water to escape as vapor before the final "heat hardening" takes place.

To top off the appeal, the polymerization of furfuryl alcohol can be made to run with almost any partner containing reactive hydrogen. Thus, furfuryl alcohol can be reacted with formaldehyde to give a series of thermosetting resins featuring more methylene bridges than are obtained by the resinification of furfuryl alcohol alone. Such resins are commonly used for the bonding of wood.

The most important application of furan resins from furfuryl alcohol lies in the foundry industry. Sand bonded with clay is usually used for molds, but for cores and for more stringent molding applications, a stronger more rigid chemical binder is required. One of these chemical binders is furan resin made from furfuryl alcohol. 100 parts by weight of sand are mixed with 20 to 30 parts by weight of an aqueous acid catalyst, and to this mixture is then added a quantity of 1 to 2 parts by weight of a suitably compounded furfuryl alcohol resin to give a mold material solidifying to the desired strength in the desired period of time. Depending on the catalyst used, there are three different processes [102]:

(1) The "nobake process" using an "active catalyst" by which the resin is self-setting at room temperature.

(2) The "warmbox process" using a "passive catalyst" by which the resin is cured at 250 to 350 °F (121 to 177 °C).

(3) The "hotbox process" using a "latent catalyst" by which the resin is cured at 425 to 525 °F (218 to 274 °C).

Historically, the "nobake process" came first, and the "hotbox process" second, while the "warmbox process" is the latest version. The "warmbox process" was introduced as the high temperatures of the "hotbox process" cause high costs for heating and corebox maintenance.

In the foundry applications, one of the advantages of furfuryl alcohol resins over previously dominating phenol/formaldehyde resins lies in the fact that furfuryl alcohol resins do not release formaldehyde which is a serious health hazard for the workers in the foundry. The release of formaldehyde is due to a thermal decomposition reaction. Phenol reacts with formaldehyde in forming methylene ether bridges, and at elevated temperatures these methylene ether bridges degenerate to methylene bridges in liberating formaldehyde as shown below:

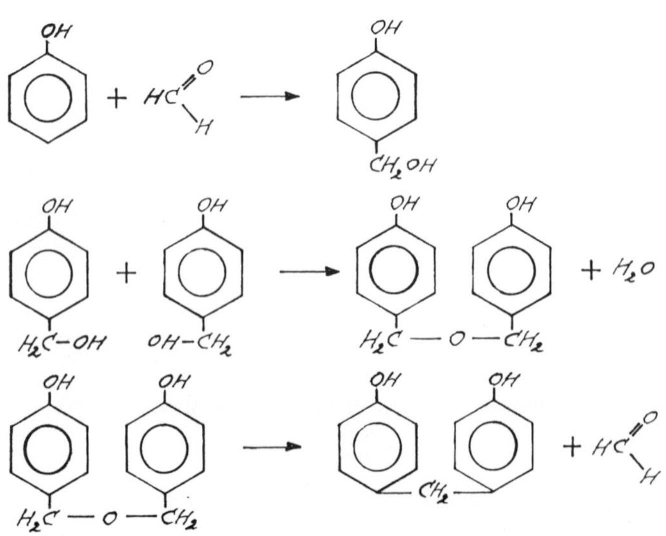

28.3. Description of a Resin Plant

Inasmuch as polycondensation reactions require both a catalyst and heat, starting and terminating the process is greatly facilitated in that

(a) the catalyst can be readily mixed with the principal reactants at a low temperature,

(b) the reaction can be readily started by heating the reaction mixture, and

(c) the reaction can be readily terminated by neutralizing the catalyst.

With polycondensation being exothermic, heat must be withdrawn during the reaction proper to maintain a constant temperature. After completion of the polymerization, excess water (water introduced by diluted reactants and water liberated by the process) is simply removed by vacuum distillation.

A typical plant for the production of resins from furfural or furfuryl alcohol is illustrated in Figure 101. The process is carried out in a batchwise fashion. From the tanks 1 to 3, the principal reactants and the catalyst are fed into a reactor 4 featuring a mixer 5 and capable of being heated as well as cooled. The reaction is started by raising the temperature. With the mixture thus rendered exothermic, the heat released is continuously removed by cooling to keep the process under control. When the desired degree of polymerization is attained, a neutralizing agent from tank 6 is added to bring the reaction to an end. After cooling, the reactor is submitted to an underpressure by means of vacuum pump 7, and heating is resumed to remove excess water by distillation. The vapors are liquefied in condenser 8, the condensate being collected in tank 9. As soon as the product in the reactor has reached the required consistency, the heat input is stopped and the vacuum discontinued before pump 10 conveys the finished resin to the weighing station 11 for loading in barrels.

References

[100] A. J. Norton, Ind. Eng. Chem. 40 (1948) 236-238.

[101] I. S. Goldstein and W. A. Dreher, Ind. Eng. Chem. 52 (1960) 57-58.

[102] R. H. Kottke and A. E. Bloomquist, AFS Transactions 86 (1978) 215-220.

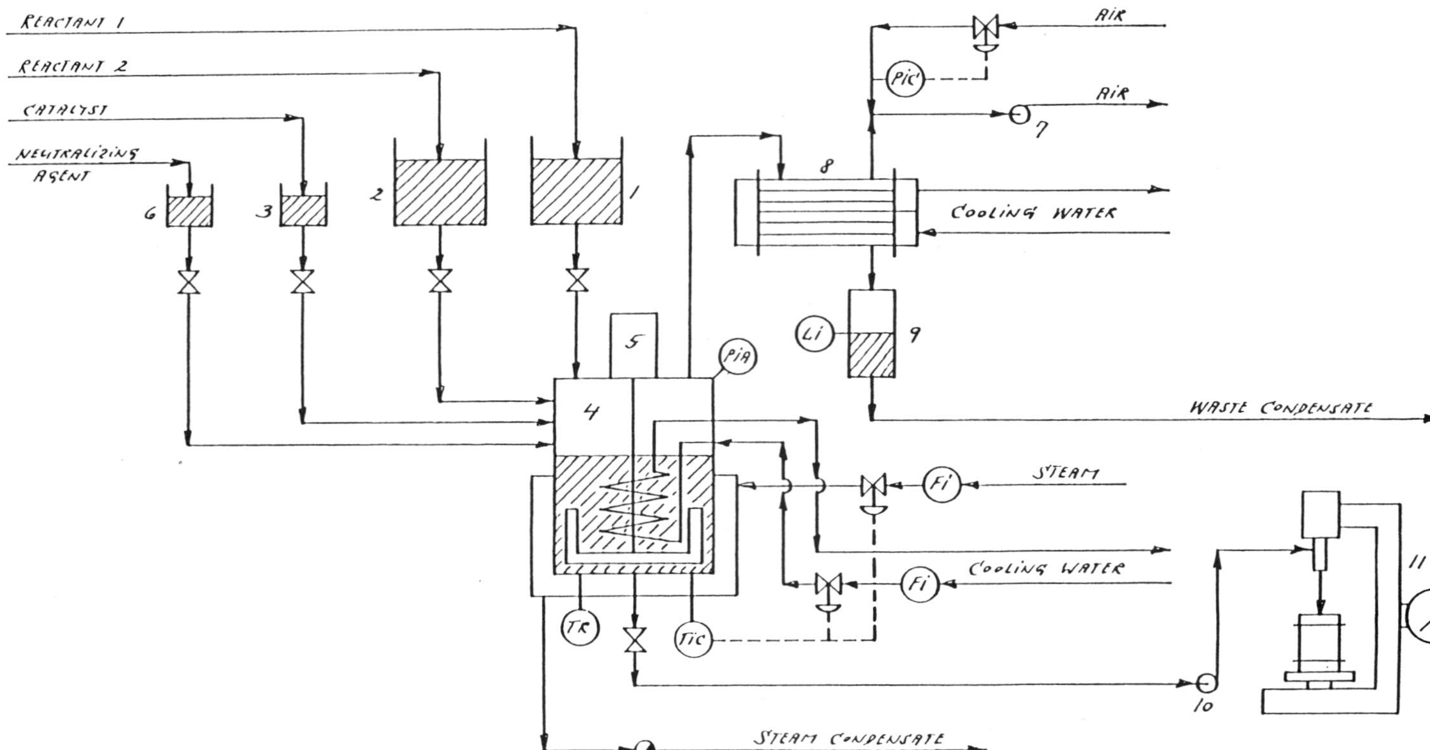

Figure 101. Plant for the Production of Resins.

29. Tetrahydrofurfuryl Alcohol

Unlike furfuryl alcohol, tetrahydrofurfuryl alcohol cannot be made successfully with a copper chromite catalyst as the latter has essentially no effect on the furan ring, its action being limited to the reduction of the carbonyl group.

Tetrahydrofurfuryl alcohol is made commercially from furfural by a high-pressure hydrogenation at 1000 to 1500 psi (67.889 to 101.833 ATM) and 170 to 180 °C, using a mixture of copper chromite and RANEY nickel as the catalyst [103]:

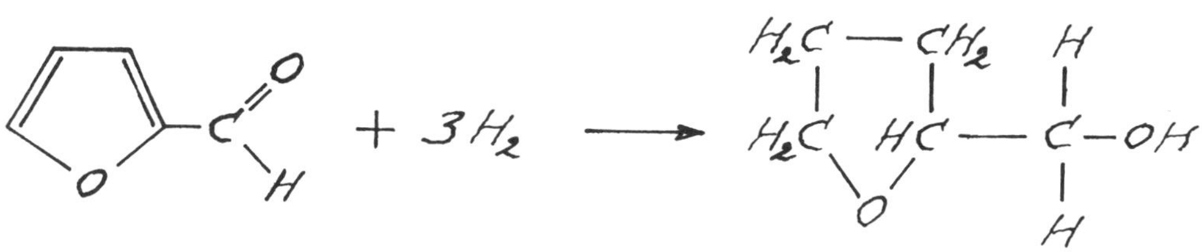

Under the given conditions, the yield ranges from 70 to 80 percent, the by-product being 1,5-pentanediol. Higher yields up to 90 percent can be obtained at lower temperatures, between 100 and 125 °C, but this requires unreasonably longer reaction times. The RANEY nickel is made by leaching a nickel-aluminum alloy with caustic.

With a batch operation under the given conditions, the hydrogenation is completed in 45 to 90 minutes, the shorter time referring to the higher pressure. Continuous production can be carried out in a plant as shown in Figure 78 (page 153).

Tetrahydrofurfuryl alcohol does not seem to have any direct application, but it is needed as a stepping stone for the production of dihydropyran [104] as shown in the next chapter.

References

[103] B. H. Wojcik, Ind. Eng. Chem. 40 (1948) 210-216.

[104] O. W. Cass, Ind. Eng. Chem. 40 (1948) 216-219.

30. Dihydropyran

Dihydropyran (DHP), having an atmospheric boiling point of 86 °C and a very limited solubility in water (3 g in 100 g of water at 20 °C), can be produced from tetrahydrofurfuryl alcohol by dehydration and ring expansion over an aluminum oxide catalyst at 300 to 350 °C [105]:

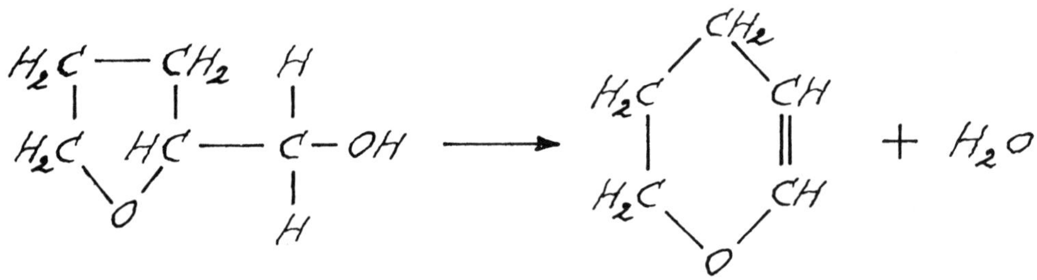

The mechanism of this astounding reaction is not known, but the driving force is certain to be the difference in ring strain [106].

The double bond of dihydropyran is highly reactive, adding widely differing materials such as hydrogen, water, chlorine, hydrogen chloride, alkyl hypochlorites, phosgene, alcohols, glycols, and organic acids. Consequently, dihydropyran is a very important intermediate for a large number of syntheses.

References

[105] O. W. Cass, Ind. Eng. Chem. 40 (1948) 216-219.

[106] T. H. Lowry and K. Schueller Richardson, Mechanismen und Theorie in der Organischen Chemie, Verlag Chemie, Weinheim, 1980.

31. Maleic Acid

With a yield in the order of 75 percent, maleic acid can be made from furfural by a catalytic vapor phase oxidation with air at 270 °C [107]. The overall reaction is

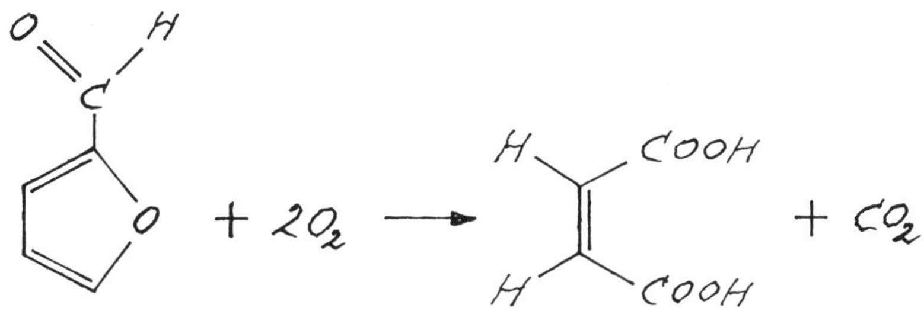

but at the reaction temperature the real conversion is

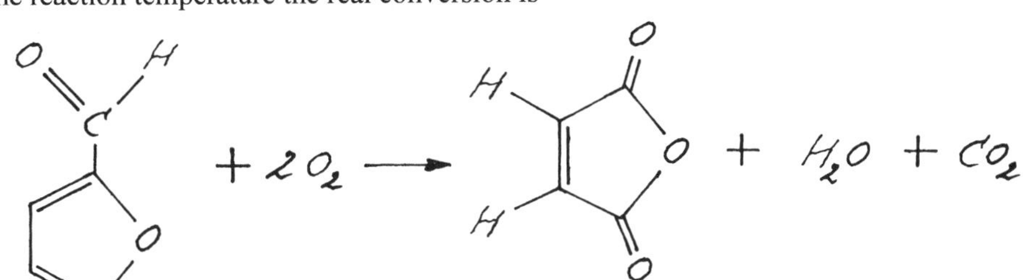

MALEIC ANHYDRIDE

so that the real product is maleic anhydride, but inasmuch as at room temperature any water readily converts maleic anhydride to the acid, and inasmuch as water is one of the oxidation products, it is maleic acid which is obtained as the product. A small by-product is formaldehyde, believed to be due to a further oxidation of the maleic anhydride in the presence of water [108]:

$$C_4H_2O_3 + O_2 + H_2O \rightarrow\rightarrow 2\ HCHO + 2\ CO_2$$

Maleic acid forms colorless prismatic crystals having a melting point of 130.5 °C. It exhibits a high solubility in water, amounting to 78.8 g per 100 ml of water at 25 °C, and 392.6 g per 100 ml of water at 97.5 °C.

Dissolved in water, maleic acid is a very strong organic acid. Its first dissociation constant at 25 °C is $K_1 = 1.42 \times 10^{-2}$ mole/liter ($pK_1 = 1.83$) as compared to $K = 1.77 \times 10^{-4}$ mole/liter ($pK = 3.75$) for formic acid.

Maleic acid, which does not exist in nature, is an important chemical for the production of alkyd resins (resins made by heating dicarboxylic acids, such as maleic acid, with polyvalent alcohols, such as glycerol), and for the synthesis of dienes.

To obtain a high yield, it is important to use the right catalyst and the right material for containing the catalyst. With vanadium pentoxide in pyrex tubes, the yield is only 25 percent of the input furfural, but with a vanadium pentoxide/molybdenum trioxide/iron molybdate catalyst in nickel tubes, the yield is in the order of 75 percent. Interestingly, the best yields are obtained when the catalyst, prepared from appropriate ammonium salts, is cured with air at 300 °C in situ, and when it is then used directly at the reaction temperature of 270 °C without allowing it to cool down.

Instead of using furfural as the input material, it is also possible to use furan, furoic acid, or furfuryl alcohol [109], but of course this has the drawback of requiring an additional step from furfural. It is noted, however, that furan and furoic acid have the advantage of not forming polymers.

On the gas side, it is also possible to use oxygen instead of air, but this has no yield benefit and can cause explosions [109] so that this alternative can be ruled out.

A plant for the production of maleic acid from furfural is shown schematically in Figure 102. Furfural is fed into a vaporizer system comprising a packed column 1, a circulation pump 2, and a heater 3 energized by hot water to maintain the furfural temperature at 45 °C. Below the packing of column 1, a dosed quantity of air is introduced. In the countercurrent system of air flowing upwards and liquid furfural flowing downwards, the air gets saturated with the vapor pressure of furfural at 45 °C. The resulting mixture of air and furfural vapor passes a demister pad 4 and a superheater 5 before it enters a reactor 6 carrying the catalyst in nickel tubes. The catalytic reactor is maintained at a temperature of 270 °C by means of a molten salt circuit. As the reaction is exothermic, the molten salt flowing around the reactor tubes must withdraw heat from the system. The gaseous mixture of reaction products enters a scrubber system comprising a packed column 7, a circulation pump 8, and a cooler 9. The pump circulates water through the cooler 9 and onto the packing of column 7 where it meets a countercurrent of the gaseous products, from which maleic anhydride is dissolved as maleic acid, while excess air and carbon dioxide liberated by the reaction are vented. A portion of the scrubber stream is fed into an evaporator consisting of a cyclone 10, a circulation pump 11, and a heater 12 energized by steam. The evaporator produces a waste vapor stream of water, formaldehyde, and carbon dioxide, as well as a concentrate fed into a

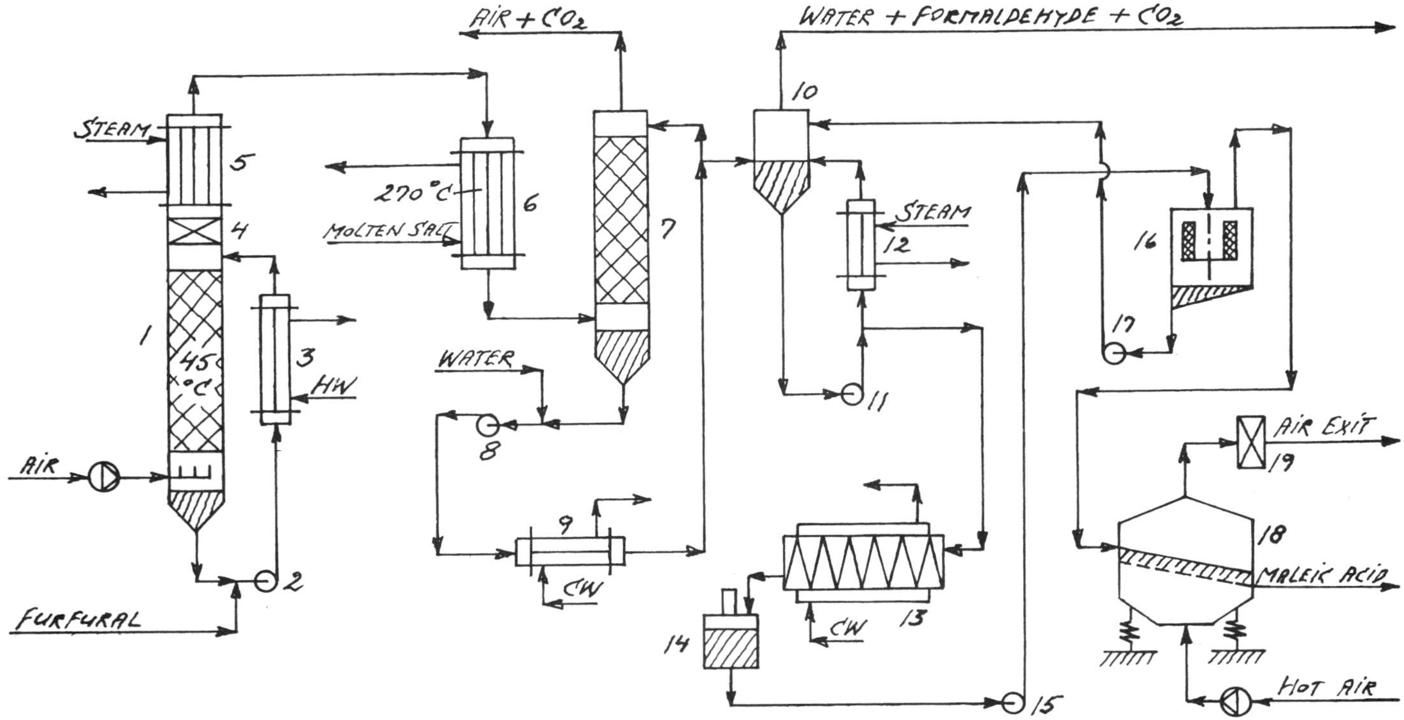

Figure 102. Plant for the Production of Maleic Acid from Furfural.

cooled crystallizer 13. The slurry of crystals is collected in tank 14 from where pump 15 takes it into a filter centrifuge 16. Pump 17 conveys the drained mother liquor back into cyclone 10, while the crystal cake is introduced into a vibratory drier 18 operated with hot air. A filter 19 retains fines.

References

[107] E. R. Nielsen, Ind. Eng. Chem. 41 (1949) 365-368.

[108] W. V. Sessions, J. Amer. Chem. Soc. 50 (1928) 1696-1698.

[109] N. A. Milas and W. L. Walsh, J. Amer. Chem. Soc. 57 (1935) 1389-1393.

32. Methylfuran

Methylfuran, more precisely 2-methylfuran, also known as sylvan, a colorless liquid having an atmospheric boiling point of 63 °C, can be produced from furfural in very much the same fashion as furfuryl alcohol, but at a higher temperature.

It was pointed out already in the chapters 17.1 and 17.3 that 2-methylfuran appears as an unwanted by-product when furfuryl alcohol is made from furfural by the vapor phase process at 135 °C, using a copper chromite catalyst, and that the relative quantity of 2-methylfuran increases when the reaction temperature is raised to compensate the gradual reduction of catalyst activity.

The reaction producing 2-methylfuran is shown below:

If 2-methylfuran is to be made intentionally, it was found that the reaction temperature should be 250 °C (instead of 135 °C for furfuryl alcohol), and that the copper chromite catalyst should be mixed with activated charcoal [110]. As the reaction is highly exothermic, good care must be exercised to prevent an inactivation of the catalyst by overheating.

Although the reaction given above calls for a molar hydrogen to furfural ratio of only 2 : 1, it is recommended to run the process at a ratio between 6 : 1 and 7 : 1. This means that in an atmospheric process the furfural in the vaporizer should have a temperature of 104 °C as at this temperature the furfural has a vapor pressure of 100 torr, so that the molar hydrogen to furfural ratio is (760 – 100)/100 = 6.6. Under these conditions, the yield of 2-methylfuran is 92.5 percent. Apart from the inevitable reaction water, the principal by-product is furfuryl alcohol.

Small production units employ 4 inch diameter brass tubes 16 inches long, heated by electric resistor ribbons, while large production units require shell and tube reactors heated by a molten salt circuit. Organic heat exchange media are not recommended as at 250 °C they would undergo a fairly rapid thermal degradation.

Reference

[110] L. E. Schniepp, H. H. Geller, and R. W. von Korff, J. Amer. Chem. Soc. 69 (1947) 672-674.

33. Pyrolysis of Furfural

Contrary to common belief, when submitted to nothing but thermal loading, furfural is an extremely stable compound. No gases are formed when furfural is heated to 275 °C in a sealed tube for a period of three hours. After such treatment, almost all of the furfural is recoverable by distillation, leaving only a tiny amount of tarry residue.

Using a flow test and a contact time of 20 seconds, 565 °C was found to be the initial decomposition temperature of furfural [111]. Other data obtained by the flow test are compiled in Figure 103.

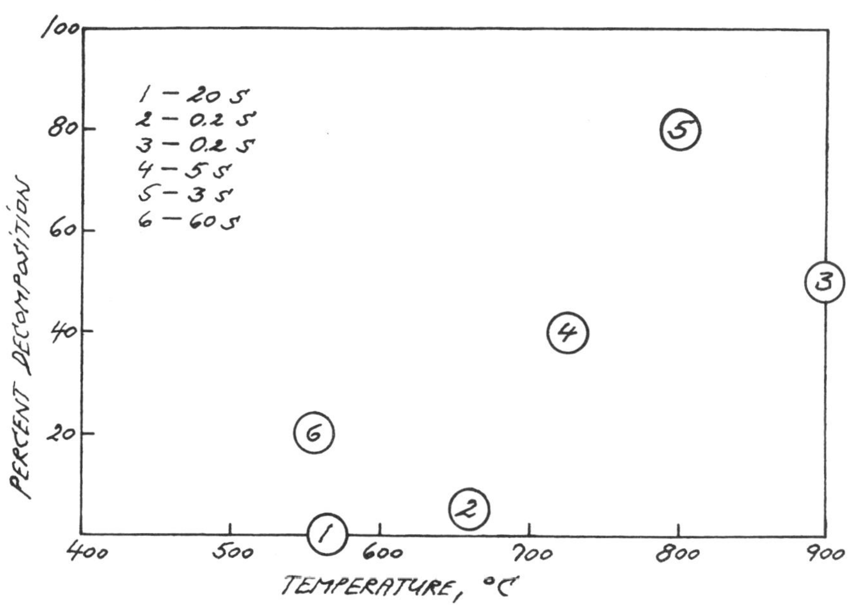

Figure 103. Pyrolysis Data for Furfural as Obtained by a Flow Test.

Noteworthily, at a contact time of 0.2 seconds, only 5 percent of the furfural decomposes at 660 °C. At such elevated temperatures, furfural decomposes to furan and carbon monoxide, in complete analogy to benzaldehyde which decomposes to benzene and carbon monoxide. The best conditions for the production of furan by pyrolysis of furfural were found to be 725 °C and a contact time of five seconds. Such a process gives a furan yield of 16.5 percent.

As all pyrolysis reactions, the decomposition of furfural to furan and carbon monoxide is accompanied by other minor reactions. Of the gases formed, only 80 percent is carbon monoxide, other gases identified being 10 percent hydrogen as well as small quantities of carbon dioxide, butadiene, propadiene, ethylene, propylene, acetylene, methylacetylene, and cyclopropene. As the quantities of carbon dioxide and butadiene are roughly equal on a molar basis, it is believed that a part of the furfural reacts with pyrolytically liberated hydrogen:

$$\text{Furfural} + H_2 \rightarrow\rightarrow \text{Butadiene} + CO_2$$

On the side of liquid products, apart from the furan, small quantities of benzene, toluene, and phenol are formed.

Reference

[111] C. D. Hurd, A. R. Goldsby, and E. N. Osborne, J. Amer. Chem. Soc. 54 (1932) 2532-2536.

APPENDICES

Subjects of a more specialized nature are treated in the following supplementary chapters. To set them apart from the principal text, they are listed by letters rather than numbers, with uppercase letters referring to properties and property measurements, while lowercase letters deal with selected theoretical topics.

A. Properties of Furfural

Molecular weight: 96.082 g/mole

Boiling point (760 mm Hg): 161.7 °C

Freezing point: -36.5 °C

Specific gravity d^{25}_{25}: 1.1610

Refractive index n_D^{20}: 1.52608

Spectroscopic polarity index E_T^N: 0.426 (measured by C. Reichardt)

ΔH_f^0(liq): -48.20 kcal/mole = -201.67 kJ/mole

ΔH_f^0(g): -36.10 kcal/mole = -151.04 kJ/mole

Viscosity at 25 °C: 1.494 cP

Flash point (closed cup): 59 °C

Ignition temperature: 315 °C

Solubility in water at 20 °C: 8.3 g per 100 ml of water

Solubility in water as a function of temperature as shown in Figure 104.

Azeotrope with water at 1 ATM: 65 % by weight of water

Boiling point 97.85 °C

Vapor/liquid equilibrium diagram of the furfural/water system at 1 ATM as shown in Figure 105.

Vapor/liquid equilibrium diagram of the furfural/water system at various pressures as shown in Figure 106 (range from 0 to 10 % furfural by mole).

Toxicity: LD_{50} (oral) for a rat:..127 mg/kg

LD_{50} (oral) for a mouse:..400 mg/kg

LD_{50} (oral) for a guinea pig:..541 mg/kg

LD_{50} (oral) for a dog:..2300 mg/kg

LD_{50} (intravenous) for a mouse:..152 mg/kg

LD_{50} (intravenous) for a dog:..250 mg/kg

LC_{50} in air (1 hour inhalation) for a rat:..1037 ppm

Vapor pressure as a function of temperature as shown in Figure 107.

Furfural vapor is irritating to the mucous membranes, but the low volatility reduces the risk of

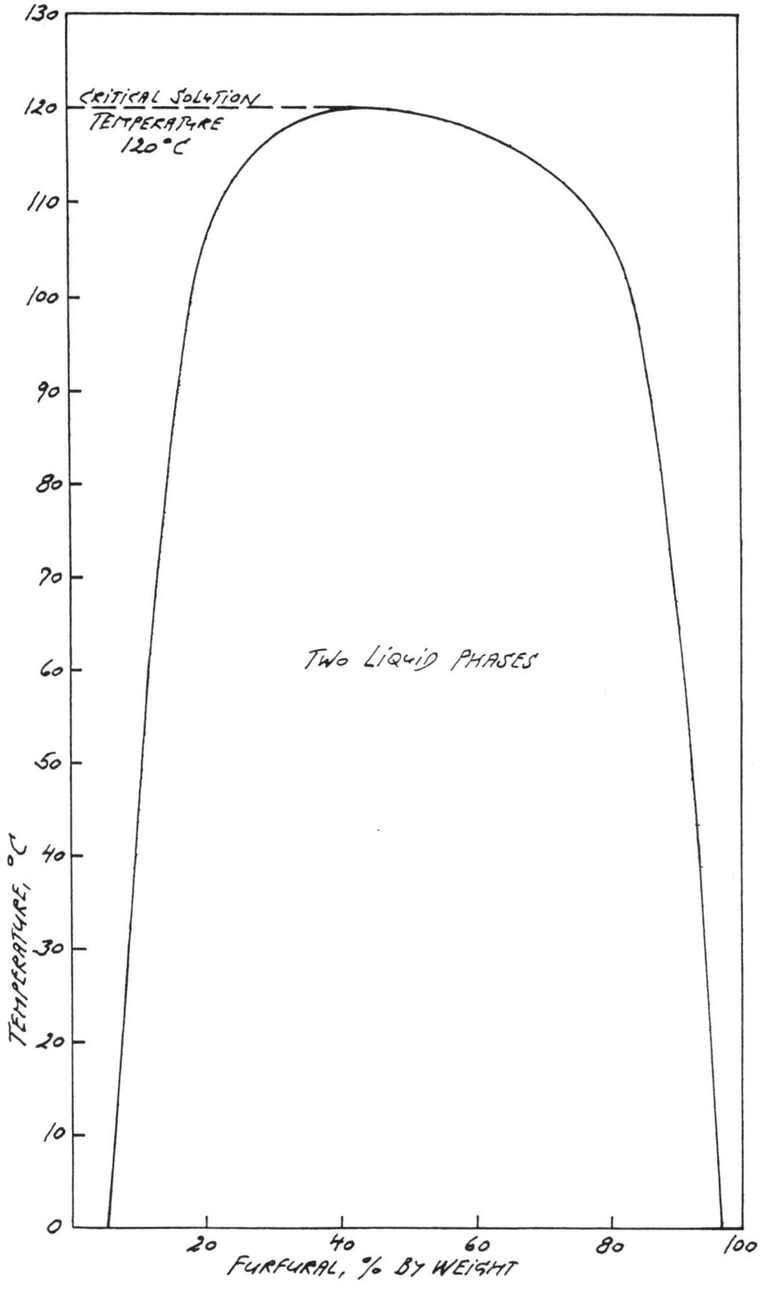

Figure 104. The Solubility of Furfural in Water as a Function of Temperature. Unlimited Solubility above 120 °C.

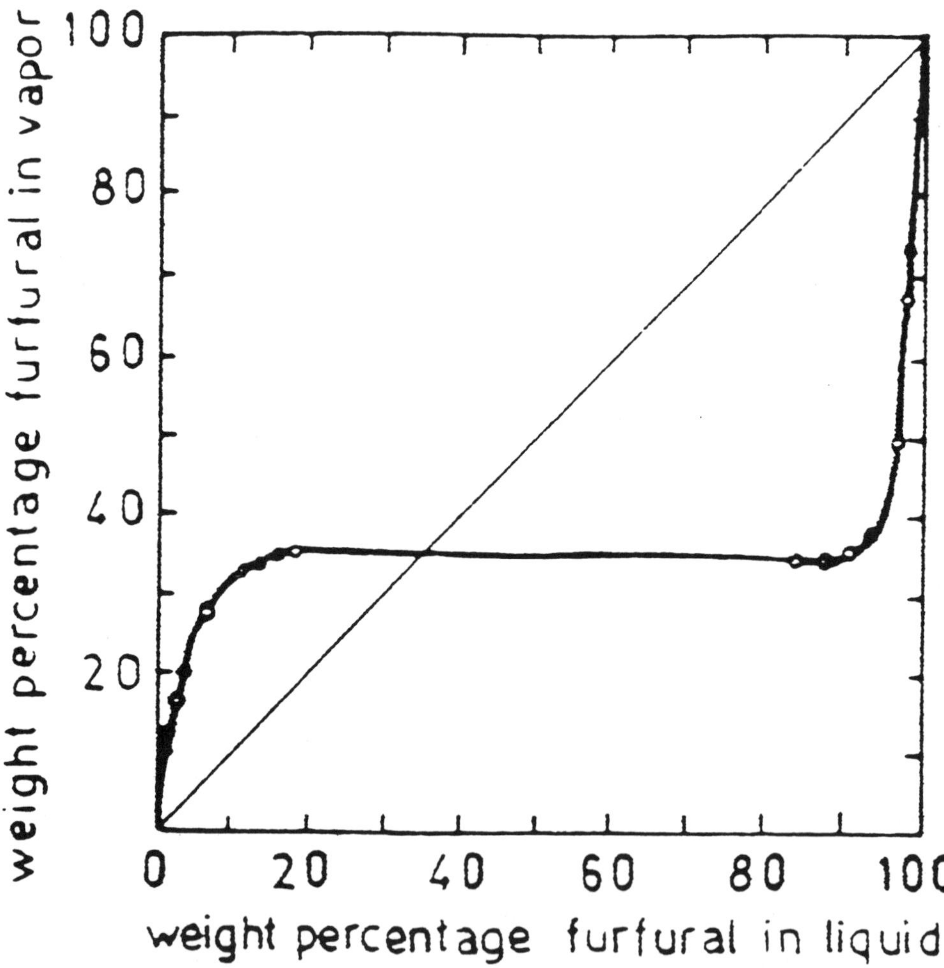

Figure 105. The Vapor/Liquid Equilibrium Diagram of the Furfural/Water System at 1 ATM.

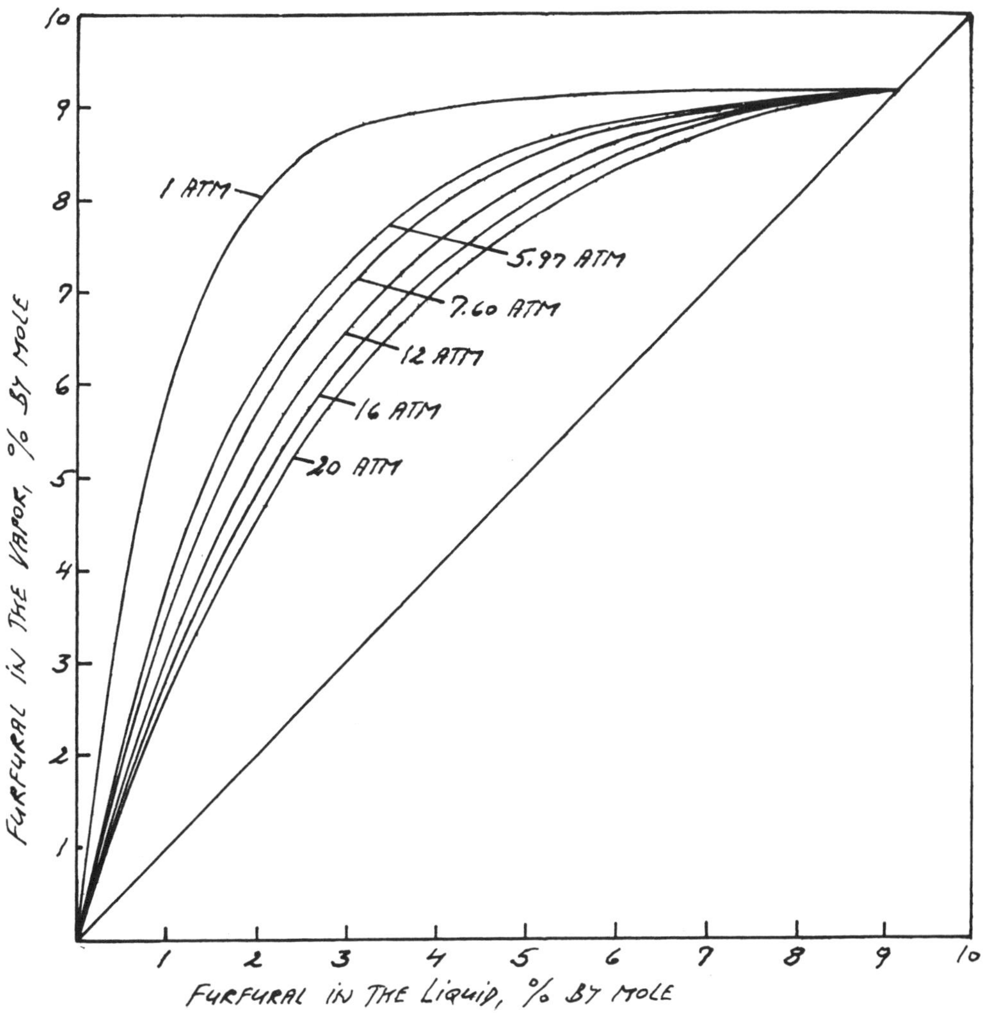

Figure 106. The Vapor/Liquid Equilibrium Diagram of the Furfural/Water System at Various Pressures (range from 0 to 10 % furfural by mole).

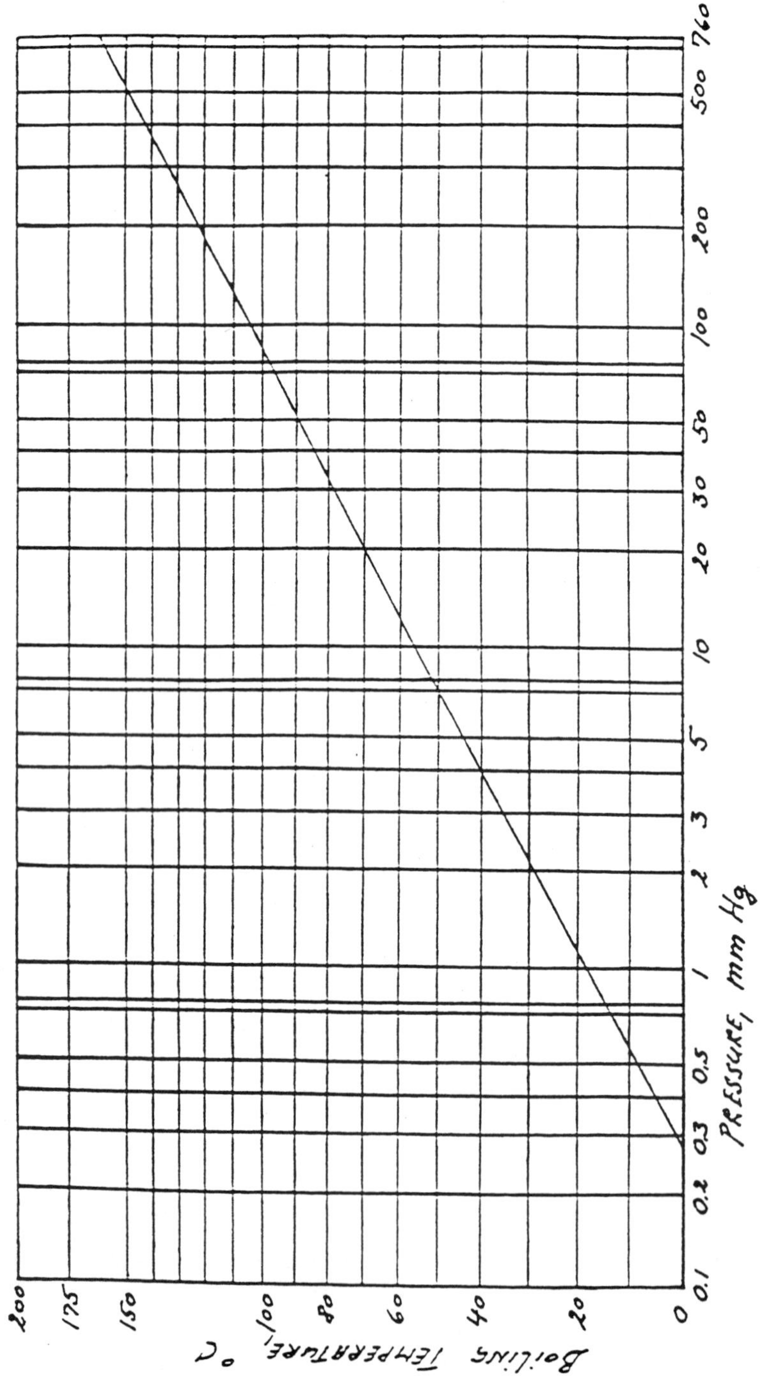

Figure 107. The Vapor Pressure of Furfural as a Function of Temperature.

exposure. Furfural is metabolized rapidly and excreted in the urine as furoylglycine, with a smaller portion (0.5 to 5 %) as 2-furanacrylurie acid.

Furfural was found to penetrate the skin of rabbits, but with no fatal effect up to 500 mg/kg. Several drops of furfural in standard eye tests with rabbits cause significant irritation but no irreversible damage.

B. Properties of Furfuryl Alcohol

Molecular weight: 98.098 g/mole

Boiling point (760 mm Hg): 169.35 °C

Freezing point: -14.6 °C

Specific gravity $d^{25}{}_{25}$: 1.1296

Refractive index $n_D{}^{25}$: 1.4850

Spectroscopic polarity index $E_T{}^N$: 0.605 (measured by C. Reichardt)

$\Delta H_f{}^0$(liq): -66.02 kcal/mole = -276.23 kJ/mole

$\Delta H_f{}^0$(g): -50.62 kcal/mole = -211.79 kJ/mole

Viscosity at 25 °C: 5.0 cP

Flash point (closed cup): 77 °C

Ignition temperature: 391 °C

Unlimited solubility in water

Azeotrope with water at 1 ATM: 80 % by weight of water

Boiling point 98.5 °C

Vapor pressure as a function of temperature:

t, °C	40	60	80	100	120	140
p, torr	1.81	6.33	20.33	53.46	128.01	271.83

A violent explosion occurs when furfuryl alcohol is polymerized with concentrated mineral acids such as HCl, H_2SO_4, or HNO_3.

Toxicity: LD_{50} (oral) for a mouse:....................160 mg/kg

LD_{50} (oral) for a rat:....................275 mg/kg

LD_{50} (intravenous) for a rabbit:....................650 mg/kg

LC_{50} in air (1 hour inhalation) for a rat:....................592 ppm

Furfuryl alcohol is irritating to the eyes and to the skin. The occupational exposure limit is 50 ppm.

C. Properties of Furan

Molecular weight: 68.072 g/mole

Boiling point (760 mm Hg): 31.36 °C

Freezing point: -85.6 °C

Specific gravity d^{15}_{4}: 0.9444

Refractive index n_D^{20}: 1.42157

Spectroscopic polarity index E_T^N: 0.164 (measured by C. Reichardt)

ΔH_f^0(liq): -14.91 kcal/mole = -62.38 kJ/mole

ΔH_f^0(g): -8.30 kcal/mole = -34.73 kJ/mole

Flash point (closed cup): -35.5 °C

Solubility in water at 25 °C: 1.0 g per 100 ml of water

Toxicity: LD_{50} (intravenous) for a mouse:..7 mg/kg

LD_{50} (intravenous) for a rat:..5200 mg/kg

LC_{50} in air (1 hour inhalation) for a rat:..3464 ppm

Exposure to furan produces a narcotic effect. The low boiling point requires that adequate ventilation be provided in areas where furan is handled. Liquid furan can be absorbed by the skin. People with circulatory disorders, abnormal liver conditions, or chronic gastrointestinal complaints should not work with furan. Symptoms of furan exposure include fatigue, headache, and gastrointestinal disturbances.

D. Properties of Tetrahydrofuran

Molecular weight: 72.104 g/mole

Boiling point (760 mm Hg): 66 °C

Freezing point: -108.5 °C

Specific gravity $d^{20}{}_4$: 0.886

Refractive index $n_D{}^{20}$: 1.4073

Spectroscopic polarity index $E_T{}^N$: 0.207 (measured by C. Reichardt)

$\Delta H_f{}^0$(liq): -51.67 kcal/mole = -216,19 kJ/mole

$\Delta H_f{}^0$(g): -44.02 kcal/mole = -184.18 kJ/mole

Flash point (closed cup): -22 °C

Ignition temperature: 260 °C

Limited solubility in water

Azeotrope with water at 1 ATM: 5.3 % by weight of water

Boiling point 63.5 °C

Toxicity: Tetrahydrofuran rates as one of the least toxic solvents, but it has a narcotic effect similar to that of diethyl ether. It gives rise to skin and mucous membrane irritation. As it dissolves the upper layers of the skin, hand protection should be provided to prevent contact. The occupational exposure limit is 200 ppm.

E. Properties of Diacetyl

Molecular weight: 86.088 g/mole

Boiling point (760 mm Hg): 88 °C

Freezing point: -2.5 °C

Specific gravity $d^{20}{}_4$: 0.9831

Refractive index $n_D{}^{18.5}$: 1.3933

Spectroscopic polarity index $E_T{}^N$: 0.454 (measured by C. Reichardt)

$\Delta H_f{}^0$(liq): -87.45 kcal/mole = -365.89 kJ/mole

$\Delta H_f{}^0$(g): -78.20 kcal/mole = -327.19 kJ/mole

Flash point (closed cup): 26 °C

Solubility in water at 15 °C: 25 g per 100 ml of water

Azeotrope with water at 1 ATM: 10.71 % by weight of water
Boiling point 78.5 °C

Azeotrope with ethanol at 1 ATM: 53 % by weight of diacetyl
Boiling point 73.9 °C

Azeotrope with propanol at 1 ATM: 75 % by weight of diacetyl
Boiling point 85.5 °C

Vapor pressures: At –70 °C (solid): 0.01 torr
At –30 °C (solid): 1.0 torr
At 0 °C (liquid): 13 torr

F. Properties of 2,3-Pentanedione

Molecular weight: 100.114 g/mole

Boiling point (760 mm Hg): 112 °C

Freezing point: -52 °C

Specific gravity d^{15}_{4}: 0.955

Refractive index n_D^{19}: 1.4081

Flash point (closed cup): 18 °C

Solubility in water at 25 °C: 6.67 g per 100 ml of water

Azeotrope with water at 1 ATM: Boiling point 86 °C

G. Properties of Acetoin

Molecular weight: 88.104 g/mole

Boiling point (760 mm Hg): 143.6 °C

Freezing point: +15 °C

Specific gravity d^{15}_{4}: 1.002

Refractive index n_D^{15}: 1.4194

Unlimited solubility in water

Azeotrope with water at 1 ATM: 85 % by weight of water

Boiling point 99.87 °C

Vapor pressures: At 36.5 °C: 11 torr

At 49.2 °C: 20 torr

At 51 °C: 23 torr

H. Properties of Acetic Acid

Molecular weight: 60.052 g/mole

Boiling point (760 mm Hg): 118.1 °C

Freezing point: +16.6 °C

Specific gravity d^{20}_4: 1.049

Refractive index n_D^{20}: 1.37182

Spectroscopic polarity index E_T^N: 0.648 (measured by C. Reichardt)

ΔH_f^0(liq): -116.4 kcal/mole = -487.02 kJ/mole

ΔH_f^0(g): -103.97 kcal/mole = -435.01 kJ/mole

Viscosity at 20 °C: 1.183 cP

Flash point (closed cup): 40 °C

Ignition temperature: 430 °C

Unlimited solubility in water

No azeotrope with water at 1 ATM

Dissociation constant at 25 °C: 1.845 x 10^{-5} mole/liter

Acetic acid vapor in concentrations above 50 ppm causes irritations of mucous membranes. For this reason, the maximum admissible work place concentration is 10 ppm. The smell threshold lies between 1 and 5 ppm.

Liquid acetic acid in concentrations above 80 percent has the same destructive effect on the human skin as concentrated hydrochloric acid. Glacial acetic acid can cause perforation of the esophagus.

Vapor pressure as a function of temperature:

t, °C	20	40	60	80	100
p, torr	11.80	34.06	88.50	202.56	417.52

I. Properties of Formic Acid

Molecular weight: 46.026 g/mole

Boiling point (760 mm Hg): 100.7 °C

Freezing point: +8.4 °C

Specific gravity d^{15}_4: 1.22647

Refractive index n_D^{20}: 1.37137

Spectroscopic polarity index E_T^N: 0.728 (measured by C. Reichardt)

ΔH_f^0(liq): -97.8 kcal/mole = -409.20 kJ/mole

ΔH_f^0(g): -86.67 kcal/mole = -362.63 kJ/mole

Viscosity at 20 °C: 1.804 cP

Flash point (closed cup): 48 °C

Ignition temperature: 480 °C

Unlimited solubility in water

Azeotrope with water at 1 ATM: 22.6 % by weight of water

Boiling point 107.2 °C

Dissociation constant at 20 °C: 1.77×10^{-4} mole/liter

Toxicity: LD_{50}(oral) for a rat:......................1830 mg/kg

LD_{50}(oral) for a mouse:......................1076 mg/kg

LC_{50} in air (1 hour inhalation) for a rat:......................7.4 g/m^3

Formic acid causes irritation of the eyes and of the mucous membranes. It can also cause corneal opacity. The occupational exposure limit is 9 mg/m^3.

Vapor pressure as a function of temperature:

t, °C	20	39.89	59.98	79.93
p, torr	33.68	85.52	193.47	397.18

J. Properties of Difurfural (5,5'-diformyl-2,2'-difuran)

Molecular weight: 190.148 g/mole

Melting point: 264 °C

Yellow crystals

Soluble in furfural and acetic acid

Insoluble in chloroform, methylene chloride, diethyl ether, benzene, dioxane, acetone, and ethyl acetate.

K. Properties of Xylose

Molecular weight: 150.130 g/mole

Melting point: 153 ^{o}C

White crystals

Specific gravity $d^{20}{}_4$: 1.525

Refractive index $n_D{}^{20}$: 1.517

Solubility in water at 20 ^{o}C: 117 g per 100 ml of water

L. Properties of Tetrahydrofurfuryl Alcohol

Molecular weight: 102.130 g/mole

Boiling point (760 mm Hg): 178 °C

Freezing point: Below –80 °C

Specific gravity d^{20}_4: 1.0495

Refractive index n_D^{19}: 1.4502

Viscosity at 25 °C: 6.24 cP

Flash point (closed cup): 83.9 °C

Ignition temperature: 282 °C

Unlimited solubility in water

Toxicity: LD_{50} (oral) for a mouse:..........2300 mg/kg

LD_{50}(oral) for a rat:..........2500 mg/kg

LD_{50}(oral) for a guinea pig:..........3000 mg/kg

Massive doses of tetrahydrofurfuryl alcohol vapor produce a narcotic effect. Tetrahydrofurfuryl alcohol is slightly irritating to the human skin, and severely irritating to the eyes of albino rabbits. When young adult albino rabbits were exposed to air containing 4700 ppm of tetrahydrofurfuryl alcohol for a period of 4 hours, no deaths occurred during the subsequent 14-day observation period, and body weight gains were within normal limits.

M. Properties of Dihydropyran

Molecular weight: 72.104 g/mole

Boiling point (760 mm Hg): 84.5 °C

Freezing point: -70 °C

Specific gravity d^{20}_4: 0.923

Refractive index n_D^{20}: 1.4380

Flash point (closed cup): -17.8 °C

Solubility in water at 25 °C: 1.63 g per 100 ml of water

Toxicity: LD_{50}(oral) for a rat: 256 mg/kg

Dihydropyran is harmful if absorbed by the skin, inhaled, or ingested. Its vapor or mist is irritating to the eyes, to mucous membranes, and to the upper respiratory tract, but a maximum admissible work place concentration has not been established as yet.

N. Properties of Furoic Acid

Molecular weight: 112.082 g/mole

Melting point: 133 °C

White crystals

Solubility in water at 15 °C: 3.57 g per 100 ml of water

Triple point pressure: 10.3 torr

O. Properties of Methylfuran

Molecular weight: 82.098 g/mole

Boiling point (760 mm Hg): 63.2 °C

Freezing point: -88.7 °C

Specific gravity $d^{20}{}_4$: 0.913

Refractive index $n_D{}^{20}$: 1.4320

Flash point (closed cup): -27 °C

Solubility in water at 25 °C: Smaller than 0.3 g per 100 ml of water

Toxicity: LD_{50}(oral) for a rat: 167 mg/kg

LC_{50} in air (1 hour inhalation) for a rat: 1485 ppm

Symptoms of overexposure are nausea and temporary lowering of the blood pressure.

P. Properties of 5-Methyl Furfural

Molecular weight: 110.108 g/mole

Boiling point (760 mm Hg): 187 °C

Specific gravity d^{18}_{4}: 1.1072

Solubility in water at 20 °C: 3.3 g per 100 ml of water

Q. Properties of 2-Furyl Methyl Ketone

Molecular weight: 110.108 g/mole

Boiling point (760 mm Hg): 173 °C

Freezing point: +33 °C

R. Properties of Furan Dialdehyde

Molecular weight: 124.068

Melting point: 110 °C

White lamellar crystals

Dipole moment: 3.46 D

Solubility in water at 17 °C: 1.0 g per 100 ml of water

Solubility in acetone at 17 °C: Approximately 10 g per 100 ml of acetone

Insoluble in the light fraction ("ether fraction") of gasoline.

Slightly soluble in hexane, cyclohexane, ligroin, carbon tetrachloride, diethyl ether, and benzene.

Well soluble in hot water, hot ethanol, dimethylsulfoxide, and methylisobutyl ketone.

S. Explosion Limits in Air at 760 mm Hg and 20 °C

Furfural: 2.1 – 19.3 % by mole

Furfuryl alcohol: 1.8 – 16.3 % by mole

Furan: 2.3 – 14.3 % by mole

Tetrahydrofuran: 1.5 – 6.0 % by mole

Tetrahydrofurfuryl alcohol: 1.5 – 9.7 % by mole

Acetic acid: 4.0 – 17.0 % by mole

Acetaldehyde: 4.0 – 57 % by mole

Acetone: 1.6 – 13.0 % by mole

Methanol: 6.0 – 36.5 % by mole

Ethanol: 2.6 – 18.9 % by mole

Hexane: 1.2 – 7.4 % by mole

Hydrogen: 4.0 – 75 % by mole

T. Spectroscopic Polarity

In this book, use has been made of a spectroscopically determined "polarity index" called E_T^N, where E stands for "energy", T stands for "transition", and N stands for "normalized". The polarity index characterizes the polarity of organic compounds relative to water, the latter being given the polarity index 1.000. All organic compounds have polarity indices smaller than that of water, the only known exception being 1,1,1,3,3,3-hexafluoro-2-propanol, for which $E_T^N = 1.068$.

Worldwide, this polarity index has found extensive applications in distillation, extraction, and organic synthesis, and for this reason, where available, it was included in the property lists given in the preceding chapters.

The polarity index was invented by C. Reichardt, professor of organic chemistry at the University of Marburg/Germany, and author of a rather famous book on "SOLVENTS AND SOLVENT EFFECTS IN ORGANIC CHEMISTRY" (VCH Verlagsgesellschaft, Weinheim, 1988). In this book, the E_T^N values are listed for hundreds of compounds.

The effect underlying the E_T^N concept can be explained as follows: When a small quantity of a special dye ("REICHARDT's dye") is dissolved in the liquid to be characterized, the maximum absorption peak of this dye undergoes a shift depending on the polarity of the liquid in which it is dissolved. This is illustrated in Figure 108. It depicts the absorption spectrum of REICHARDT's dye in ethanol, acetonitrile, and 1,4-dioxane. In ethanol, the maximum absorption peak is at 550 nm. This means that green light is absorbed, so that in ethanol the dye assumes the complementary color of green, which is purple. In acetonitrile, the maximum absorption peak is at 622 nm. This means that orange light is absorbed, so that in acetonitrile the dye assumes the complementary color of orange, which is blue-green. This effect is called "solvatochromism" (= change of color in different solvents). It is due to the fact that the polarity of the solvent molecules changes the excitation energy (electronic transition energy) of the dye, so that inversely a measurement of the excitation energy (measurement of the frequency of the absorption peak) can be used to determine the polarity of various solvents.

Table 4 gives the E_T^N values of some important solvents in the order of decreasing

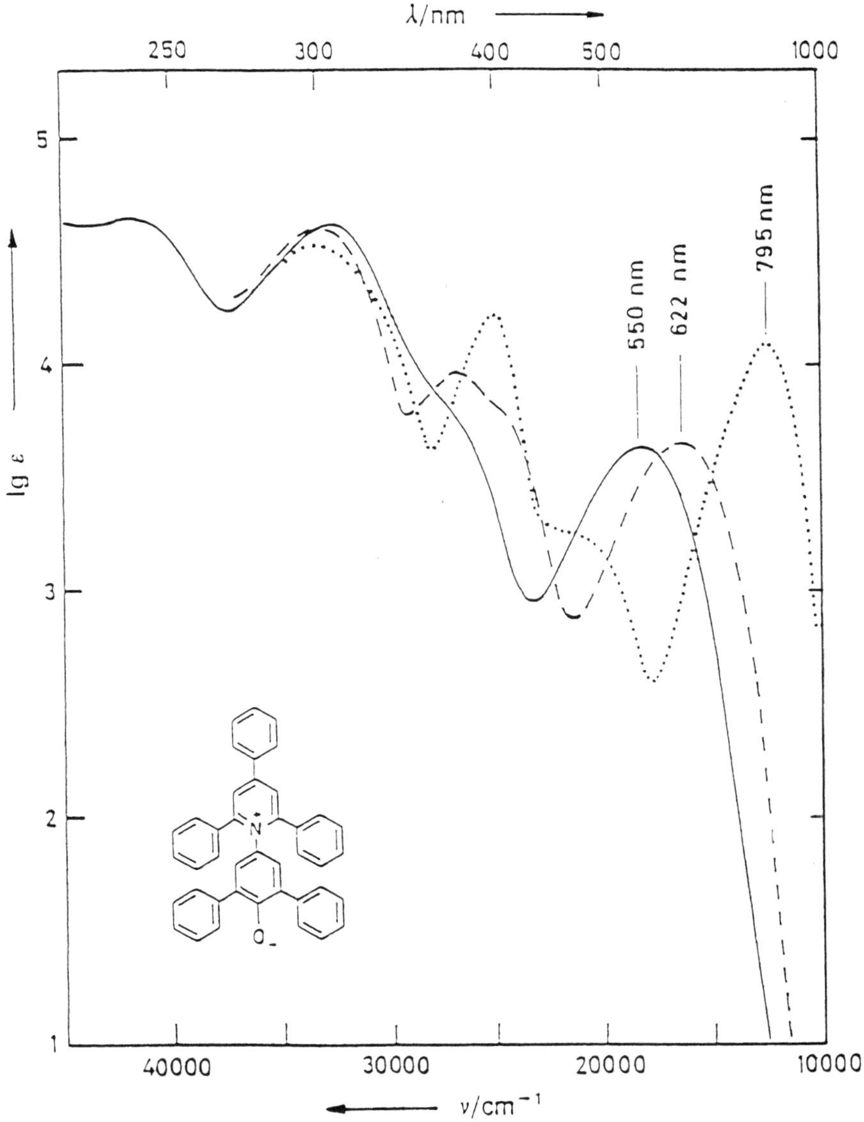

Figure 108. The Absorption Spectrum of "REICHARDT's Dye" [2,6-diphenyl-4-(2,4,6-tri-phenyl-1-pyridinio)phenoxide] at 25 °C. The formula of the dye is shown in the lower left corner of the diagram.

———————— spectrum in ethanol

— — — — spectrum in acetonitrile

---------------- spectrum in 1,4-dioxane

Table 4. E_T^N Values of some Important Solvents in the Order of Decreasing Polarity.

Water	1.000
Formamide	0.799
Methanol	0.762
Ethanol	0.654
Acetic Acid	0.648
Nitromethane	0.481
Acetonitrile	0.460
Diacetyl	0.454
Dimethyl sulfoxide	0.444
Furfural	0.426
Aniline	0.420
N,N-Dimethylformamide	0.404
Acetone	0.355
Nitrobenzene	0.324
Chloroform	0.259
Ethyl acetate	0.228
Tetrahydrofuran	0.207
1,4-Dioxane	0.164
Diethyl ether	0.117
Benzene	0.111
Carbon disulfide	0.065
n-Hexane	0.009

polarity. Furfural is seen to be more polar than aniline and N,N-dimethylformamide, and less polar than dimethyl sulfoxide and acetonitrile.

A comprehensive table of E_T^N values as given in professor Reichardt's book permits a judicious selection of volatility modifiers in distillation as well as an appraisal of solvents for liquid/liquid extraction. Consequently, for such applications, it is essential that the E_T^N value be known.

U. Pentosan Determination

The analytical determination of the pentosan content of a raw material is based on converting the pentosan to furfural, and on then measuring the furfural thus obtained. Obviously, a prerequisite for this procedure is a 100 % conversion of the pentosan to furfural, which means that furfural losses by resinification (reaction of furfural with itself) and by condensation (reaction of furfural with intermediates of the pentose-to-furfural conversion) must be completely avoided. This can be accomplished by using the method described by Hughes and Acree [112], where the raw material containing the pentosan is digested in boiling 12 % hydrochloric acid saturated with sodium chloride. The atmospheric boiling point of 12 % aqueous hydrochloric acid is approximately 108 °C, and the saturation with sodium chloride raises the boiling point to 110 °C. Any furfural formed in this system is instantly transferred into the vapor phase, where loss reactions cannot take place. The theoretical explanation for this transfer process on the basis of the respective phase diagram was given in chapter 7.

A suitable apparatus for the pentosan analysis along these lines is shown in Figure 109. This setup consists of a distillation flask 1 equipped with a steam inlet 2, a thermometer 3, a feed funnel 4, a condenser 5, and a metering cylinder 6 sealed by a trap 7. The trap consists of small RASCHIG rings and water. The distillation flask is charged with 20 g of sodium chloride, with a quantity of raw material corresponding to approximately 0.15 g of pentosan, and with 100 ml of 12 % aqueous hydrochloric acid (density 1.06 g/cc). This mixture is rapidly heated, and shortly before boiling, at approximately 104 °C, the needle valve 8 is opened to introduce some steam from a pressure cooker operated at 115 °C (1268 torr). Then the temperature is raised to boiling (110 °C), with the heat input adjusted so as to produce approximately 100 ml of distillate in 15 minutes. To the extent as evaporation takes place from the still, fresh 12 % hydrochloric acid is added from the feed funnel in intervals of five minutes to restore the initial level, and this procedure is continued until the distillate is free of furfural. Thereafter, the water from the trap (containing some furfural) is added to the distillate. The total quantity of furfural in the resulting mixture can then be determined in various ways.

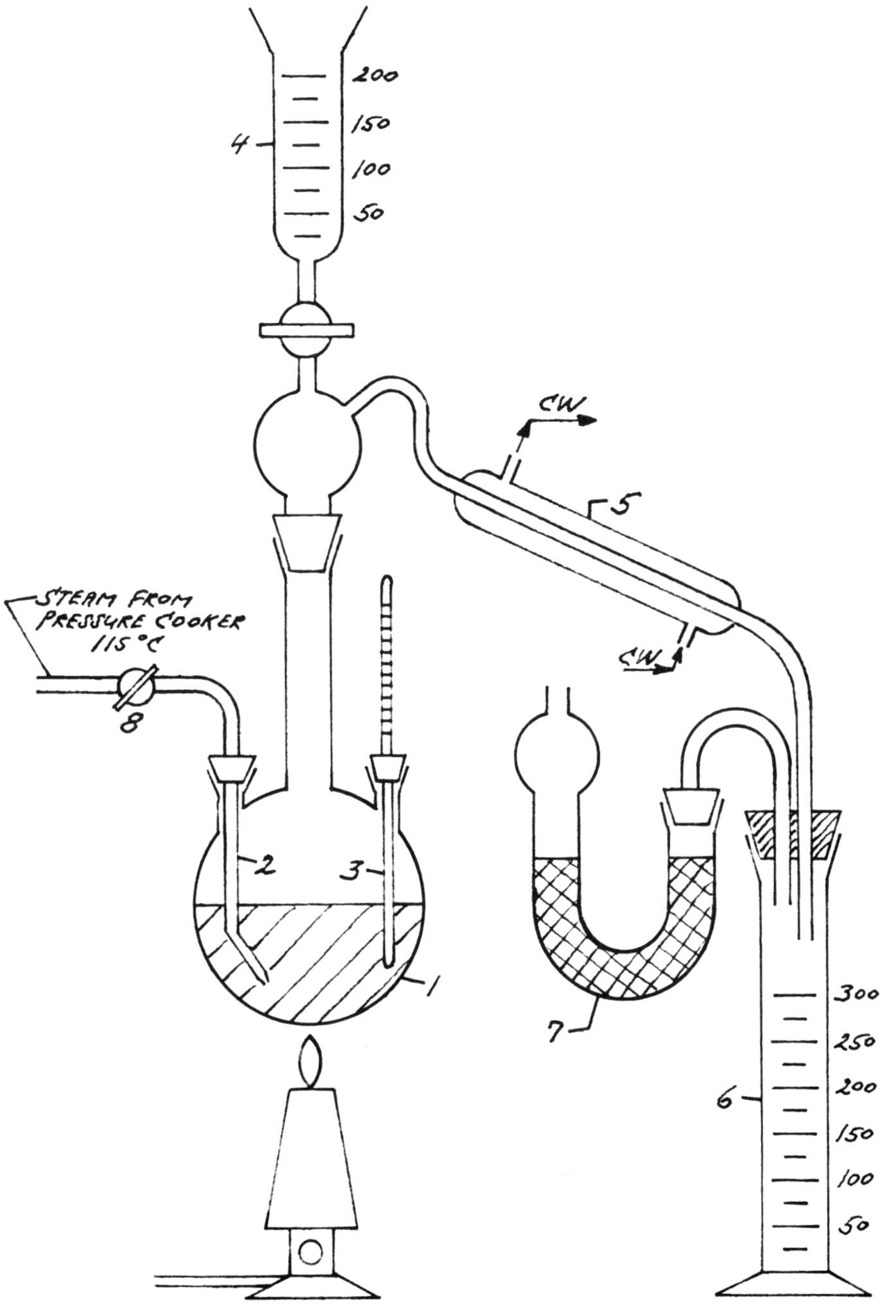

Figure 109. Laboratory Setup for the Determination of Pentosan.

One such way is a precipitation of the furfural with barbituric acid to form furfural barbituric acid [113]. To this end, 0.5 g of barbituric acid is dissolved in 25 ml of 12 % aqueous hydrochloric acid, and this solution is added to the mixture containing the furfural. After a standing period of 18 hours to permit the crystallization to take place, the precipitate is separated by a dried and weighed glass filter, washed twice with distilled water, and drained by a vacuum. Then the filter is dried at 130 °C until a constant weight is reached.

If W is the weight of the dried precipitate, and if V is the volume of the solution after addition of the precipitating agent, then the quantity of furfural obtained from the raw material is

$$FU = (M_{FU}/M_{FBA})[W + \lambda V]$$

where M_{FU} and M_{FBA} are the molecular weights of furfural and furfural barbituric acid, respectively, and where λ is the solubility of furfural barbituric acid in 12 % aqueous hydrochloric acid. The second term in the brackets accounts for the tiny quantity of furfural barbituric acid lost in the drained liquid. $M_{FU}/M_{FBA} = 0.46606$, and $\lambda = 0.0000122$ g/ml, so that

$$FU = 0.46606\ [W + 0.0000122\ (g/ml)\ V]$$

and inasmuch as according to the stoichiometry (chapter 2.1, page 3) 132.114 g of pentosan yield 96.082 g of furfural, it follows that each gram of furfural produced in the test corresponds to 1.37501 g of pentosan.

Needless to say, the same test procedure can be employed to determine the pentose content of a solution, in which case 150.130 g of pentose yield 96.082 g of furfural, so that each gram of furfural produced in the test corresponds to 1.56252 g of pentose.

References

[112] E. E. Hughes and S. F. Acree, Journal of Research of the National Bureau of Standards 21 (1938) 327-336.

[113] A. Beythien and W. Diemair, Laboratoriumsbuch für den Lebensmittelchemiker, Verlag Gisela Liedl, München, 1970.

V. Methyl Pentosan Determination

As discussed in chapter 11, the methyl pentosan content of the raw material is of considerable interest as it determines the level of 5-methyl furfural and furyl methyl ketone in the final furfural. Both 5-methyl furfural and furyl methyl ketone are unwanted by-products requiring a sizable expense if they are to be removed from the furfural, and if they exceed the critical level of 1 %, the furfural thus loaded loses the attribute of being a “fine chemical”, which is an important aspect for pricing the product.

The procedure for determining the methyl pentosan content of the raw material is based on the fact, discovered by Schorger [114], that methyl furfural phloroglucide is soluble in hot ethanol while furfural phloroglucide is not, so that the two can be separated.

Two grams of comminuted 40-mesh dry raw material and 100 ml of 12 % aqueous hydrochloric acid are charged into the distillation flask of a setup as shown in Figure 109 of the preceding chapter, and the distillation procedure is followed as in the pentosan determination. The total furfural distillate, including the liquid of the trap, is then mixed with 40 ml of a filtered phloroglucinol (1,3,5-benzenetriol) solution prepared beforehand by heating 11 grams of phloroglucinol in 1500 ml of 12 % hydrochloric acid. The resulting mixture soon turns greenish black, and after standing for 16 hours, solid furfural phloroglucide and solid methyl furfural phloroglucide will have settled to the bottom of the beaker. In the unlikely event that a drop of the supernatant liquid gives a pink color with aniline acetate paper, the precipitation is not complete. In that case, some more phloroglucinol solution is added and the beaker allowed to stand for another 16 hours to complete the precipitation. The precipitate is separated by a dried and weighed glass filter, washed with 150 ml of distilled water, and drained by a vacuum. Then the filter is dried at 130 ^{o}C until a constant weight is reached. This yields a weight W_1. The filter with its precipitate is then placed into a beaker filled with 95 % ethanol, and this beaker is heated in a water bath of 60 ^{o}C for ten minutes. This dissolves the methyl furfural phloroglucide while leaving the furfural phloroglucide behind. The ethanol solution is decanted, and the procedure is repeated four or five times until the alcohol is colorless. The beaker and its content is then dried for two hours in an oven. Weighing and deduction of the filter and beaker weights yields a weight W_2.

Calculating W_1 minus W_2 gives the weight of methyl furfural phloroglucide, from which the amount of methyl pentosan in the original sample can be computed, with 1 mg of methyl furfural phloroglucide corresponding to 0.61867 mg of methyl pentosan.

Reference

[114] A. W. Schorger, Ind. Eng. Chem. 9 (1917) 556-560.

a. The Entropy Effect in Furfural Loss Reactions

In In the chapters 6 and 7, it was discussed in detail that in furfural reactors any furfural in the vapor phase is "safe", i.e. incapable of undergoing loss reactions, but that any furfural in solution can react with itself and with intermediates of the xylose-to-furfural conversion as the solution contains hydrogen ions capable of "specific acid catalysis" as well as neutral molecules capable of "general acid catalysis".

Inasmuch as the furfural loss reactions lead to larger molecules, they are analogous to polymerization, with its buildup of oligomer and polymer molecules from monomer molecules. In representing an increase in "order", this buildup of larger molecules causes a decrease in entropy, so that the "entropy change of the reaction" ΔS_R is a negative quantity.

Reactions can proceed to any significant extent only when their "change of free energy"

$$\Delta G_R = \Delta H_R - T\Delta S_R$$

is negative, where ΔH_R is the change in enthalpy, and T is the absolute temperature. Thus, with ΔS_R being negative, so that $-T\Delta S_R$ is a positive quantity, to be significant, polymerization reactions must have a negative ΔH_R, which means that they must be exothermic. In addition, it is noted that with increasing temperatures the driving force ΔG_R of the reaction diminishes as it becomes less negative, on account of the positive nature of $-T\Delta S_R$, and at a critical temperature T_c, where ΔG_R becomes zero, so that

$$\Delta H_R - T_c\Delta S_R = O$$

or

$$T_c = \Delta H_R/\Delta S_R$$

the reaction essentially ceases. (Note that both ΔH_R and ΔS_R are negative quantities.)

In polymerization technology, the temperature T_c is called the "ceiling temperature" of the reaction [115] as it represents the "ceiling" up to which the polymerization proceeds significantly, whereas above this temperature it becomes insignificant.

The same reasoning applies to the loss reactions in furfural reactors: Above a critical temperature T_c, there will be no significant loss reactions, and on the way to this temperature, consideration must be given to the fact that the equilibrium constant K of any reaction is related to its free energy change ΔG_R [116] by the extremely important equation

$$\Delta G_R = - RT \ln K$$

leading to

$$K = \exp(-\Delta G_R/RT)$$

which shows that

(a) at $\Delta G_R < 0$ the equilibrium constant K is huge, so that loss reactions are highly significant, whereas

(b) at $\Delta G_R > 0$ the equilibrium constant K becomes tiny, so that loss reactions are insignificant.

In other words, as the temperature approaches T_c, the furfural loss reactions diminish, so that the yield increases, and eventually, at a sufficiently high temperature, these loss reactions become insignificant, so that the furfural yield approaches 100 percent.

This is illustrated in Figure 110 [117] referring to results obtained with acidified xylose solutions in sealed ampoules. As can be seen, the furfural yield increases steadily with increasing temperature, although, noteworthily, in the sealed ampoule process there is absolutely no removal of furfural from the scene of the reaction.

It is a tragedy that the first industrial furfural process, described in the introduction, had to be carried out in a total absence of these facts, and that its temperature was limited by the low pressure capacity of old reactors from an abandoned cereal process, so that severe loss reactions and correspondingly low yields became a trademark of the furfural industry from the very start.

In the light of all the facts now available from many independent sources, new furfural processes, as the SUPRATHERM and STAKE processes, aim at the increased yields obtainable at high temperatures, even without removal of the furfural from the scene of the reaction. Although this leads to somewhat uncomfortable high pressures, it is certainly a correct route towards higher yields, based on a fundamental principle of thermodynamics, and in hindsight the circumstances at the birth of the furfural industry must be deplored.

References

[115] B. Vollmert, Polymer Chemistry, Springer-Verlag, New York, 1973.

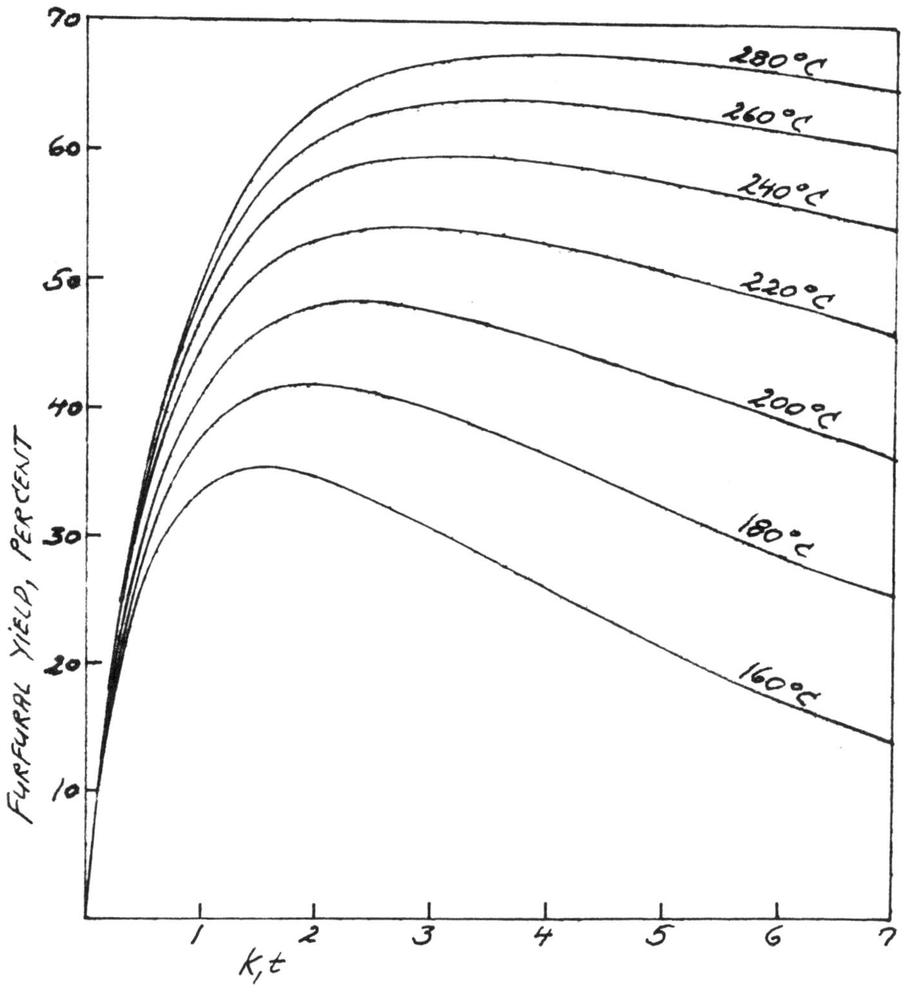

Figure 110. Dependence of the Furfural Yield on Temperature and Time as obtained by a "Sealed Ampoule Process" with an Initial Xylose Concentration of 0.666 mole/liter (100 g/liter).

[116] G. M. Barrow, Physical Chemistry, McGraw-Hill Book Company, New York, 1966.
[117] D. F. Root, J. F. Saeman, J. F. Harris, and W. K. Neill, Forest Products Journal 9 (1959) 158-165.

b. The "Temperature Compensation" of Acidity

In chapter 5, it was shown that the rate constant for the reaction of xylose disappearance has the form

$$k_1 = \alpha\, c_H \exp(-\Delta E/RT)$$

which means that the rate of this reaction increases linearly with the hydrogen ion concentration c_H, and exponentially with the absolute temperature T. Thus, acidity can be compensated by temperature, and vice versa. In other words, a lowering of the acidity by Δc_H can be compensated in raising the temperature by ΔT, the correlation being

$$\alpha(c_H - \Delta c_H) \exp[-\Delta E/\{R(T + \Delta T)\} = \alpha\, c_H \exp[-\Delta E/RT]$$

which leads to

$$\Delta c_H/c_H = 1 - \exp[(-\Delta E/RT)\{(\Delta T/T)/(1 + \Delta T/T)\}]$$

For the reaction of xylose disappearance, the quantity

$$\Delta E/R = 16894\ ^{\circ}K$$

so that for any absolute temperature T the value of $\Delta c_H/c_H$ can be plotted versus $\Delta T/T$. It is more instructive, however, to choose a reference temperature ϑ_R, in °C, and to plot

$$1 - (\Delta c_H/c_H) = (c_H - \Delta c_H)/c_H$$

versus the corresponding temperature ϑ, in °C as $(c_H - \Delta c_H)/c_H$ is the fraction to which the acidity can be reduced when the temperature is raised from ϑ_R to ϑ. For $\vartheta_R = 180$ °C, such a plot is shown in Figure 111, where the ordinate is logarithmic. As can be seen, when the temperature is raised from 180 °C to 230 °C, the acidity can be reduced to 2.45 percent of the value at 180 °C if the reaction rate is to be kept unchanged. Inversely, at equal reaction rate, the acidity must be grossly increased if the temperature is lowered. This explains why the batch process of QUAKER OATS (chapter 10.1), at 153 °C, requires a monstrous acidity for many hours, whereas the STAKE process (chapter 10.7.2), at 230 °C, can operate without any foreign acid addition with a residence time in the order of only six minutes.

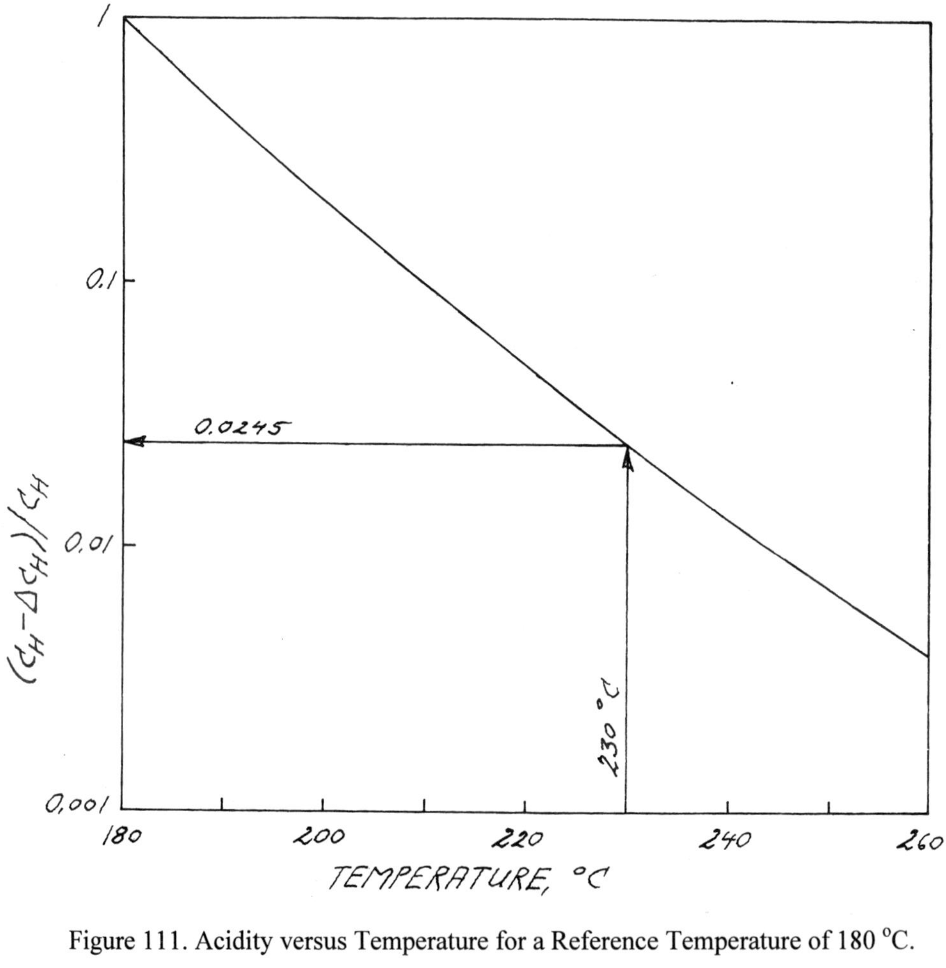

Figure 111. Acidity versus Temperature for a Reference Temperature of 180 °C.

c. The Corrosion Debacle in Extracting Furfural with Chloroform

In 1992, a rather unusual furfural plant was built. With a front end according to the AGRIFURANE process described in chapter 10.2, the back end was designed as shown in Figure 112. The filtered reactor condensate containing 5 % furfural, 1.7 % acetic acid, 0.17 % formic acid, and various low boilers was introduced into an extraction tower 1 fed with chloroform at the top. On the way downwards, the heavy chloroform (density 1.498 g/cc at room temperature) picked up the furfural, and in view of the poor solubility of chloroform in water, it formed a chloroform/furfural extract at the bottom. This extract entered a distillation column 2 removing the chloroform as the head fraction. From a buffer tank 3, this chloroform was recycled to the extraction tower 1. The sump fraction of the distillation column 2 consisted of furfural, polymers, waxes, and some low boilers. This fraction was introduced into a distillation column 4, which yielded a head fraction of low boilers, a side stream of furfural, and a sump fraction of polymers and waxes.

The raffinate of the extraction tower 1 consisted of water, acetic acid, formic acid, and a small quantity of chloroform. This fraction was fed into a distillation column 5 producing a head fraction of chloroform joining the other chloroform in tank 3. The sump fraction of column 5 consisted of water, acetic acid, and formic acid. This stream entered an extraction tower 6. A mixture of a high-boiling amine (ALAMINE 336 produced by HENKEL) and a high-boiling hydrocarbon (SHELLSOL produced by SHELL), introduced into the bottom of tower 6, picked up the carboxylic acids to form an extract phase at the top. The raffinate, being water, was discharged from the bottom.

The extract of tower 6 entered a distillation column 7 recovering the amine/hydrocarbon mixture as the sump fraction, to be recycled into the extractor 6, while the head fraction of column 7, a mixture of acetic acid and formic acid, was separated in the distillation column 8.

Apart from the fact that this design had the disadvantage of using a high-boiling extractant for the carboxylic acids, so that various undefined high-boiling impurities accumulated in the extractant, thus necessitating its frequent replacement, the use of chloroform proved to be a disaster as any stainless steel in contact with this medium was

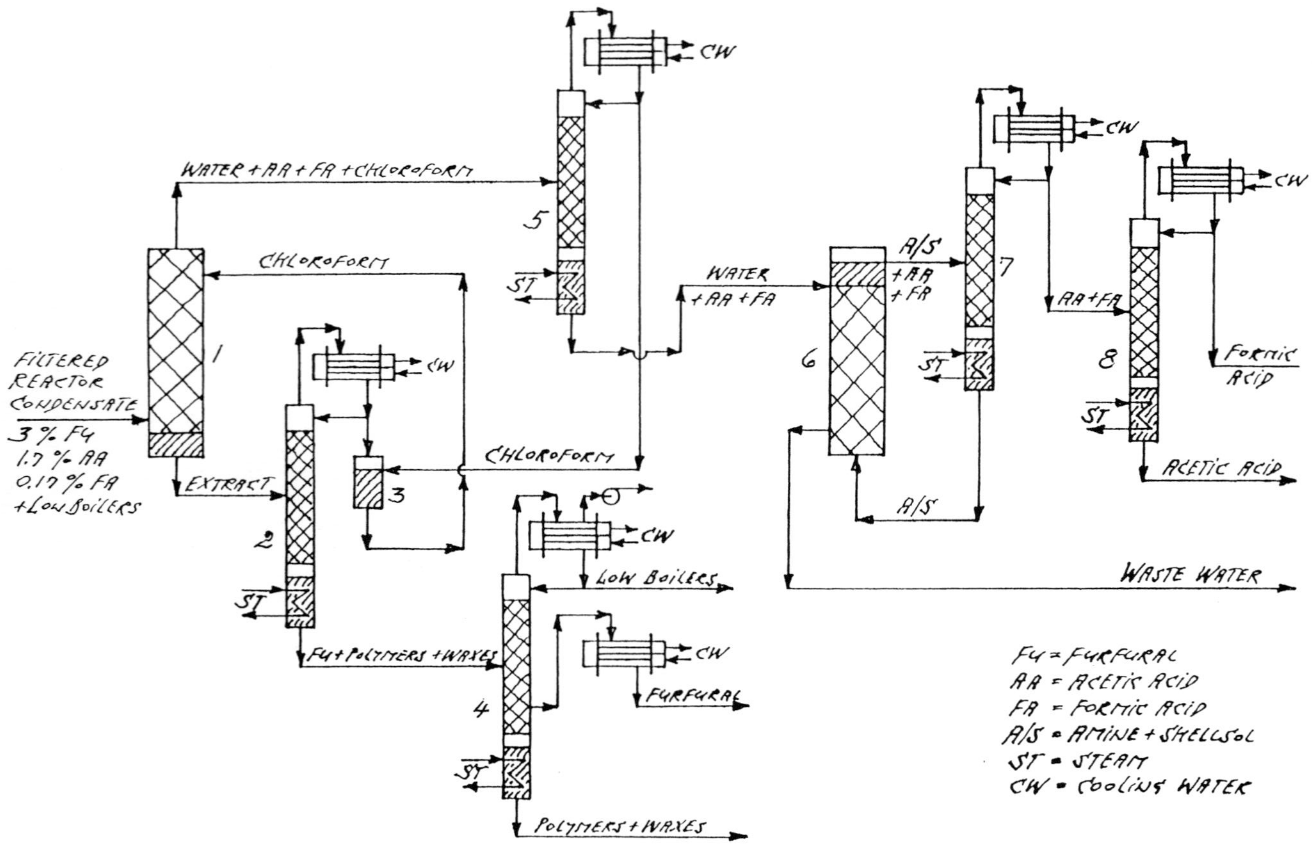

Figure 112. Furfural Extraction with Chloroform.

doomed by corrosion.The reason for this debacle is a reaction of chloroform with oxygen to form phosgene ($COCl_2$) and hydrochloric acid:

$$CHCl_3 + 0.5\ O_2 \rightarrow\rightarrow COCl_2 + HCl$$

This reaction falls into the category of "autoxidation" defined as the slow low-temperature oxidation of organic compounds by O_2 via a radical chain mechanism, based on the fact that the oxygen molecule is a diradical

and that the C-H bond in chloroform is very weak, on account of the high electronegativity of Cl. The reaction mechanism is shown below:

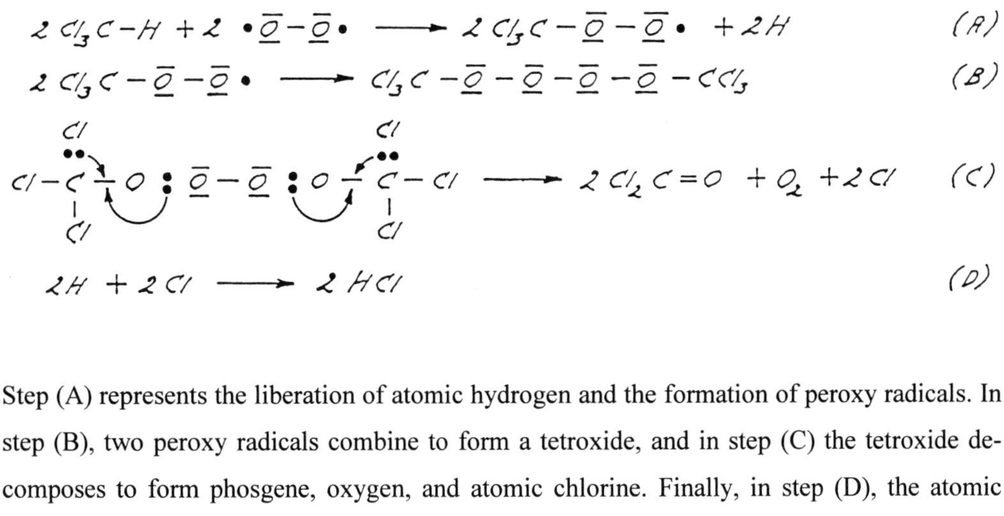

Step (A) represents the liberation of atomic hydrogen and the formation of peroxy radicals. In step (B), two peroxy radicals combine to form a tetroxide, and in step (C) the tetroxide decomposes to form phosgene, oxygen, and atomic chlorine. Finally, in step (D), the atomic chlorine of (C) reacts with the atomic hydrogen of (A) to form hydrochloric acid.

Unfortunately, in the design of Figure 112, the chloroform is also in contact with water. This amplifies the formation of hydrochloric acid as water reacts with phosgene to yield additional HCl:

$$COCl_2 + H_2O \rightarrow\rightarrow CO_2 + 2\ HCl$$

Combination of these reactions shows that in the presence of water each oxidized chloroform molecule yields three molecules of HCl.

The hydrochloric acid thus generated severely attacks normal stainless steel by "pit corrosion" and "stress fissure corrosion". In the plant shown in Figure 112, severe corrosion became apparent in the extraction tower 1 and particularly in the distillation column

3 where an elevated temperature accelerated the corrosion process. In view of these gruesome facts, the plant had to be abandoned.

The phenomenon of severe corrosion by chloroform has been well known for many decades [118].

Reference

[118] Lexikon der Korrosion, Volume 2, Mannesmannröhren-Werke, Düsseldorf, 1970.

d. Corrosion in the Extractive Distillation of Diacetyl

As described in chapter 16.6.1, extractive distillation of diacetyl requires the use of aqueous sulfuric acid as volatility modifier throughout most of the column. Although the process calls for an acid strength of only 0.3 % by weight, with a column made of stainless steel type 316 L this led to considerable corrosion within a year of operation. However, it was found possible to completely eliminate this problem by adding a trace of nitric acid.

It was known from the literature [119] that at 145 °C (a temperature not reached anywhere in the extractive distillation of diacetyl), the corrosion rate of the German stainless steel type 1.4301 (18 % Cr, 10 % Ni) in an aqueous mixture of

0.12 % H_2SO_4 by volume = 0.22008 % H_2SO_4 by weight, and

0.02 % HNO_3 by volume = 0.03004 % by weight

was "far below 0.1 mm/a".

On the basis of this information, it was estimated that 0.05 % by weight of HNO_3 added to the 0.3 % by weight of H_2SO_4 would be sufficient to prevent corrosion of 316 L stainless steel at temperatures slightly below 100 °C.

The nitric acid was added to the reboiler since HNO_3 forms a low-boiling azeotrope with diacetyl, so that the HNO_3 went upwards with the diacetyl, thereby protecting the entire column. After this simple measure, the corrosion problem was a specter of the past.

The reason for the protective action is known to lie in the fact that the nitric acid, even in a very small concentration, enforces passivation of the stainless steel. Passivation is based on the formation of a thin layer of chromium (III) oxide, which brings the attack to a virtual halt:

$$Cr + HNO_3 \rightarrow\rightarrow 0.5\ Cr_2O_3 + NO + 0.5\ H_2O \qquad (1)$$

The driving force of this reaction as expressed by the change of free enthalpy is –106.32 kcal/mole = -444.84 kJ/mole, which is a gigantic value, giving this reaction a huge thermodynamic preference over the competitive sulfuric acid reaction

$$Cr + 1.5\ H_2SO_4 \rightarrow\rightarrow Cr^{3+} + 1.5\ SO_4^{2-} + 1.5\ H_2 \qquad (2)$$

where the change of the free enthalpy is a mere –10.15 kcal/mole = -42.47 kJ/mole.

Contrary to (1), in reaction (2) the chromium goes into solution, thus causing a continuing corrosion loss. Another fundamental difference lies in the gaseous reaction products, as in (1) there is only an extremely brief burst of NO, while in (2) there is a continuous formation of hydrogen.

The "oxidizing power" of a corroding medium depends on the chemical substances in the medium, on their concentration, on the reactions they can undergo, and on the temperature. The relation expressing the "oxidizing power" as a function of these variables is called "NERNST's Law":

$$\varepsilon = \varepsilon_0 + (RT/nF) \ln [\Pi\ c_{ox}/\Pi\ c_{red}] \qquad (3)$$

where ε is the "oxidizing power" (expressed in volt)

ε_0 is the "oxidizing power" for unit concentrations (1 mole/liter) and 25 °C (tabulated in handbooks),

R is the universal gas constant,

T is the absolute temperature,

n is the number of electrons exchanged in the reaction,

F is FARADAY's constant = 23.0618 kcal/volt,

Π means "product of",

c_{ox} means mass action concentrations of the reagents in the oxidation reaction, and

c_{red} means mass action concentrations of the reagents in the reduction reaction, but at the exclusion of water, the concentration of which, by international convention, is already included in the tabulated ε_0 values.

On this basis, determining the "oxidizing power" of 0.05 % aqueous nitric acid at 97 °C requires the following input:

(a) The "redox reaction" to be applied, from handbook tables, is

$$NO_3^- + 4H^+ + 3e \rightarrow\rightarrow NO + 2\ H_2O \qquad (4)$$

where e stands for electrons.

(b) For reaction (4), handbook tables give ε_0 = +0.96 volt.

(c) From reaction (4), it is seen that n = 3.

(d) From the oxidation side (left-hand side) of (4), it is seen that

$$\Pi\ c_{ox} = [NO_3^-]\ [H^+]^4$$

where

$$[NO_3^-] = [H^+] = [HNO_3] = 7.934 \times 10^{-3} \text{ mole/liter.}$$

(e) $\Pi\ c_{red} = [NO]$

where according to (4)

$[NO] = [NO_3^-]$

The evaluation of (3) with this input results in

$\varepsilon(0.05\ \%\ HNO_3) = +0.754$ volt.

By comparison, the “oxidizing power” of 0.3 % sulfuric acid at 97 °C is found to be

$\varepsilon(0.3\ \%\ H_2SO_4) = -0.008$ volt.

Normal stainless steel (18 % Cr, 10 % Ni) is passive when the “oxidizing power” of the medium lies between +0.075 volt and +1.150 volt as shown in Figure 113 [120], and for stainless steels containing some molybdenum, the “passivity range” is even wider. Thus, from the given calculation, it can be seen that 0.3 % H_2SO_4, with $\varepsilon = -0.008$ volt, lies well outside the passivity range, in the range of active corrosion, whereas 0.05 % HNO_3, with $\varepsilon = +0.754$ volt, lies well within the passivity range. In other words, the addition of very little HNO_3 takes the system out of the “range of active corrosion” into the “range of passivity” where corrosion is halted by a protective layer of chromium (III) oxide.

References

[119] Lexikon der Korrosion, Volume 2, page 120, Mannesmannröhren-Werke, Düsseldorf, 1970.

[120] Lexikon der Korrosion, Volume 1, page 30, Mannesmannröhren-Werke, Düsseldorf, 1970.

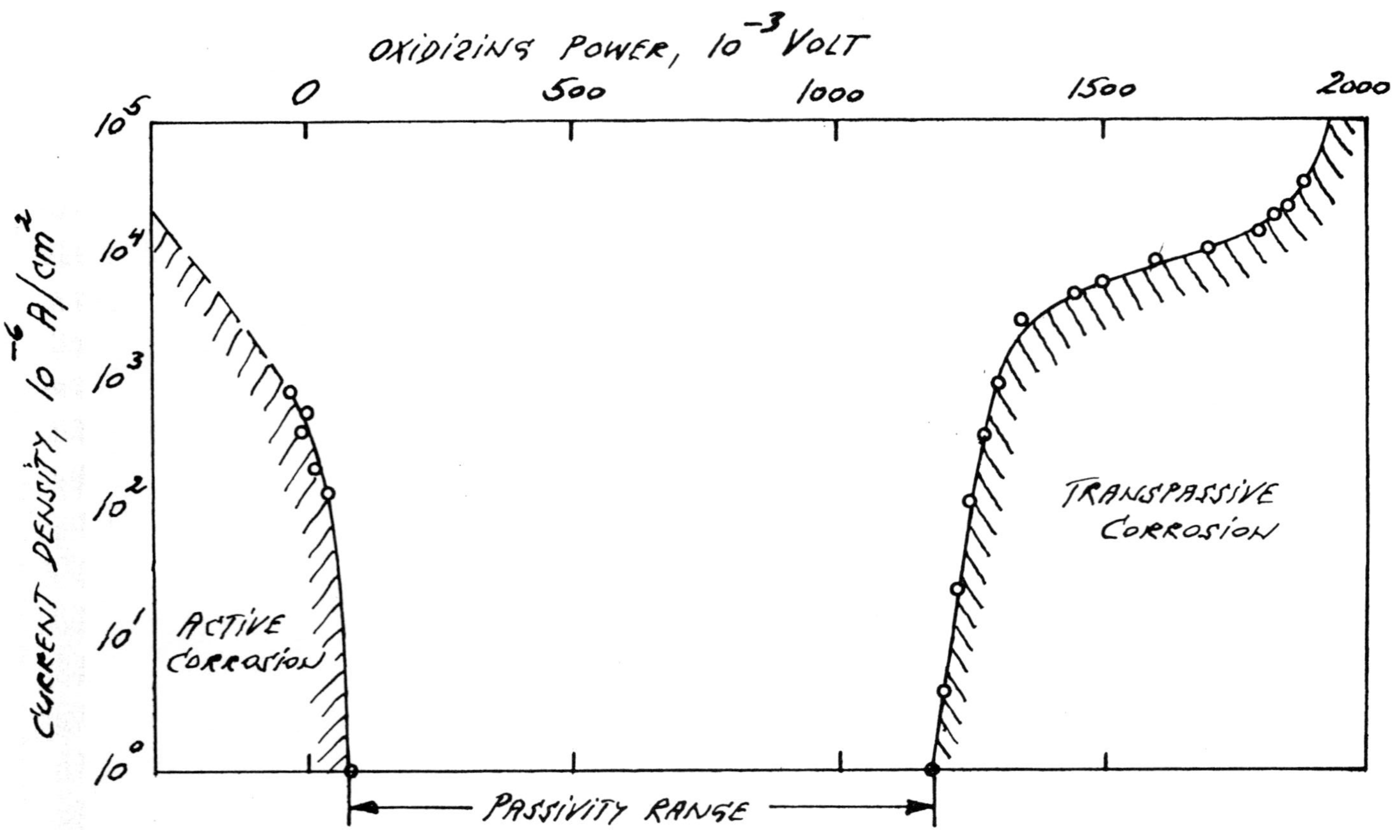

Figure 113. The Passivity Range of Normal Stainless Steel (18 Cr/10 Ni).

e. Corrosion in the Extraction of Acetic Acid and Formic Acid

In the conventional process of extracting acetic acid and formic acid from the waste water of the furfural process by means of ethyl acetate as shown in Figure 51 (page 106), particular attention must be paid to the final distillation column where acetic acid is separated from formic acid. In the sump and the reboiler of this column, corrosion of stainless steel is severe because of the high temperature caused by high-boiling impurities, and in the head of this column the corrosion of stainless steel is also severe as this is where the formic acid concentration is very high, with formic acid being much more agressive than acetic acid.

When the final column is run at atmospheric pressure, even a highly alloyed stainless steel such as UDDEHOLM 904 L containing 20 % Cr, 25 % Ni, 4.5 % Mo, 1.5 % Cu, 0.4 % Si, and 1.7 % Mn was found to be unsatisfactory. In one known case, the column was made of glass, but a less expensive solution of the problem is running the column at reduced pressure to lower all temperatures. As the rate of any chemical reaction, the corrosion rate contains the factor

$$\exp(-\Delta E/RT)$$

where ΔE is the activation energy, R is the universal gas constant, and T is the absolute temperature. Consequently, plotting the logarithm of available corrosion data versus 1/T (or more practically versus 1000 ^{o}K/T) yields a straight line. This is illustrated in Figure 114 referring to the corrosion of normal stainless steel by concentrated formic acid. As can be seen, a lowering of the temperature from the atmospheric boiling point of approximately 101 ^{o}C to 75 ^{o}C as effected by a pressure reduction from 760 to 330 mm Hg diminishes the corrosion rate by a factor in the order of 0.03, thus increasing the life expectancy of the column by a factor of 33.

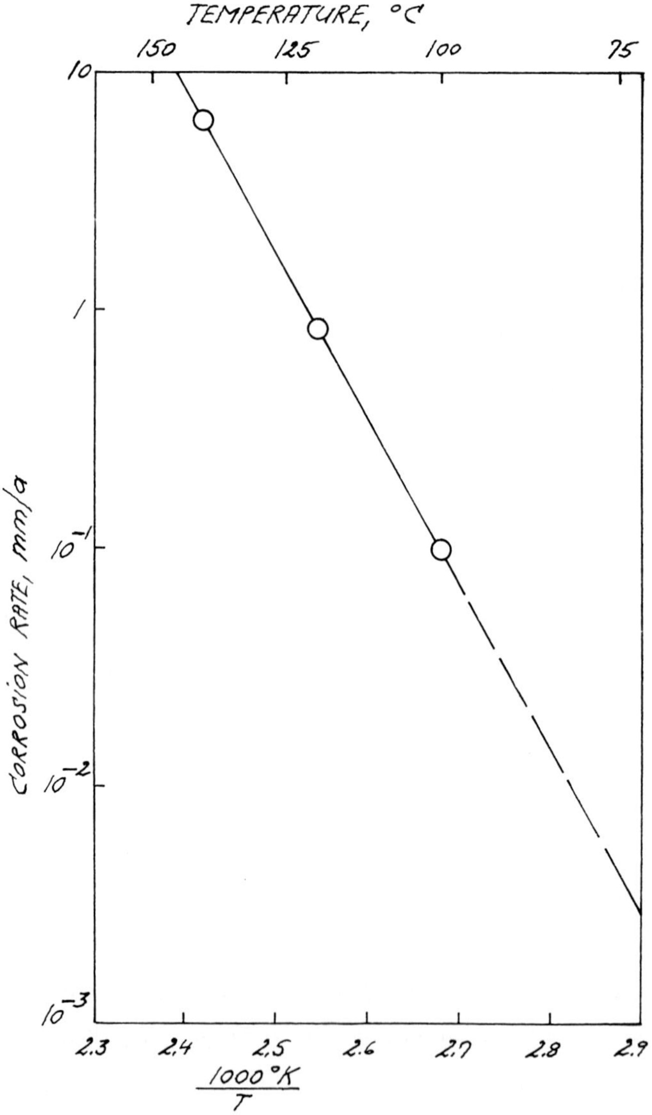

Figure 114. The Corrosion Rate for Normal Stainless Steel (18 % Cr, 8 % Ni) in Concentrated Formic Acid as a Function of Temperature.

f. Neutralization of Raw Furfural

When the raw furfural produced by the decanter of the azeotropic column turns out to be acid, due to an insufficient number of trays between the feed inlet and the side stream withdrawal, it is customary to submit this raw furfural (the heavy phase of the decanter) to a "neutralization" with sodium carbonate. This, however, is problematic, the principal cause being the fact that the dielectric constant of furfural amounts to a mere 48 percent of the value for water (38.0 versus 78.5 at 25 °C). Thus, raw furfural, exhibiting a typical furfural content of 92 % by weight, will have a dielectric constant only half as great as that of water. This has far-reaching consequences.

The dielectric constant of a solvent reflects its capacity to cause particles of opposite electrical charge to separate from one another. In a solvent with a high dielectric constant, e.g. water, a minimum amount of work is required to separate a positively charged ion from one with a negative charge. By contrast, a much greater amount of energy must be expended in accomplishing this process in a solvent with a low dielectric constant. Consequently, the dielectric constant determines the degree of dissociation. While in water the dissociation of acetic acid

$$CH_3COOH \rightarrow\rightarrow H^+ + CH_3COO^-$$

has the dissociation constant

$$[H^+]\,[CH_3COO^-]/[CH_3COOH] = K = 1.76 \times 10^{-5} \text{ mole/liter},$$

the situation is quite different in a medium of lower dielectric constant. In methanol, having a dielectric constant of only 32.6, similar to that of raw furfural, the dissociation constant of acetic acid is [121]

$$K = 3.0 \times 10^{-10} \text{ mole/liter}.$$

From this, it can be estimated that the dissociation constant of acetic acid in raw furfural is in the order of 10^{-9} mole/liter, i.e. roughly 1/20000 of the value in water. In other words, while the dissociation of acetic acid in water is already poor, it is almost nonexistent in raw furfural. This changes the course of neutralization.

Dissolved in water, sodium carbonate hydrolyzes according to

$$Na_2CO_3 + H_2O \rightarrow\rightarrow 2\,Na^+ + HCO_3^- + OH^- \qquad (1)$$

and when such a solution is used to neutralize acetic acid, the hydroxyl ions of (1) react with the available hydrogen ions of the acid to form water:

$$CH_3COOH + Na_2CO_3 + H_2O \rightarrow\rightarrow 2\,H_2O + CH_3COO^- + 2\,Na^+ + HCO_3^- \quad (2)$$

When the few hydrogen ions of the poorly dissociated acid are thus consumed by hydroxyl ions, the dissociation equilibrium of the acid is disturbed, so that the acid keeps dissociating in order to maintain this equilibrium. As a result, the number of CH_3COOH molecules diminishes, while the number of CH_3COO^- ions increases, until no more CH_3COOH molecules are left, which corresponds to total neutralization. In other words, in the course of neutralization, the ratio of CH_3COOH molecules to CH_3COO^- ions, i.e. the ratio

$$[CH_3COOH]/[CH_3COO^-] \quad (3)$$

decreases. According to the definition of the dissociation constant K

$$[H^+] = K\,[CH_3COOH]/[CH_3COO^-] \quad (4)$$

Thus, for various ratios (3) and with the known dissociation constant, the hydrogen ion concentration can be calculated, which means that the degree of neutralization, expressed by (3), can be related to the pH value observed. For instance, when (3) is 90/10, corresponding to a 10 % neutralization, equation (4) yields

$$[H^+]_{90/10} = K \times 9$$

In raw furfural, where a dissociation constant $K = 10^{-9}$ mole/liter applies, this amounts to

$$[H^+]_{90/10} = 9 \times 10^{-9} \text{ mole/liter}$$

whereas in water, where the dissociation constant is 1.76×10^{-5} mole/liter, it yields

$$[H^+]_{90/10} = 9 \times 1.76 \times 10^{-5} \text{ mole/liter} = 1.584 \times 10^{-4} \text{ mole/liter}$$

Hence, 10 % neutralization is reached when in raw furfural

$$pH = 8.05$$

while in water the same neutralization is reached already at

$$pH = 3.80$$

On this basis, the degree of neutralization can be plotted as a function of the pH value as illustrated in Figure 115, where the neutralization in raw furfural is compared to that in water. As can be seen, neutralization in raw furfural requires much higher pH values than equivalent neutralization in water.

As the neutralization of raw furfural is commonly carried out to a pH of 9, it can be concluded from Figure 115 that under such circumstances only 50 percent of the acid is neutralized whereas in water a pH of 9 would effect 100 % neutralization.

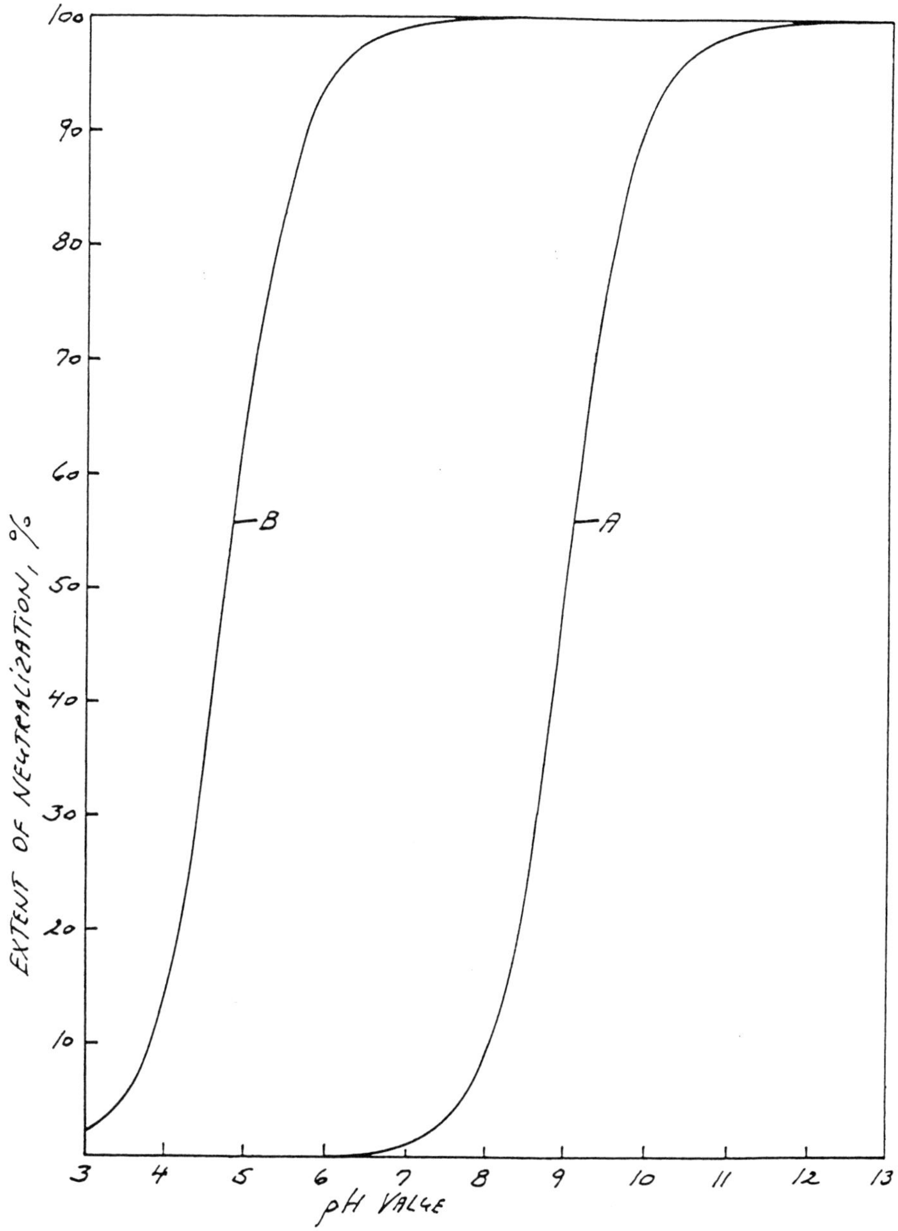

Figure 115. Neutralization Characteristics of Acetic Acid.

Curve A: In raw furfural.

Curve B: In water.

In this connection, it is noted that in the decanter of the azeotropic column the concentrations of acetic acid in the light aqueous phase and the heavy organic phase are almost equal. This can be seen from the respective distribution diagram shown in Figure 116 [121]. Thus, the phase separation in the decanter is unfortunately no help for obtaining a raw furfural of low acidity.

Reference

[121] A. E. Skrzec and N. F. Murphy: Ind. Eng. Chem. 46 (1954) 2245-2247.

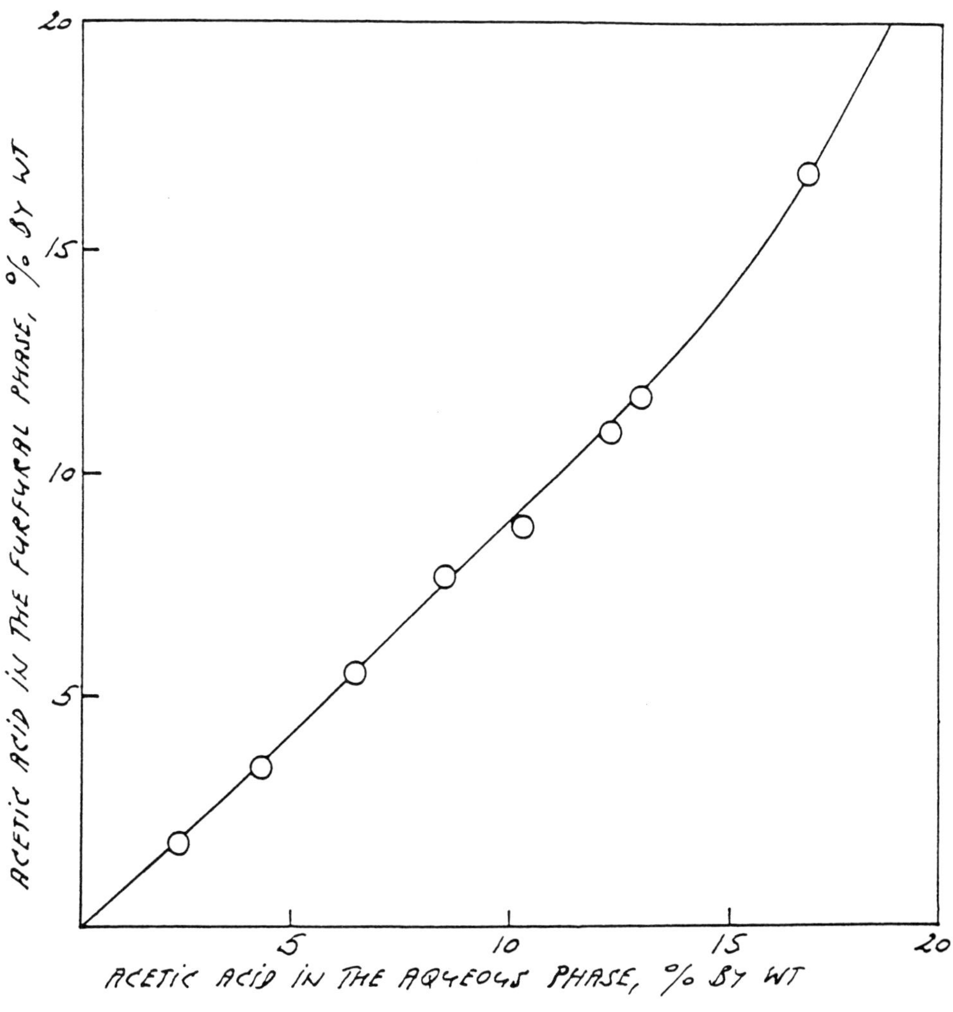

Figure 116. The Distribution of Acetic Acid between Water and Furfural at 27 °C.

g. Distillation Measures against the Acidity of Raw Furfural

The preceding chapter has shown that acids arriving in the side stream of the azeotropic column of a furfural plant end up in the raw furfural, thus requiring a respective neutralization. Inasmuch, however, as preventing a disease is better than healing it, it is decidedly preferable to prevent the acids from getting up there in the first place.

To deal with this problem, consideration must be given to the vapor/liquid equilibria in the water/acetic acid and the water/formic acid systems.

The vapor/liquid equilibrium in the water/acetic acid system at 760 mm Hg is shown in Figure 117. As can be seen, in the range of very small acetic acid concentrations (i.e. at water concentrations close to 100 %) as encountered in the product stream of furfural reactors, the equilibrium "curve" is a straight line. On this basis, the distillation effects above the feed plate as far as acetic acid is concerned are illustrated in Figure 118. With water being more volatile than acetic acid, the column portion above the feed inlet produces an increase of the water concentration, i.e. a decrease of the acetic acid concentration.

Inasmuch as almost all of the acetic acid withdrawn with the side stream is refluxed with the light aqueous phase of the decanter, the rectification conditions for acetic acid are very close to total reflux. Hence, in very good approximation, the "operating line" for the graphical stage determination procedure can be taken to be the diagonal as actually done in Figure 118. The result, converted from % by mole to % by weight, and from the number of theoretical stages n_{th} to the number of real stages n_r by setting $n_r = n_{th}/0.7$, is depicted in Figure 119. As can be seen, reducing the acetic acid concentration to 1 % of its feed value requires 15 trays above the feed inlet.

Formic acid can be dealt with in an analogous fashion. The vapor/liquid equilibrium diagram for the water/formic acid system is shown in Figure 120, where in analogy to the case of acetic acid the range of low formic acid concentrations (high water concentrations) is represented by a straight line. On this basis, the distillation effects above the feed plate as far as formic acid is concerned are illustrated in Figure 121, and the result is summarized in Figure 122. Comparison with Figure 119 reveals that formic acid drops off much more sharply than acetic acid, the reason being that for formic acid the slope of the

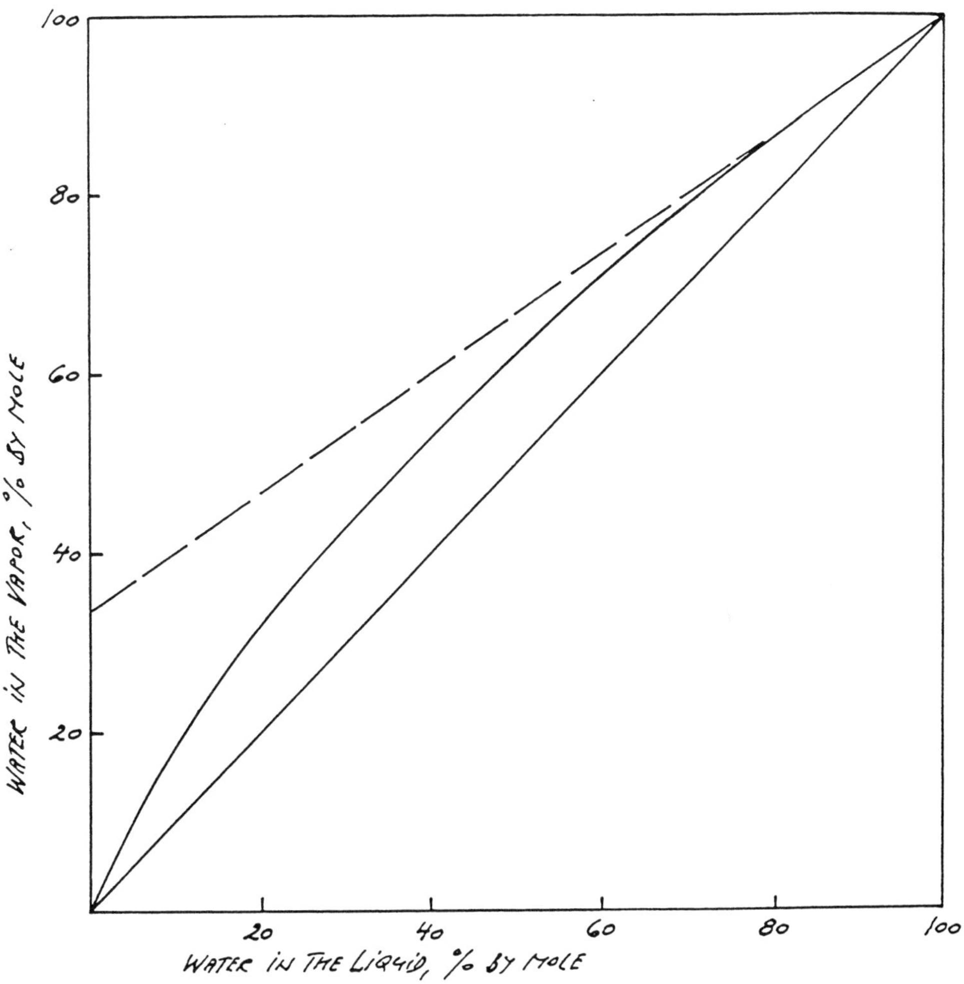

Figure 117. Vapor/Liquid Equilibrium Diagram for the Water/Acetic Acid System at 760 mm Hg.

The dashed line represents the linear approximation for low acetic acid concentrations (high water concentrations).

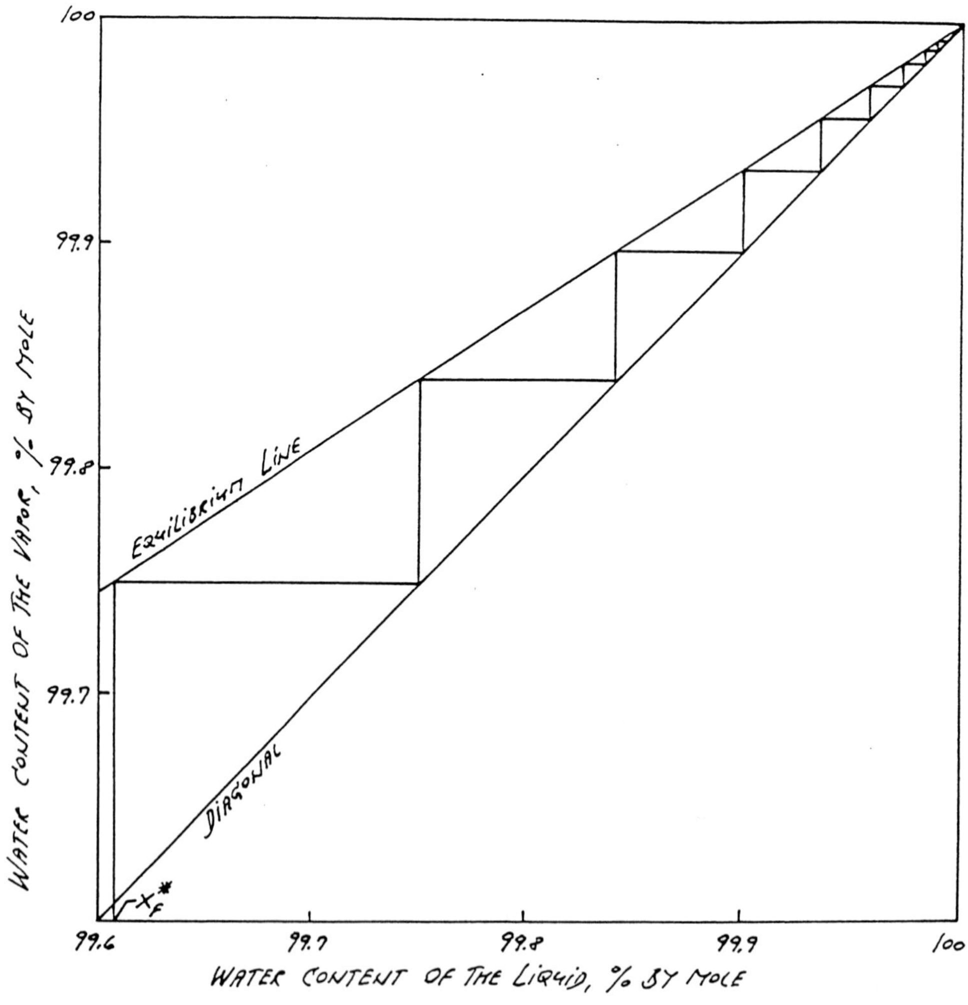

Figure 118. Acetic Acid Distillation Effects above the Feed Tray.

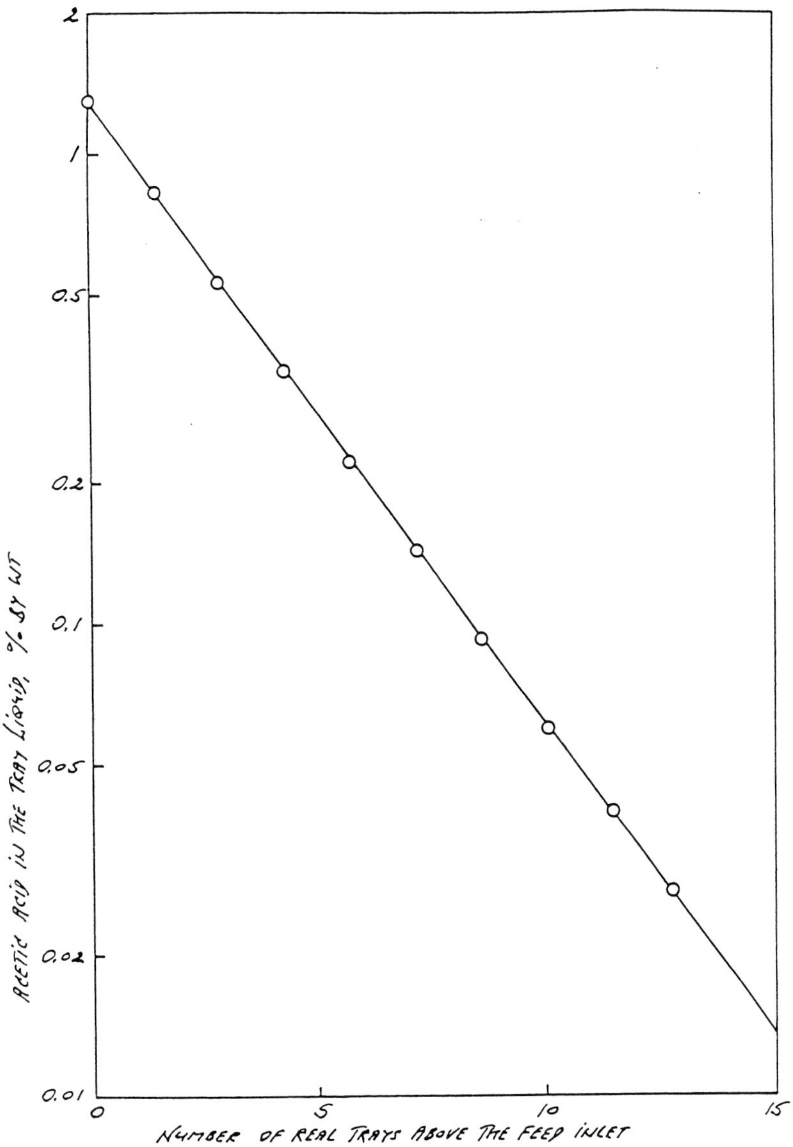

Figure 119. Reduction of Acetic Acid in the Side Stream as a Function of the Number of Real Trays above the Feed Inlet.

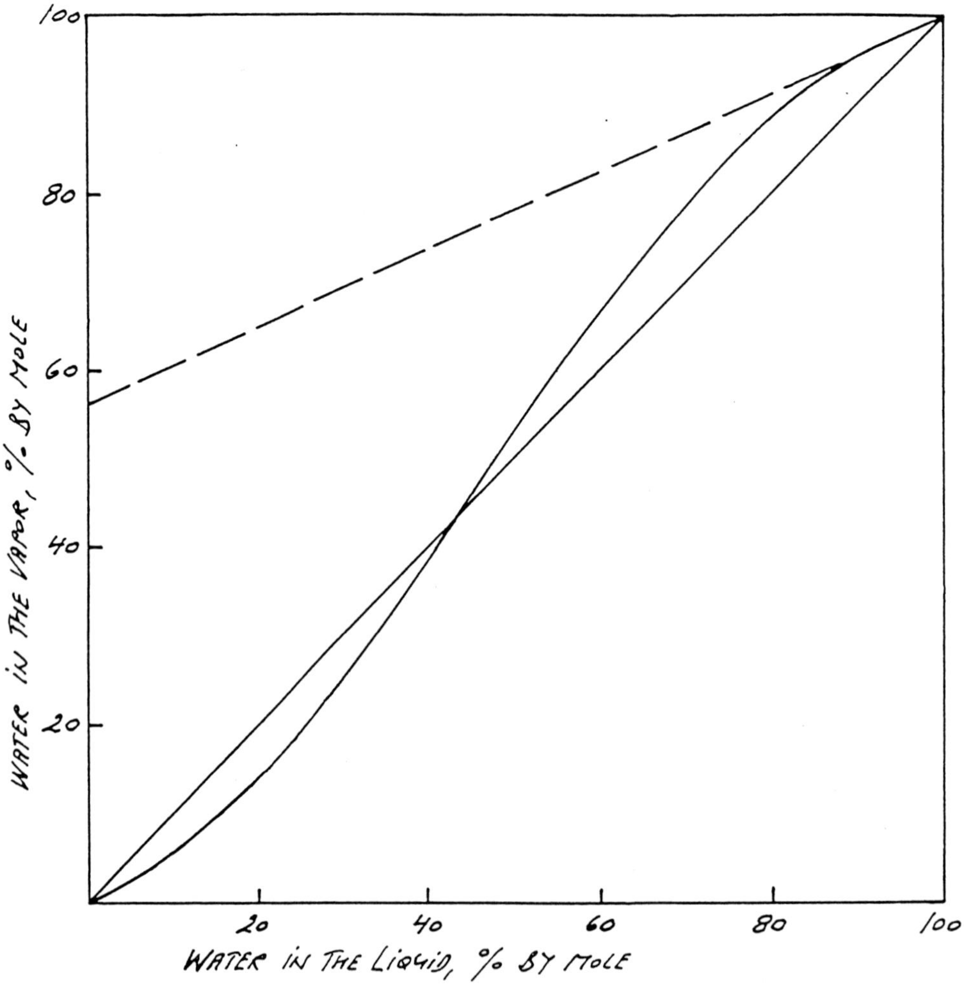

Figure 120. Vapor/Liquid Equilibrium Diagram for the Water/Formic Acid System at 760 mm Hg.

The dashed line represents the linear approximation for low formic acid concentrations (high water concentrations).

The azeotrope boils at 107.6 °C (higher than both components).

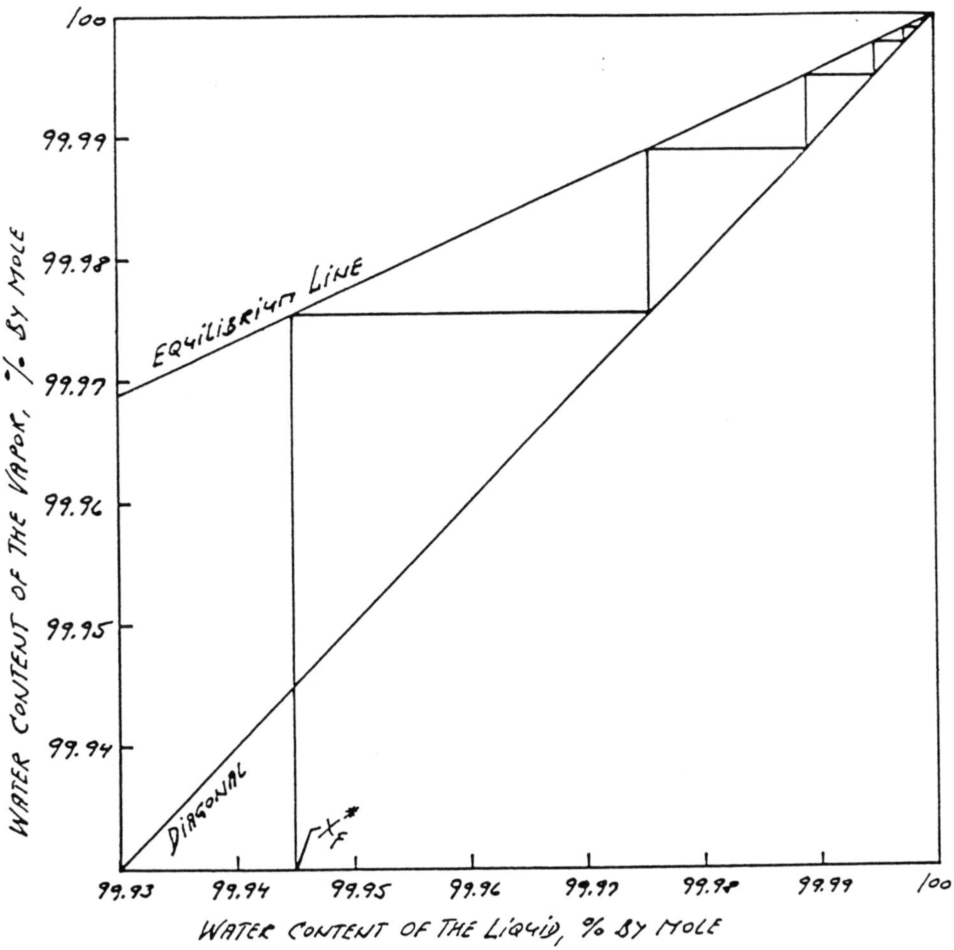

Figure 121. Formic Acid Distillation Effects above the Feed Tray.

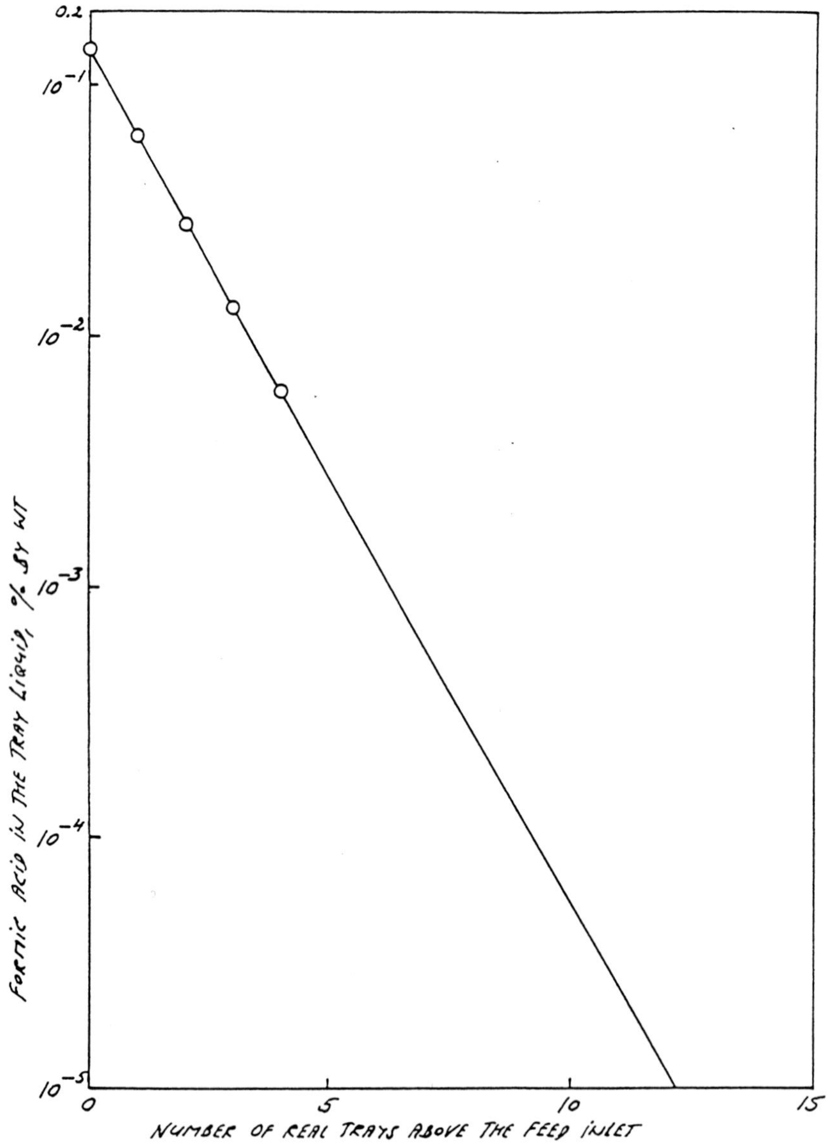

Figure 122. Reduction of Formic Acid in the Side Stream as a Function of the Number of Real Trays above the Feed Inlet.

equilibrium line in the range considered is significantly smaller than in the case of acetic acid. Hence, formic acid is only a minor problem, so that the column design must be based on acetic acid. For a good column design, it is customary to use 15 trays between the feed inlet and the side stream withdrawal point. It is most unwise to use much less as this results in an increased acidity of the raw furfural, thus requiring a massive neutralization.

h. Flashing of Residues

Inasmuch as all industrial furfural processes are carried out at elevated pressures, they all involve a depressurization (flashing) of the residue. In the case of processes where the raw material is steam-stripped to the point of exhaustion, the flashing of the residue yields very little if any furfural, but in the case of non-stripped single pass processes, such as the SUPRATHERM and STAKE processes, the flashing of the residue is an important part of the overall process in that it yields a vapor stream containing most of the furfural produced, and a residue still containing some furfural in its liquid phase. This is illustrated schematically in Figure 123.

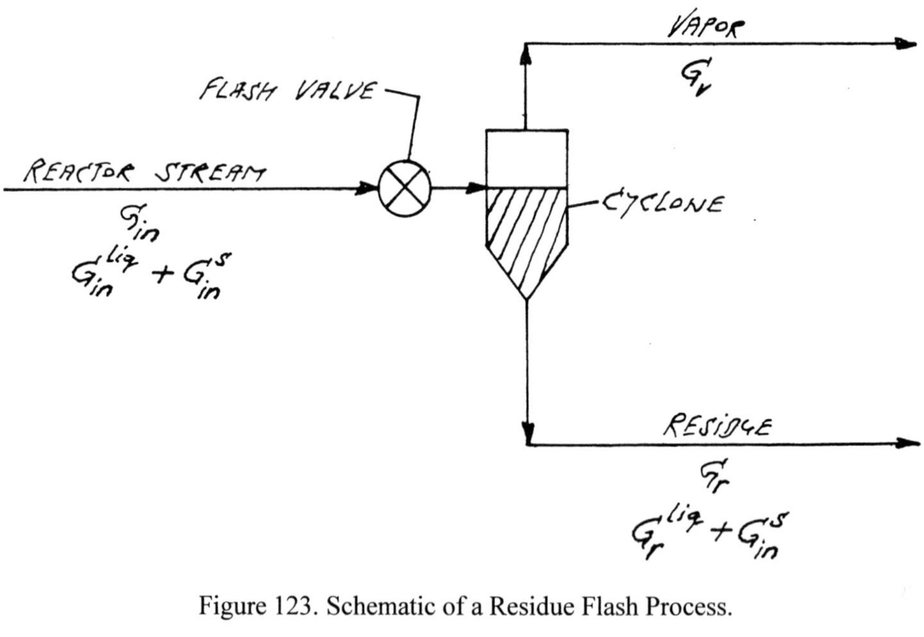

Figure 123. Schematic of a Residue Flash Process.

After passing a flash valve, the reactor stream G_{in} enters a cyclone held at atmospheric or subatmospheric pressure, where it is split into a vapor stream G_v and a residue stream G_r. The reactor stream consists of a liquid portion G_{in}^{liq} and a solid portion G_{in}^{s}, and analogously the residue stream consists of a liquid portion and the same solids as the reactor stream.

To design or appraise any such process, it is necessary to accurately compute the relative magnitudes of the two output streams and their concentrations as a function of the characteristics of the incoming stream. The mass flow rates of the vapor stream and the residue stream can be determined from a mass balance and an enthalpy balance.

The mass balance is

$$G_{in} = G_v + G_r \tag{1}$$

and the enthalpy balance is

$$G_{in}^{liq}\, i_{in}^{liq} + G_{in}^{s}\, i_{in}^{s} = G_v\, i_v + G_r^{liq}\, i_r^{liq} + G_{in}^{s}\, i_r^{s} \tag{2}$$

where the i terms are the specific enthalpies (kcal $kg^{-1}\,{}^{o}C^{-1}$ or kJ $kg^{-1}\,{}^{o}C^{-1}$).

Since

$$G_r^{liq} = G_{in}^{liq} - G_v \tag{3}$$

equation (2) results in

$$G_{in}^{liq}\, i_{in}^{liq} + G_{in}^{s}(i_{in}^{s} - i_r^{s}) = G_v\, i_v + i_r^{liq}(G_{in}^{liq} - G_v) \tag{4}$$

leading to

$$\boxed{G_v = \frac{G_{in}^{liq}\left(i_{in}^{liq} - i_r^{liq}\right) + G_{in}^{s}\left(i_{in}^{s} - i_r^{s}\right)}{i_v - i_r^{liq}}} \tag{5}$$

On the other hand, combination of the equations (3) and (5) results in

$$\frac{G_r^{liq}}{G_{in}^{liq}} = 1 - \frac{\left(i_{in}^{liq} - i_r^{liq}\right) + \dfrac{G_{in}^{s}}{G_{in}^{liq}}\left(i_{in}^{s} - i_r^{s}\right)}{i_v - i_r^{liq}} \tag{6}$$

so that all mass streams are found.

The furfural balance yields

$$FU_{in} = \xi_v\, G_v + \xi_r^{liq}\, G_r^{liq} \tag{7}$$

where FU_{in} is the incoming mass of furfural,

ξ_v is the mass fraction of furfural in the vapor, and

ξ_r^{liq} is the mass fraction of furfural in the liquid phase of the residue.

The furfural concentration in the vapor and in the liquid phase of the residue are governed by the vapor/liquid equilibrium in the cyclone. With this equilibrium depending on the pressure as shown in Figure 106 (page 237), flashing into a cyclone maintained at a reduced pressure differs from flashing into an atmospheric cyclone in that the concentration ratio of the furfural in the vapor to the furfural in the liquid phase of the residue varies with the cyclone pressure.

For the small concentrations of interest in flashing furfural residues (5 % by weight of furfural in water corresponds to a mere 0.977 % by mole), this ratio can be well approximated by the initial slope of the vapor/liquid equilibrium curve. Referred to mass fractions, this slope is known as the "amplification factor κ". Its dependence on pressure is illustrated in Figure 124. As can be seen, κ increases strongly with decreasing pressure.

Against this background, with

$$\xi_v/\xi_r^{liq} = \kappa \qquad (8)$$

equations (8) and (7) give

$$\xi_v = FU_{in}/[G_v\{1 + (G_r^{liq}/\kappa G_v)\}] \qquad (9)$$

so that

$$\frac{FU_v}{FU_{in}} = \frac{1}{1 + \frac{1}{\kappa}\frac{G_r^{Liq}}{G_v}}$$

where $FU_v = \xi_v\, G_v$ is the furfural in the vapor. Thus, large values of κ (low cyclone pressures) favor bringing most of the furfural into the vapor phase.

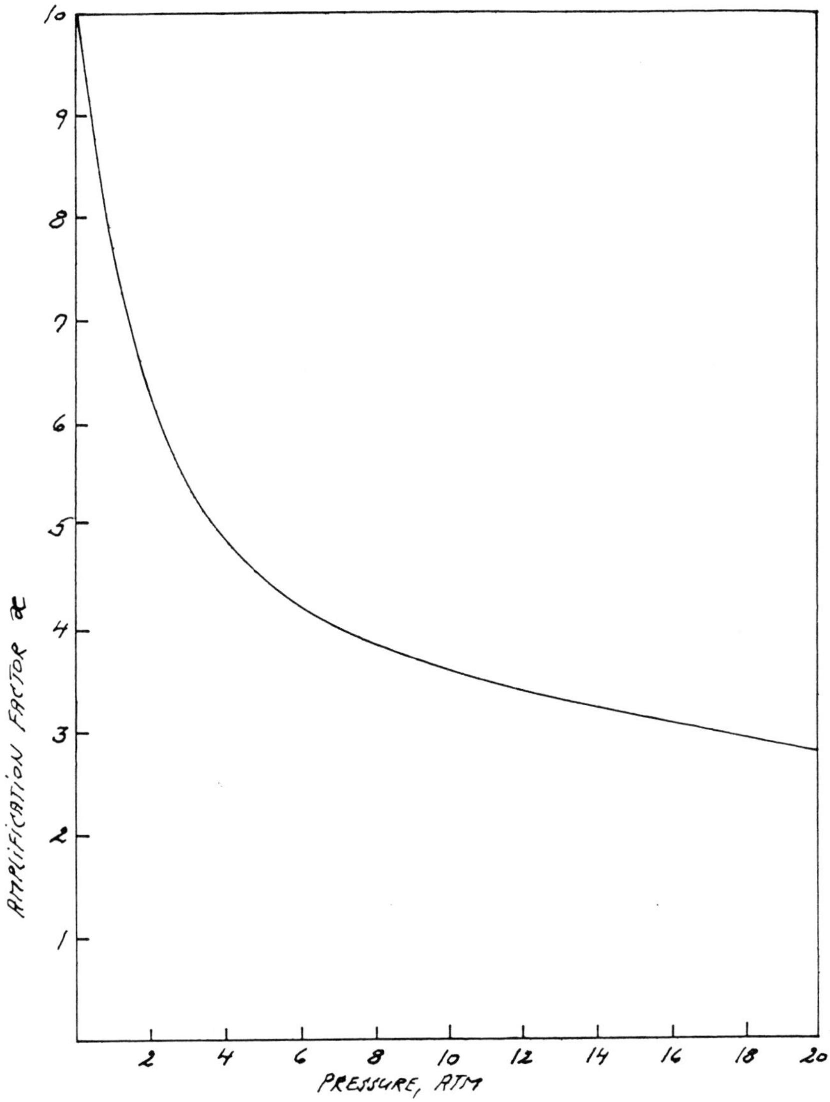

Figure 124. Dependence of the "Amplification Factor" on the Pressure in the Cyclone.

i. Operational Details of the QUAKER OATS Batch Process

A typical QUAKER OATS batch process as described in chapter 10.1 can be characterized by the following figures [122]:

Rotary digester 12 feet long by 8 feet in diameter (3.65 m long by 2.45 m in diameter). Operating temperature: 153 °C corresponding to 5.0853 ATM.

Input:

4500 lb of oat hulls with 6 % moisture:..4230 lb of dry solids
270 lb of water

100 lb of 95 % H_2SO_4 dissolved in 1200 lb of water:....................................95 lb of H_2SO_4
1205 lb of water

Sum:

4230 lb of dry solids (1916.190 kg)

1475 lb of water (668.175 kg)

95 lb of H_2SO_4 (43.035 kg)

Total: 5800 lb (2627.400 kg)

Liquid portion of the charge: 1570 lb in 5800 lb, which corresponds to 27.069 % by weight. Initial acid strength of the liquid phase: 95 lb in 1570 lb, which corresponds to 6.051 % by weight. This is the acid strength at the very start of the process. With the charge (solid and liquid) being heated by steam condensation, the condensate thus generated adds to the moisture content of the charge, thus lowering the acid strength. Only the minimum quantity of condensate due to heating the raw material can be specified, the actual condensate quantity being greater as heating the equipment (reactor walls, trunnions, etc) leads to additional condensate.

Specific heats of the charge components: Solids:..............0.370 kcal kg^{-1} $^oC^{-1}$
Water:..............1.000 kcal kg^{-1} $^oC^{-1}$
H_2SO_4:.............0.339 kcal kg^{-1} $^oC^{-1}$

Heating of the charge from 25 °C to 153 °C: Solids:..........90750.758 kcal
Water:..........85526.400 kcal
H_2SO_4:.........1867.375 kcal

Total heat requirement: 178144.533 kcal

Heat of steam condensation at 153 °C: 501.9 kcal/kg

Thus, heating of the charge to the reaction temperature generates the condensate quantity

ΔW = (178144.533/501.9) kg = 354.940 kg

Due to this "heat-up condensate", the acid strength after reaching the reaction temperature (if the equipment has the heat capacity zero and if no reactions have taken place) is found to be 4.036 % by weight (as compared to 6.051 % by weight at the start), and in reality the acid strength is even lower because of the sizable heat capacity of the equipment.

Pentosan content of the dry oat hulls: 30.5 %

Pentosan input: 4230 x 0.305 lb = 1290.15 lb (584.438 kg)

Theoretical furfural output (see chapter 2.1):

1290.15 lb x 0.72727 = 938.287 lb (425.044 kg)

Water consumption by the pentosan-to-pentose hydrolysis:

1290.15 lb x 18.016/132.114 = 175.93398 lb (79.698 kg)

Water liberated by the overall pentosan-to-furfural conversion:

1290.15 lb x 0.27273 = 351.86797 lb (159.396 kg)

Theoretical water content of the residue after the reaction:

(668.175 +354.940 + 159.396) kg = 1182.511 kg

Theoretical dry solid weight of the residue after the reaction:

(1916.190 – 584.438) kg = 1331.752 kg

Theoretical water content of the residue after the reaction:

1182.511/(1182.511 + 1331.752) = 0.47032, i.e. 47.032 % by weight

Actual time of digestion: 5 hours

Distillate produced in 5 hours: 7950 lb (3601.350 kg)

Furfural in the distillate: 5.8 % by weight, i.e. 461.1 lb (208.873 kg)

Furfural yield: 461.1 lb/938.287 lb = 0.49143, i.e. 49.143 %

At the given furfural concentration of 5.8 %, the distillate is made up of 94.2 lb of steam per 5.8 lb of furfural, so that the steam consumption during the "distillation" is 16.241 lb of steam per lb of furfural.

Furfural obtained in the distillate as referred to the moist input hulls:

Y* = 461.1 lb of furfural per 4500 lb of moist hulls = 0.10247 lb of furfural per lb of moist hulls.

After heating by steam condensation, the nominal (maximum possible) acid strength of 4.036 % by weight corresponds to a hydrogen ion concentration of 0.417 mole/liter. Thus, according to chapter 5, with T = (273.16 +153) °K = 426.16 °K, the maximum possible rate constant of pentose disappearance is found to be

$$k_1 = 0.02357 \text{ min}^{-1}$$

which leads to a minimum residence time requirement

$$\tau = 4.60517/k_1 = 195.383 \text{ minutes} = 3.256 \text{ hours}$$

Of course, due to the sizable heat capacity of the equipment, and due to heat losses to the ambient atmosphere, more condensate is generated in the charge, thus further diluting the acid, so that the actual hydrogen ion concentration is smaller and the residence time requirement correspondingly larger.

The dilution of the acid catalyst by the condensate formed in the heating process and in maintaining the reaction temperature stresses the importance of having the reactor thermally well insulated. Against this background, it must be kept in mind that a good thermal insulation, costing very little,

(1) reduces the acid consumption or the period of processing, and

(2) leads to a drier residue, thus increasing its caloric value.

Reference

[122] H. J. Brownlee, Ind. Eng. Chem. 19 (1927) 422-424.

j. Operational Details of the ROSENLEW Process

Typical operating details of the ROSENLEW process described in chapter 10.6 are summarized in Figure 125.

Reactor dimensions: 2.5 m diameter by 12 m high.

Operating temperature: 180 ^{o}C corresponding to steam of 9.88684 ATM.

After screening to remove fines, the bagasse feed stock has a moisture content of 49.1 % and a pentosan content of 25.3 % as referred to dry solids.

At an input rate of 6290 kg/h of this bagasse, the processing characteristics are as follows:

Input of dry solids:..3200 kg/h

Input of water:...3090 kg/h

Input of pentosan:...809 kg/h

Theoretical furfural output: 809 kg/h x 0.72727:..588.36 kg/h

Furfural in the product vapor:..350.01 kg/h

Furfural yield: 350.01 kg / 588.36 kg:...59.5 %

Input of superheated steam of 265 ^{o}C and 9.88684 ATM:..................................10500 kg/h

Output of product vapor:..10870 kg/h

Furfural concentration in the product vapor:...3.22 % by wt

Output of residue:..5920 kg/h

Moisture content of residue before flashing:..55.07 %

Specific steam consumption:....................................30 kg of steam per kg of furfural

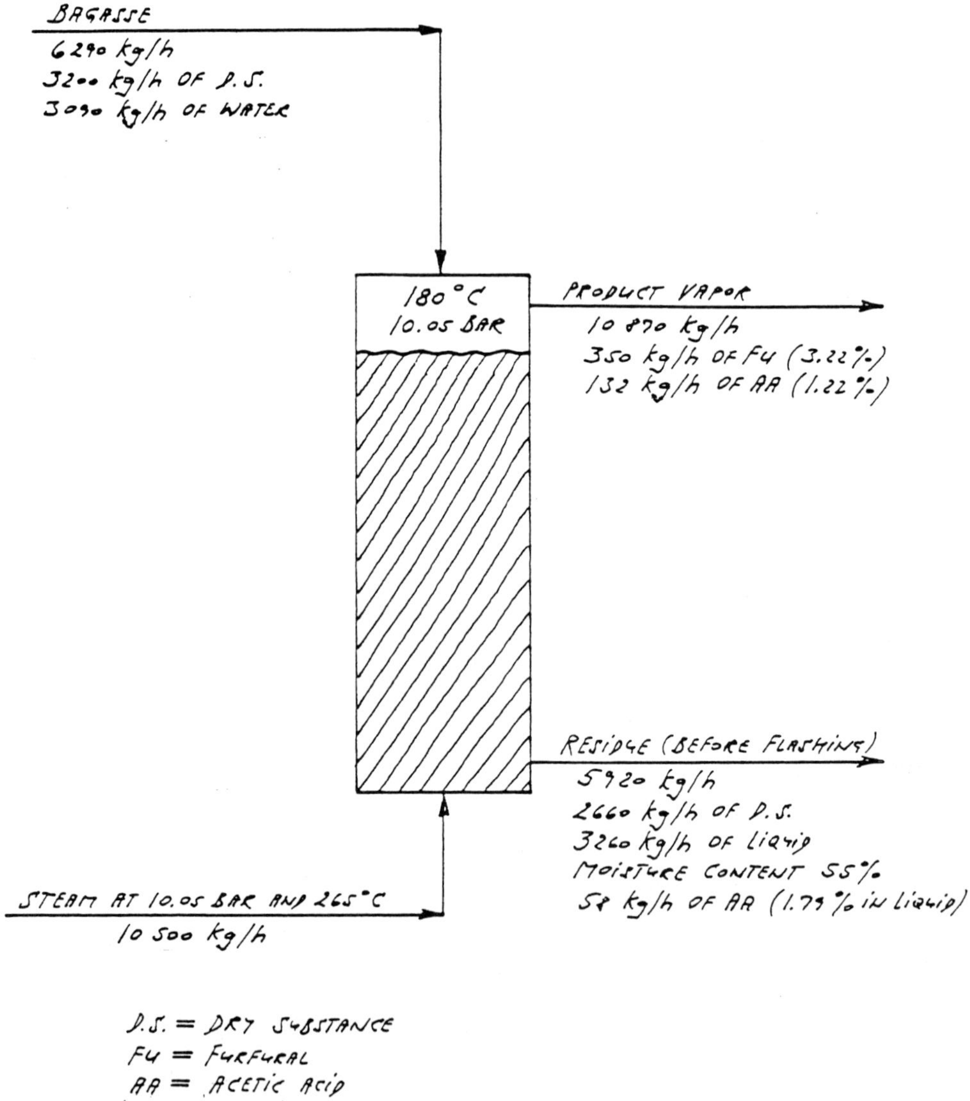

Figure 125. Mass Balance for a ROSENLEW Reactor in Typical Operation.

k. Operational Details of a ROSENLEW Distillation

The mass balance of an azeotropic column for eleven ROSENLEW reactors of 2.5 m diameter processing bagasse is shown in Figure 126. The following comments are made:

(1) The column is not a good design as the number of trays between the feed inlet and the azeotropic side stream (6 trays) is not sufficient to prevent some acetic acid of the feed stream from reaching the side stream, so that the "raw furfural" (the heavy phase of the decanter) must be submitted to a massive neutralization. For details of this point, see chapter f.

(2) The head stream (570 kg/h) containing 38 % by weight of furfural goes to an auxiliary column (not shown) where the furfural is recovered as the sump fraction. This sump fraction is the stream of 429 kg/h entering the decanter from the right-hand side. The head fraction of the auxiliary column is called "raw solvent". Specified in chapter 16.6 (page 129), it contains the low boilers as well as the diacetyl and 2,3-pentanedione by-products.

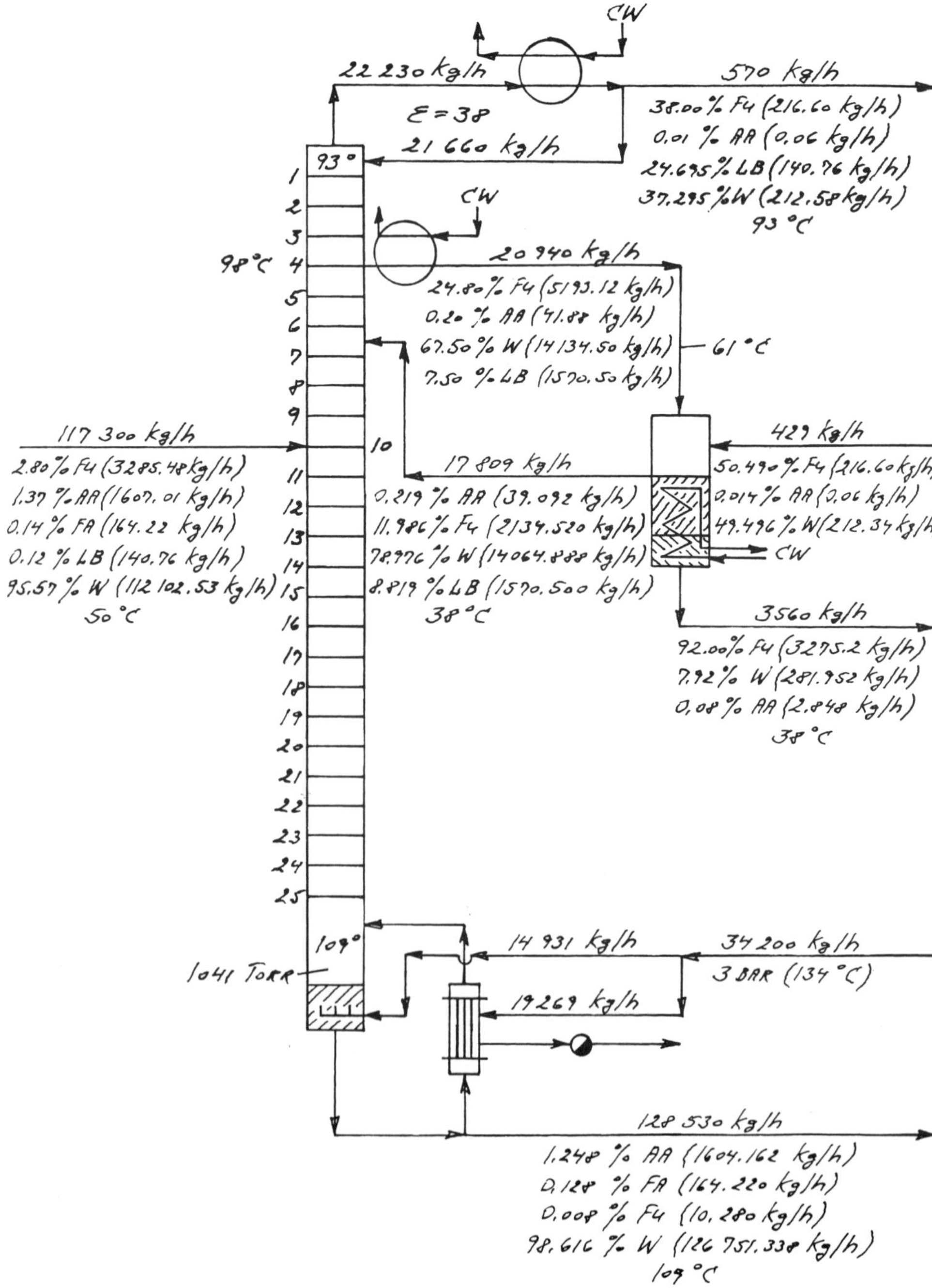

Figure 126. Mass Balance on a Distillation Column for eleven ROSENLEW Reactors.

I. Acidity Conversion Chart

Designing or appraising furfural reactors operated with sulfuric acid as the catalyst requires evaluating the rate constant k_1 for xylose disappearance treated in chapter 5. As this rate constant contains the hydrogen ion concentration as a factor, it is necessary to correlate this quantity with the technically more common acid strength in % by weight. For 25 °C, this correlation is given in Figure 127.

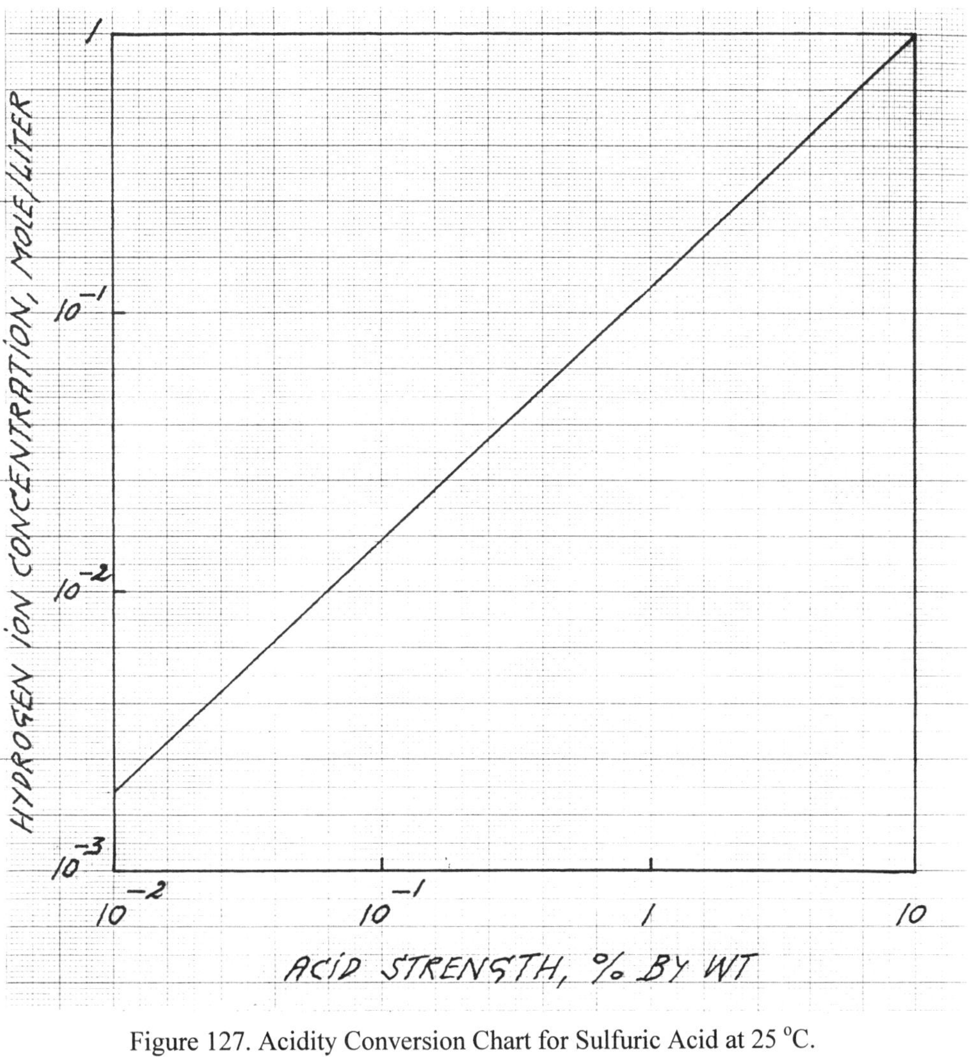

Figure 127. Acidity Conversion Chart for Sulfuric Acid at 25 °C.

As discussed already in chapter 3.1, it is noted that the available kinetics are formulated with the hydrogen ion concentration at room temperature as it existed in the laboratory before the furfural process, and not with the really active hydrogen ion concentration at the reaction temperature. Consequently, in determining the rate constant k_1, oddly enough the room temperature value of the hydrogen ion concentration as given in the graph must be used although the process takes place at a greatly elevated temperature where the hydrogen ion concentration is totally different.

m. Extraction of Vegetable Oils with Furfural

As discussed in chapter 14.1, furfural is an outstanding extractant for separating compounds with double bonds from compounds without double bonds, the reason being that the double bonds of furfural attract double bonds of other molecules, due to the fact that energy is liberated when a double bond system is enlarged.

One such application of furfural is the extraction of vegetable oils such as soybean oil to get an extract rich in double bonds, and a raffinate depleted of double bonds. The extract rich in double bonds can be used as a "drying oil" for paints and varnishes, where the double bonds react with oxygen of the ambient air to form cross-linked polymers. On the other hand, the raffinate depleted of double bonds is a more desirable product for the food industry than the nonextracted soybean oil. Thus, two advantages are obtained in a single operation.

With the rise of soybean as a major crop in the United States, there has become available a large and dependable supply of soybean oil. In the form in which it is taken from the bean, it is not a good "drying oil", but some of its constituents are of high value in this respect. On the other hand, good "drying oils" such as linseed oil, made from flax, is in relatively short supply, so that a manufacture of "drying oil" from the plentifully available soybean oil is an attractive proposition. In addition to the main goal of overcoming a shortage of "drying oils", such processing offers two ancillary advantages:

(1) The production of entirely new oils not available in nature.

(2) A control of properties to counteract the variations found in natural products.

The value of furfural in this endeavor is based on the fact that in analogy to the limited miscibility of furfural with water as shown in Figure 104 (page 235), at normal temperatures furfural is not completely miscible with vegetable oils either. Thus, when furfural is mixed with a vegetable oil at room temperature, two phases are formed, a heavy phase rich in furfural, and a light phase rich in oil, and when these two phases are analyzed, it is found that the oil constituents with double bonds have increased in the furfural phase and diminished in the oil phase. Although in a single mixer/settler stage of this type the separation effect may be small, it can be hugely increased by using a continuously operating countercurrent extraction column. In such a column, the furfural, being heavy (ρ = 1.16 g/cc), is fed into the top and flows downwards, while the oil to be refined, being light (ρ = 0.925 g/cc), is fed into the bottom and flows upwards. On its way downwards, the furfural picks up oil constituents with double bonds, so that the furfural arriving at the bottom is rich in such components, while the oil arriving at the top has lost some of these species.

A schematic of this process is shown in Figure 128 [123]. It starts in the extraction column 1 fed with furfural at the top, and with oil somewhere below the middle. The principal part of the column is packed with RASCHIG rings or BERL saddles to aid in mass transfer, and each end features a "separation chamber" without packing.

The furfural input is very much greater than the oil input, a typical ratio being 6 parts of furfural for each part of feed oil.

Contrary to customary extraction, in such a column the dispersed phase and the continuous phase change roles. In the upper portion, furfural droplets fall through a continuous oil phase, whereas in the lower portion oil droplets rise through a continuous furfural phase, with an interface between these two regions.

The furfural loaded with double bond constituents of the oil leaves the bottom of the column as the "extract". By means of pump 2, it is passed through a heater 3 and enters a vacuum evaporator 4, where at a pressure of 100 torr most of the furfural is vaporized. This vapor stream of pure furfural is liquefied in condenser 5 maintained at reduced pressure by vacuum pump 6. The condensate is collected in tank 7. The oil largely freed of furfural is taken up by pump 8 and fed into the head of a stripping column 9 operated at 50 torr. Superheated steam injected into the bottom removes any residual furfural by forming the low-boiling furfural/water azeotrope. The vapor is liquefied in condenser 10 maintained at reduced pressure by vacuum pump 11. The sump fraction, representing the "extract oil" rich in double bond constituents, is withdrawn by pump 12 and passed through a cooler 13 before going to storage and shipment.

The product of condenser 10, collected in decanter 14, forms two liquid phases, a heavy furfural phase and a light aqueous phase. The heavy furfural phase, consisting of approximately 92 % furfural and 8 % water, by weight, enters a dehydration column 15 energized by a steam coil in the bottom. The sump fraction, representing pure furfural freed of water, is withdrawn by pump 16 and fed into tank 7. The head fraction of column 15, having a composition close to the furfural/water azeotrope, joins the vapor stream of column 9 in condenser 10.

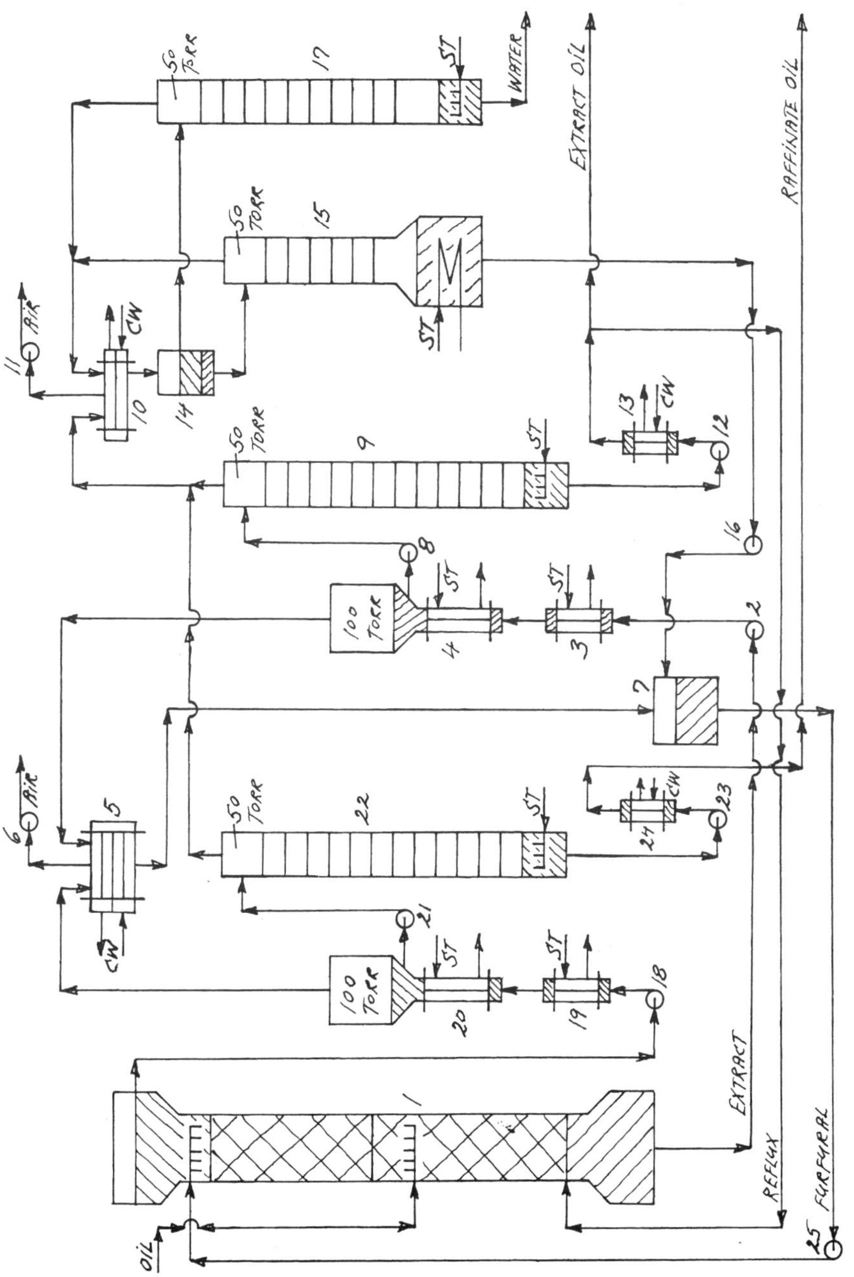

Figure 128. Extraction of Vegetable Oil with Furfural.

The light phase of decanter 14 enters the head of a distillation column 17 energized by steam injection. The sump fraction of this column is water, while the head fraction is again almost azeotropic and thus suitable for being introduced into condenser 10.

The raffinate of extraction column 1 is taken up by pump 18 and passed through a heater 19 into a vacuum evaporator 20 operated at 100 torr, where in complete analogy to the treatment of the extract in evaporator 4 most of the furfural is vaporized, to be liquefied in condenser 5. The oil largely freed of furfural is withdrawn by pump 21 and fed into the head of a stripping column 22 energized by injection of superheated steam. As in column 9, the steam removes any residual furfural by forming the low-boiling furfural/water azeotrope, so that the head fraction can join the other such streams in condenser 10. The sump fraction of column 22, representing the “raffinate oil”, is withdrawn by pump 23 and passed through a cooler 24 before going to storage and shipment.

From tank 7, pump 25 recycles the recovered furfural back into the extraction column 1.

A sizable portion of the “extract oil” is also fed back into the extraction column 1 as “reflux” as it was found that this improves the extraction of double bond constituents. This “reflux” is introduced at the bottom of the packing.

Depending on the input oil, the temperature in the extraction column is made to range between 20 and 50 °C.

The effect of the extraction process is expressed by changes of the “iodine value” I.V which is a quantitative measure for the concentration of C=C double bonds. When the oil is reacted with iodine chloride, JCl, in an appropriate solvent such as glacial acetic acid, each double bond is converted to a single bond by the addition of iodine chloride. The consumption of iodine in this “double bond elimination process” is measured by determining how much iodine is left over. To this end, potassium iodide, KJ, is added to liberate iodine, J_2, from the excess JCl, and this iodine is determined by titration with sodium thiosulfate, using starch as the indicator. In this fashion, the quantity of iodine consumed in eliminating the double bonds can be computed. The “iodine value” I.V. thus determined is the quantity, in grams, of iodine consumed by 100 grams of oil.

With the given extraction plant, a soybean oil having an I.V. of 136 can be split into an “extract oil” having an I.V. of 152 and a “raffinate oil” having an I.V. of 108.

While the “extract oil” is used as a “drying oil” for paints and varnishes, the “raffinate oil" is used for food products such as shortenings and salad oils.

What is being done with vegetable oils can also be done with animal oils such as sardine oil or dogfish liver oil. In the latter case, the "extract oil" features not only an increased I.V. but also an enrichment in vitamin A. For a normal dogfish liver oil having an I.V. of 109 and containing 17 000 units of vitamin A per gram, the "extract oil" has an I.V. of 174 and contains 82 000 units of vitamin A per gram. The yield of this fraction is in the order of 19 percent.

Reference

[123] S. W. Gloyer, Ind. Eng. Chem. 40 (1948) 228-236.

n. Furoyl Chloride

Furoyl chloride is the most important derivative of furoic acid. From a synthetic viewpoint, it is one of the most useful derivatives of furfural, because of its stability and multiplicity of reactions. In organic chemistry, furoyl chloride is used primarily for the introduction of the furoyl group into alcohols and phenols.

Furoyl chloride is made by reacting furoic acid with thionyl chloride in anhydrous benzene:

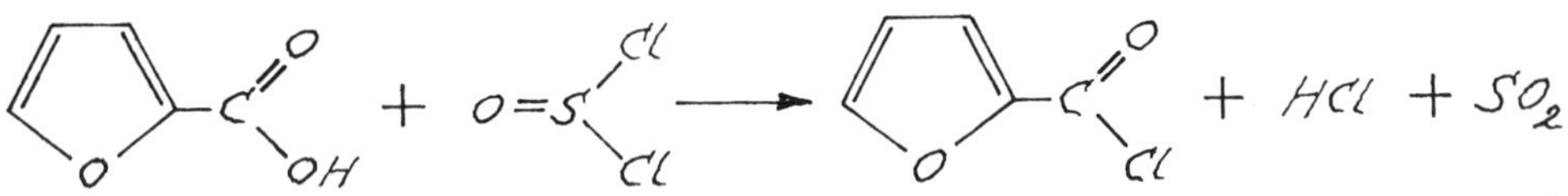

The process is carried out in a batchwise fashion as shown in Figure 129 [124]. In a reactor 1 equipped with a stirrer 2 and a reflux condenser 3, furoic acid from a bin 4 is dissolved in anhydrous benzene from a tank 5 at a ratio of 1.8 liters of benzene per kg of furoic acid, and then, after gently heating the reactor content to boiling, thionyl chloride (b.p. 12.2 °C) from a tank 6 cooled by a chiller 7 is added dropwise over a period of two days at an excess of 50 percent over the stoichiometrically required quantity. The HCl and SO_2 gases formed in the reaction are taken up by an absorption column 8 equipped with a ceramic circulation pump 9.

After the addition of the thionyl chloride, the reaction mixture is refluxed for an additional 12 hours before pump 10 passes it through a filter 11 into an atmospheric distillation column 12 energized by a steam coil. In this column, the benzene solvent and the excess thionyl chloride are boiled off, liquefied in condenser 13, and collected in tank 14.

The remaining furoyl chloride (b.p. 173 °C) is transferred into a vacuum distillation column 15 energized by a steam coil, where the furoyl chloride is obtained as the head fraction liquefied in condenser 16. The vacuum pump 17 maintains this system at a pressure of 7 mm Hg., A part of the condensate is used as reflux, while the product is collected in tank 18. The yield of furoyl chloride is in excess of 89 percent, the losses being a small fore-run collected in tank 19, a small after-run collected in tank 20, and a small quantity of a carbonaceous residue.

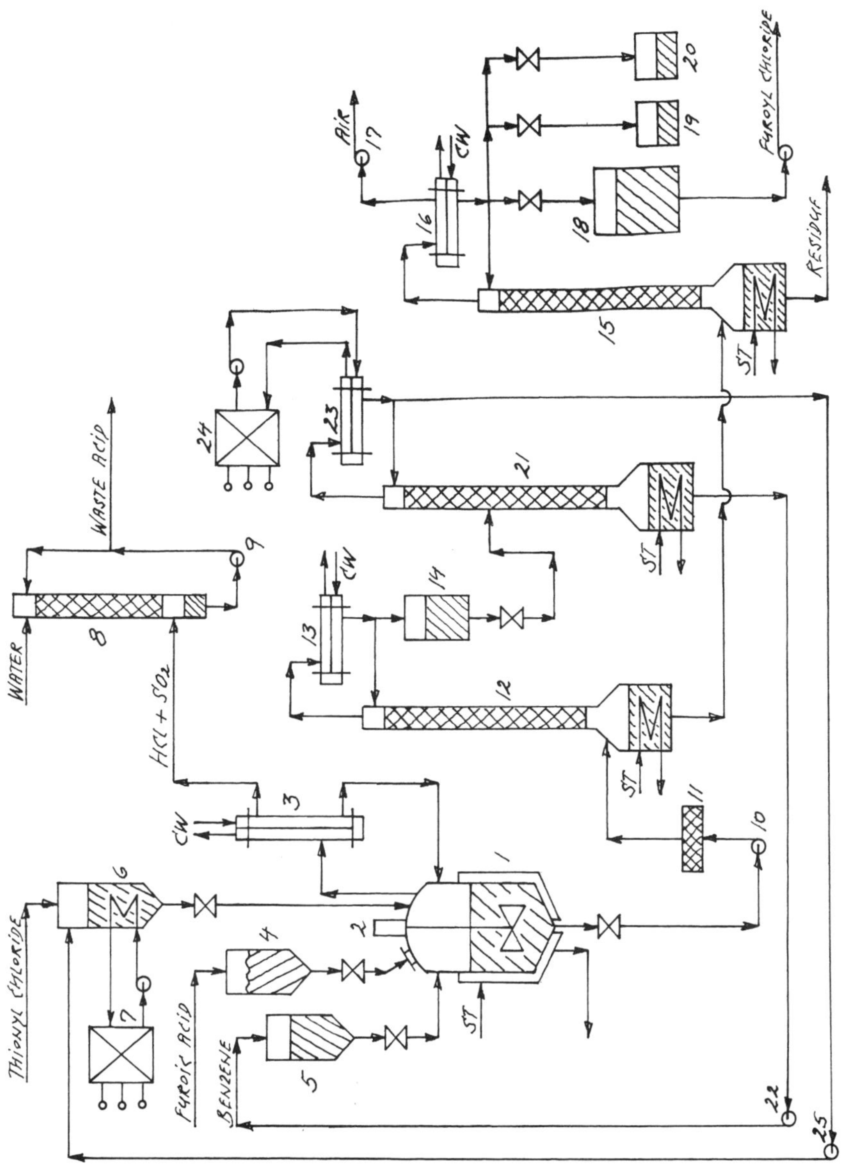

Figure 129. Plant for the Production of Furoyl Chloride.

The mixture of benzene and thionyl chloride in tank 14 is separated in a distillation column 21 heated by a steam coil. The sump fraction is benzene which is recycled back into tank 5 by means of pump 22. The head fraction of column 22 is thionyl chloride. It is liquefied in condenser 23 cooled by a chiller 24. A part of the condensate is used as reflux while the remainder is fed back into tank 6 by means of pump 25.

It is noted that furoyl chloride is a vigorous lachrymator which must be handled with care.

The reaction is completely analogous to the formation of benzoyl chloride from benzoic acid and thionyl chloride. Interestingly, however, furoyl chloride is a much more powerful lachrymator than benzoyl chloride.

Reference

[124] W. W. Hartman and J. B. Dickey, Ind. Eng. Chem. 24 (1932) 151-152.

o. Furfural as a Solvent

Having a spectroscopic polarity index $E_T^N = 0.426$, furfural is of intermediate polarity, and on account of this property, furfural is infinitely miscible with other solvents of intermediate polarity, up to E_T^N values of about 0.790, and down to E_T^N values of about 0.099. At 25 °C, this infinite miscibility with furfural includes, in alphabetical order, acetic acid, acetone, benzene, butanol, butyric acid, chloroform, ethyl acetate, ethylene glycol, formic acid, octyl alcohol, oleic acid, propionic acid, pyridine, quinoline, and toluene.

By contrast, at room temperature, furfural is not infinitely miscible with highly polar substances such as glycerol ($E_T^N = 0.812$) and water ($E_T^N = 1.000$), or with highly nonpolar substances such as hexane ($E_T^N = 0.009$).

Normally solid organic acids are only slightly soluble in furfural, some solubilities of interest being the following [125]:

	Solubility in furfural in % by weight		
	at 0°C	at 25 °C	at 40 °C
Benzoic acid	1.2	14.8	34.3
Cinnamic acid	0.6	4.1	10.9
Citric acid	0.3	3.6	9.9
Oxalic acid	3.2	4.8	9.1
Palmitic acid	1.1	1.6	10.2
Salicylic acid	1.5	11.0	28.8
Sebacic acid	0.7	0.8	2.5
Stearic acid	0.3	2.1	13.1
Succinic acid	2.0	3.0	7.0

On account of the low dielectric constant of furfural ($\varepsilon = 38$), inorganic salts are quite insoluble in fufural, their solubilities being in general smaller than 0.01 % by weight, but there are three notable exceptions:

Zinc chloride:	Solubility in furfural at 25 °C: 20.6 % by weight
Ferric chloride hexahydrate:	20.0 % by weight
Barium hydroxide octahydrate:	9.0 % by weight

Interestingly, anhydrous ferric chloride and anhydrous barium hydroxide have only very small solubilities in furfural, namely 0.55 % by weight and < 0.01 % by weight, respectively.

Petroleum ether having a boiling point range from 63 to 69 °C dissolves

3.5 % by weight of furfural at 48 °C,

81.5 % by weight of furfural at 55 °C, and

90.1 % by weight of furfural at 63 °C.

The critical solution temperature, above which the two partners are completely miscible, lies probably in the order of 70 °C.

Reference

[125] F. Trimble, Ind. Eng. Chem. 33 (1941) 660-662.

p. The Resinification Loss in Furfural Reactors

As discussed in chapter 6, in an acidified liquid non-boiling reaction medium as it exists in customary furfural reactors, there are two different losses:

(1) The loss by furfural reacting with itself (resinification loss).

(2) The loss by furfural reacting with intermediates of the xylose-to-furfural conversion (condensation loss).

The resinification loss due to furfural reacting with itself can be determined directly, easily, and accurately by studying acidified aqueous furfural solutions. Thus, the rate constant of the resinification reaction is known precisely as formulated in subchapter 6.1.

With resinification alone, without condensation, the furfural to be expected from the disappearance of xylose must be diminished by the term $k_2[FU]$ to give

$$d[FU]/dt = k_1[XY(t)] - k_2[FU(t)] \qquad (1)$$

where from xylose disappearance measurements (chapter 5) the xylose concentration as a function of time is known to be

$$[XY(t)] = [XY]_0 \exp(-k_1 t) \qquad (2)$$

and where k_2 is the rate constant of resinification as given in chapter 6.1.

Equation (1) is an inhomogeneous differential equation which can be solved by the "method of varying the constant" invented by Lagrange. The corresponding homogeneous equation

$$d[FU]/dt = -k_2[FU] \qquad (3)$$

has the solution

$$[FU(t)] = C \exp(-k_2 t) \qquad (4)$$

The inhomogeneous equation (1) can be solved by setting

$$[FU(t)] = C(t) \exp(-k_2 t) \qquad (5)$$

where C(t) is a function still to be determined. The Lagrange procedure, given in any textbook on calculus, leads to

$$C(t) = k_1[XY]_0 \{1 - \exp[-(k_1 - k_2)t\}/(k_1 - k_2) \qquad (6)$$

so that substitution of (6) in (5) gives

$$[FU(t)] = \frac{[XY]_0}{1 - \frac{k_2}{k_1}} \left\{ 1 - e^{-k_1 t \left(1 - \frac{k_2}{k_1}\right)} \right\} e^{-k_1 t \frac{k_2}{k_1}} \qquad (7)$$

Obviously, for $k_2 = 0$ (no resinification), equation (7) reduces to the time dependence of the theoretical furfural concentration as expected from the xylose disappearance, namely

$$[FU(t)]_{th} = [XY]_0 \{1 - \exp(-k_1 t)\} \qquad (8)$$

According to chapter 6.1, the ratio k_2/k_1 is a known function of the temperature, so that equation (7) can be plotted for various temperatures as parameter. In dimensionless form, this is shown in Figure 130, where the ordinate $[FU]/[XY]_0$ is the yield of a hypothetical furfural process involving resinification as the only loss.

Figure 130 illustrates the following facts:

(1) Resinification diminishes strongly with increasing temperature, so that for rough calculations of processes above 200 °C the effect of resinification can be neglected.

(2) The yield curve exhibits a maximum.

(3) With decreasing temperature, the maximum shifts towards smaller values of k_1t, and its magnitude diminishes.

Deducting equation (7) from equation (8) gives the resinification loss proper as shown in Figure 131. Here again, it is seen that the resinification loss diminishes strongly with increasing temperature.

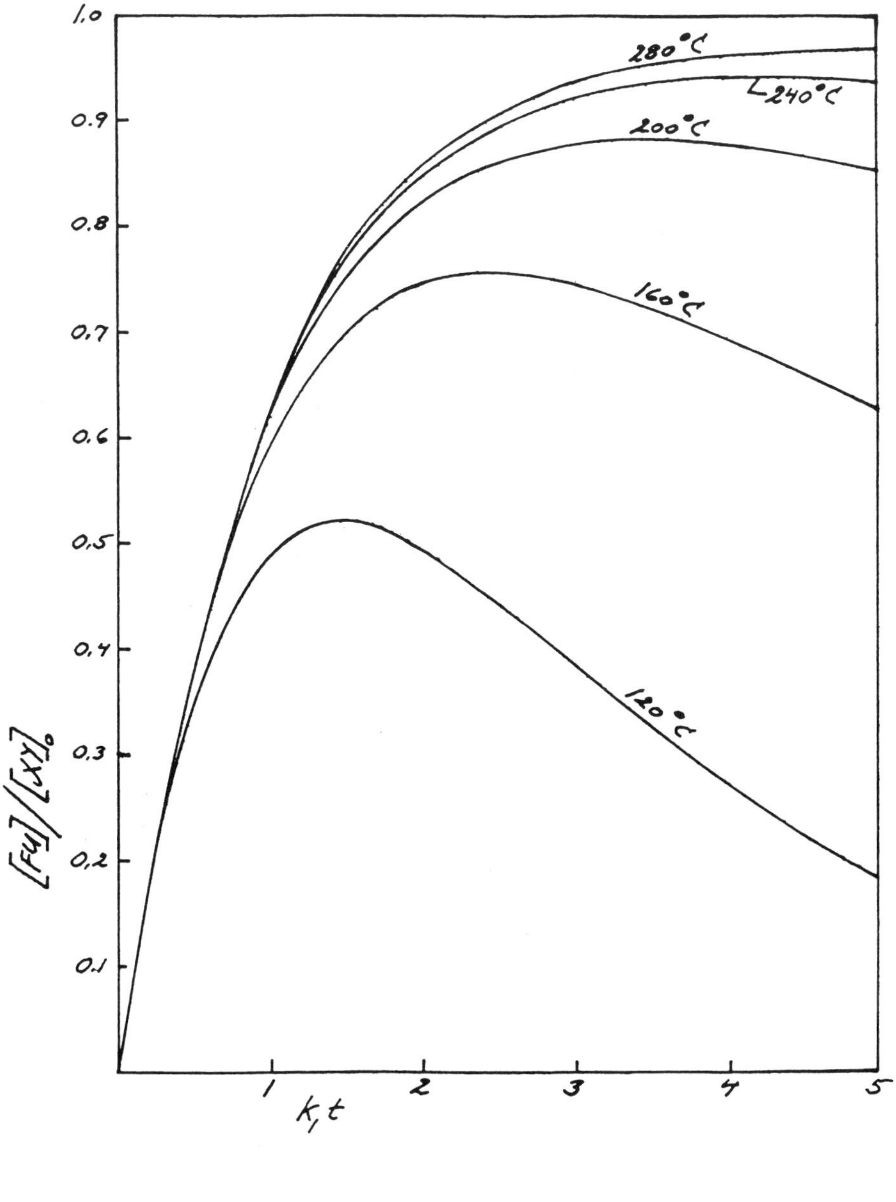

Figure 130. The Yield of a Hypothetical Furfural Process involving Resinification as the Only Loss. Graphical representation of equation (7) in dimensionless form.

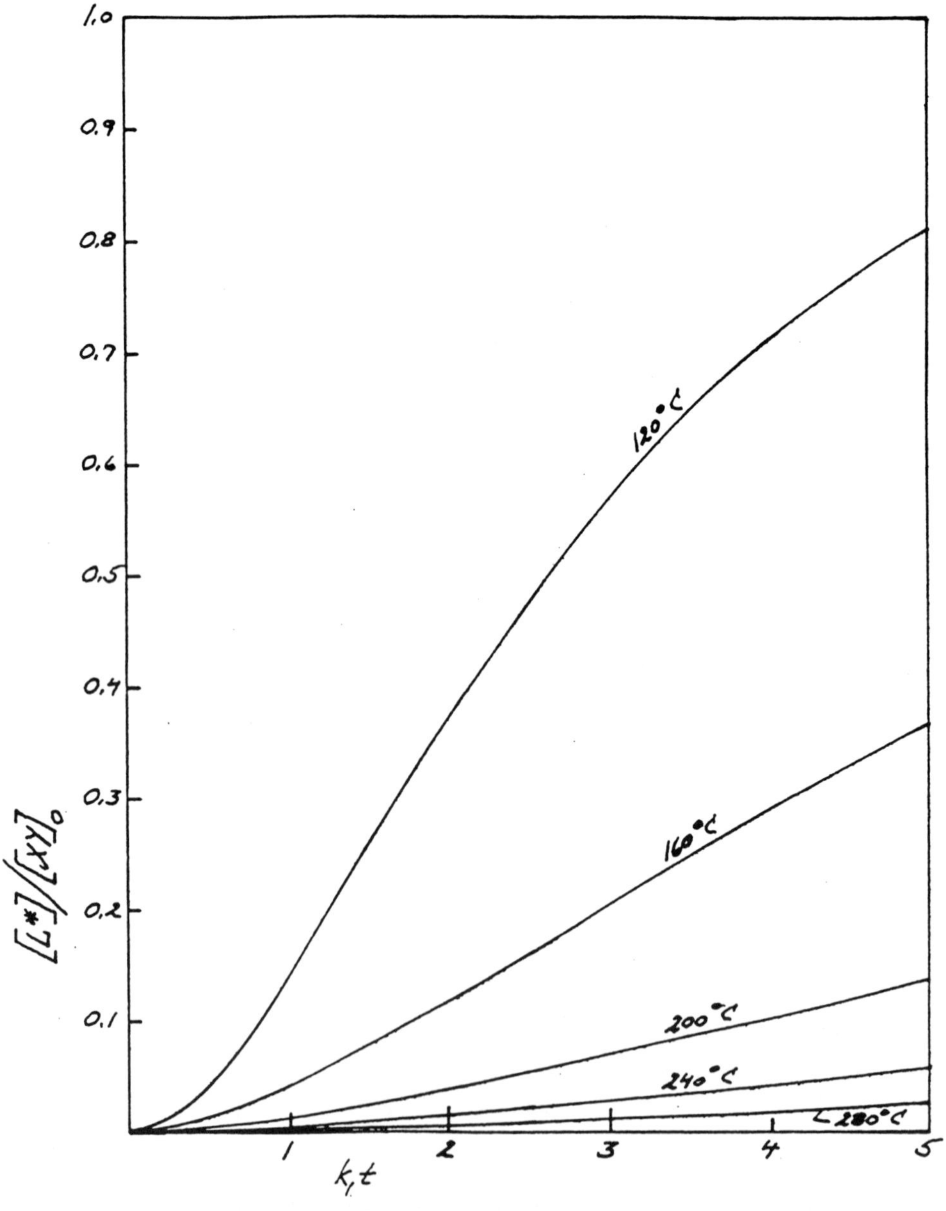

Figure 131. The Resinification Loss as a Function of Temperature in Dimensionless Form.

q. The Condensation Loss in Furfural Reactors

The loss by furfural condensation cannot be determined directly. It is found by first plotting, for the temperature of interest, the mathematically known yield of the hypothetical process with resinification as the only loss (as treated in the preceding chapter), and by then comparing this yield curve with an experimental yield curve from the literature for the same temperature. The difference between the two yield curves is the loss incurred by condensation.

For 200 °C as example, the procedure is shown in Figure 132. Curve A, representing the yield of the hypothetical process with resinification as the only loss, is obtained from equation (7) of the preceding chapter, and curve B is an experimental yield curve given in the literature [126] for 200 °C and for an initial xylose concentration of 0.666 mole/liter (100 g/liter). Experimental yield curves for other temperatures and other initial xylose concentrations are amply available in the same reference. The hatched area between the two curves A and B represents the condensation loss. To round the overall picture, the theoretical yield for the temperature considered is shown by the dashed curve C.

In subchapter 6.2, it was suggested that the condensation loss is due to one or two molecules of furfural reacting with the first intermediate of the xylose-to-furfural conversion, so that the overall rate of the condensation reactions should be

$$d[CP]/dt = k_a[FU]\,[INT] + k_b[FU]^2\,[INT] \qquad (1)$$

where [CP] stands for the concentration of the condensation products, [FU] stands for the furfural concentration, and [INT] stands for the concentration of the intermediate. As the concentration of the intermediate can be taken to be proportional to the momentary xylose concentration, equation (1) assumes the form

$$d[CP]/dt = k_a{}^*[FU(t)]\,[XY(t)] + k_b{}^*[FU(t)]^2\,[XY(t)] \qquad (2)$$

Thus, although the new rate constants $k_a{}^*$ and $k_b{}^*$ can not be determined, it is readily seen that the condensation loss increases with an increasing initial xylose concentration, so that the actual yield decreases when the initial xylose concentration increases. This is born out by the experimental yield curves shown in Figure 133, taken from the literature [126]. Hence, a furfural reactor run at a high moisture content gives a better yield than a furfural reactor run at a

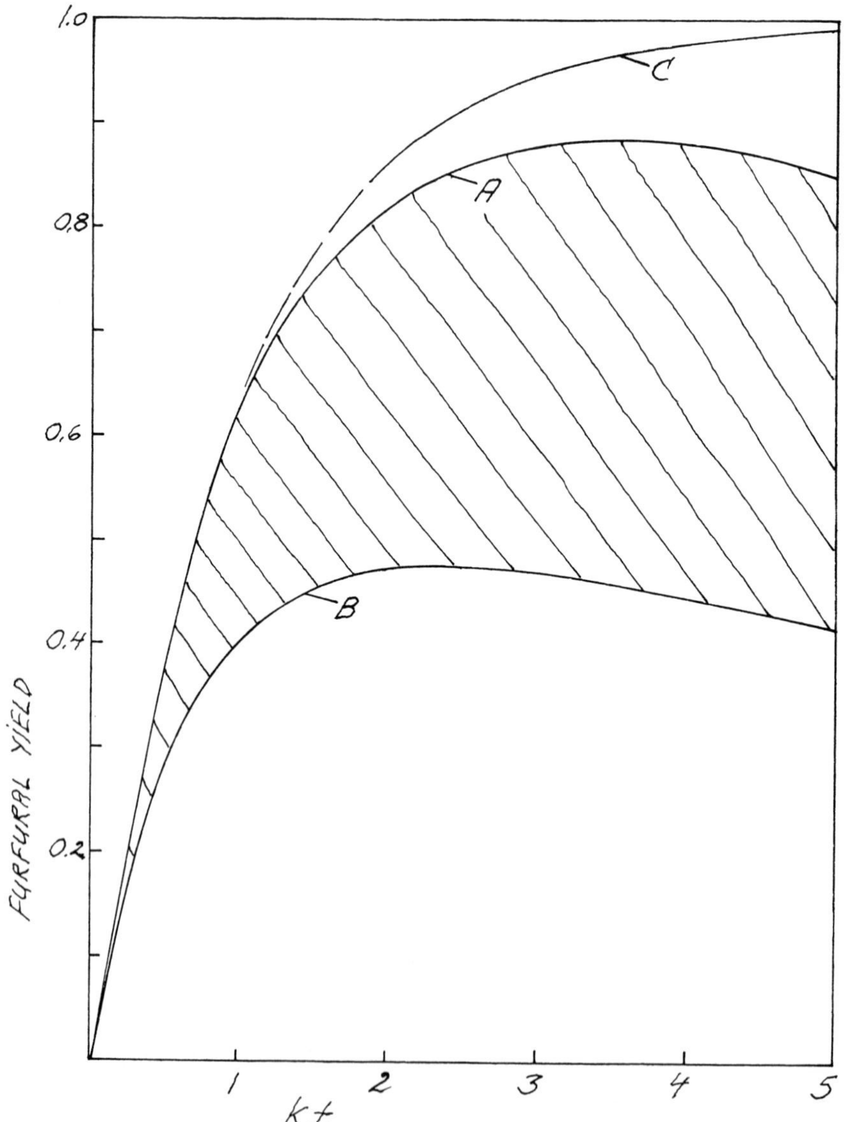

Figure 132. Graphical Representation of the Condensation Loss at 200 °C.

Curve A: Yield of a process with resinification only.

Curve B: Actual yield from experimental data [126].

Curve C: Theoretical yield.

The condensation loss is represented by the area between the curves A and B.

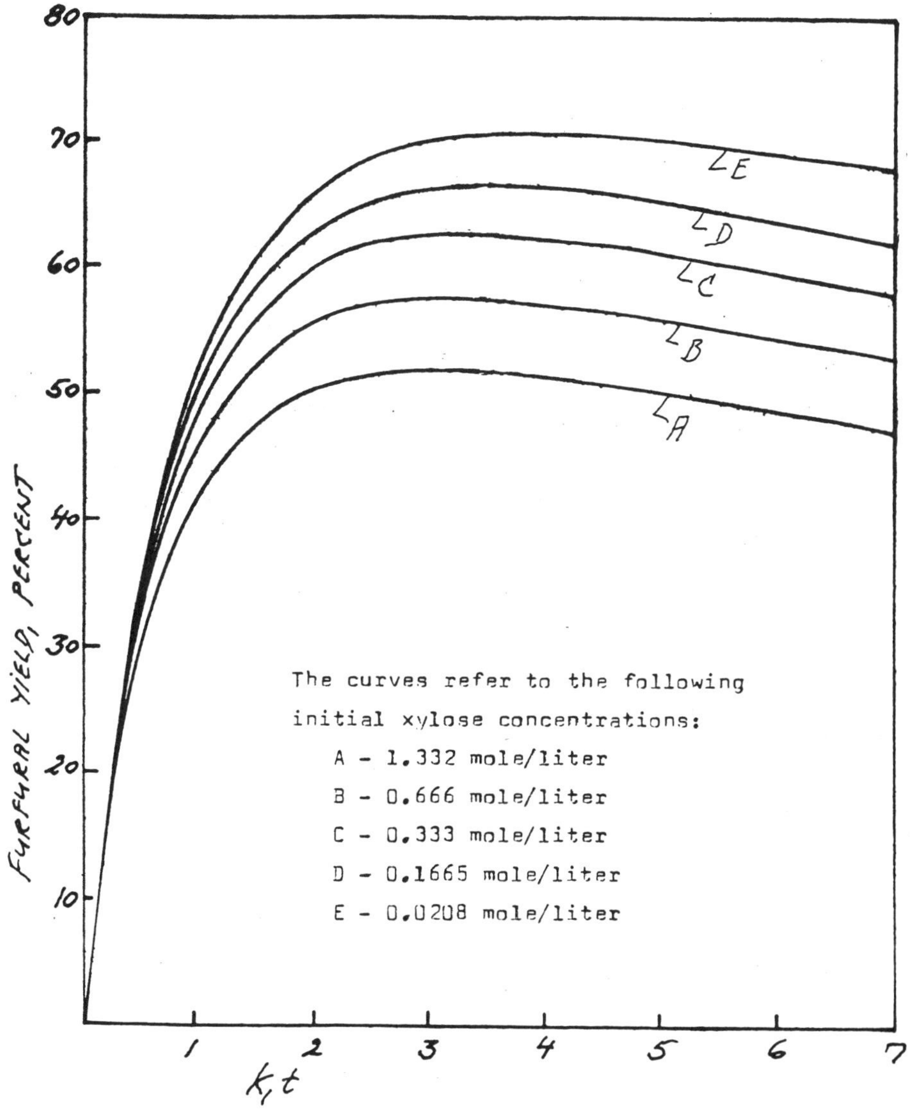

Figure 133. Experimentally Determined Yield Curves for 240 °C and Various Initial Xylose Concentrations as Parameter [126].

low moisture content. A high moisture content decreases the xylose concentration, thereby lowering the condensation loss.

By contrast, the initial xylose concentration has absolutely no effect on the resinification loss as the xylose plays no part in the respective reaction.

To obtain a comprehensive picture of the yield situation, diagrams of the type shown in Figure 132 must be drawn for various temperatures. In this fashion, it is found that the condensation loss decreases markedly with increasing temperatures. This is strong support for the high temperatures advocated in the SUPRATHERM and STAKE processes.

Reference

[126] D. F. Root, J. F. Saeman, J. F. Harris, and W. K. Neill, Forest Products Journal 9 (1959) 158-165.

r. Odd Applications

While at present the principal applications of furfural are seen to be as an extractant for lubricating oils, vegetable oils, and diesel fuel, as a fungicide, as a nematocide, and as the raw material for furfuryl alcohol to make foundry resins, other uses have been proven successfully on a multitude of fronts, and it is only a lack of research which prevents furfural and its derivatives from flooding the world in a stunning variety of different applications.

FUELS

Furfural can be and has been used as a fuel for motor cars and airplanes. Under the trade name of FURALINE, a mixture of furfuryl alcohol, methanol, and xylidine (dimethyl aniline) is used as a rocket fuel. With nitric acid as oxidizer, a mixture of furfuryl alcohol with aniline is used as the fuel for the American CORPORAL rockets and the French VERONIQUE and EMERAUDE rockets. The latter system is “hypergolic”: When the mixture of furfuryl alcohol and aniline (the “fuel”) comes into contact with the nitric acid (the “oxidizer”), it ignites by itself without external aid. Even alone, furfuryl alcohol explodes when mixed with concentrated nitric acid.

ROAD CONSTRUCTION

Somewhat related to the application of furfuryl alcohol resins “gluing” foundry sand together is the use of furfural in bituminous road construction [127]. Furfural alone, or a mixture of furfural and phenol or furfural and aniline is added to a mixture of bitumen and gravel. This addition results in an improved wetting of the gravel, thus leading to a better cohesion of the road material to give the road a longer life. On an average, only one ton of furfural per mile of road is required to produce this effect.

IMPREGNANTS

Inasmuch as furfural and furfuryl alcohol resinify with acids to form insoluble products, this reaction affords an excellent means of impregnating porous materials such as carbon or graphite artifacts with a subsequent formation of a resin in situ. This is a widespread application in the production of carbon or graphite electrodes for steel furnaces and in the production of graphite heat exchangers for the chemical industry.

BIOLOGICAL USES

In addition to the established uses of furfural as a fungicide and nematocide, furfural is a selective herbicide killing dandelions, and it is a disinfectant as effective as formaldehyde, but easier to handle and far less toxic. It can be used in poultry houses to kill lice and other parasites, it can be used to prevent infection of dehorned cattle by the screwworm fly, and a one percent aqueous solution is said to clear up infection by the parasite causing athlete's foot. Even embalming fluids for both human beings and biological specimens have been made with furfural. Certain arsenic compounds useful for preventing attack by termites and teredo (shipworm) can be carried into the wood with furfural as the solvent.

RESINS

Resins made from furfural and furfuryl alcohol excel in heat resistance and mechanical strength. For this reason, such resins are used almost exclusively to attach electric light bulbs to the brass bases, and they are commonly used to make brake linings and abrasive wheels.

EXTRACTION

Apart from its established use in the extraction of lubricating oils, vegetable oils, and diesel fuel, furfural has won an important role in the refining of wood rosin. This natural resin, obtained by extracting chips of pine tree stumps, has a dark color which must be removed before the rosin can be used in paper sizing and in the production of "resin wine" (retsina). To this end, the crude wood rosin is dissolved in petroleum ether, and this solution is extracted with furfural. In this process, the undesirable color bodies end up in the furfural extract. Furfural is also used in the refining of tall oil, a by-product of the KRAFT pulping process. The purpose of the refining is to remove undesirable rosin acids. The tall oil is first esterified with methanol, and a solution of the esterified oil in naphtha is extracted with furfural. In this process, the rosin acids end up in the furfural extract.

DIFUNCTIONAL COMPOUNDS

Work with difunctional furan compounds promises to be particularly intriguing. Many transformations have already been carried out with furan dialdehyde [128], and are likely to be applicable to difurfural as well. Linkages via nitrogen as illustrated in Figure 134 lead to 2,5-diaminofuran, a monomer for the production of furan polyamides, as well as to antibacterial and antiviral compounds. Linkages via carbon as illustrated in the Figures 135 and 136 result in compounds applied as pharmaceuticals, as insecticides, as microbicides, as fungicides, and as building blocks of special polymers. Macrocycles as shown in Figure 137

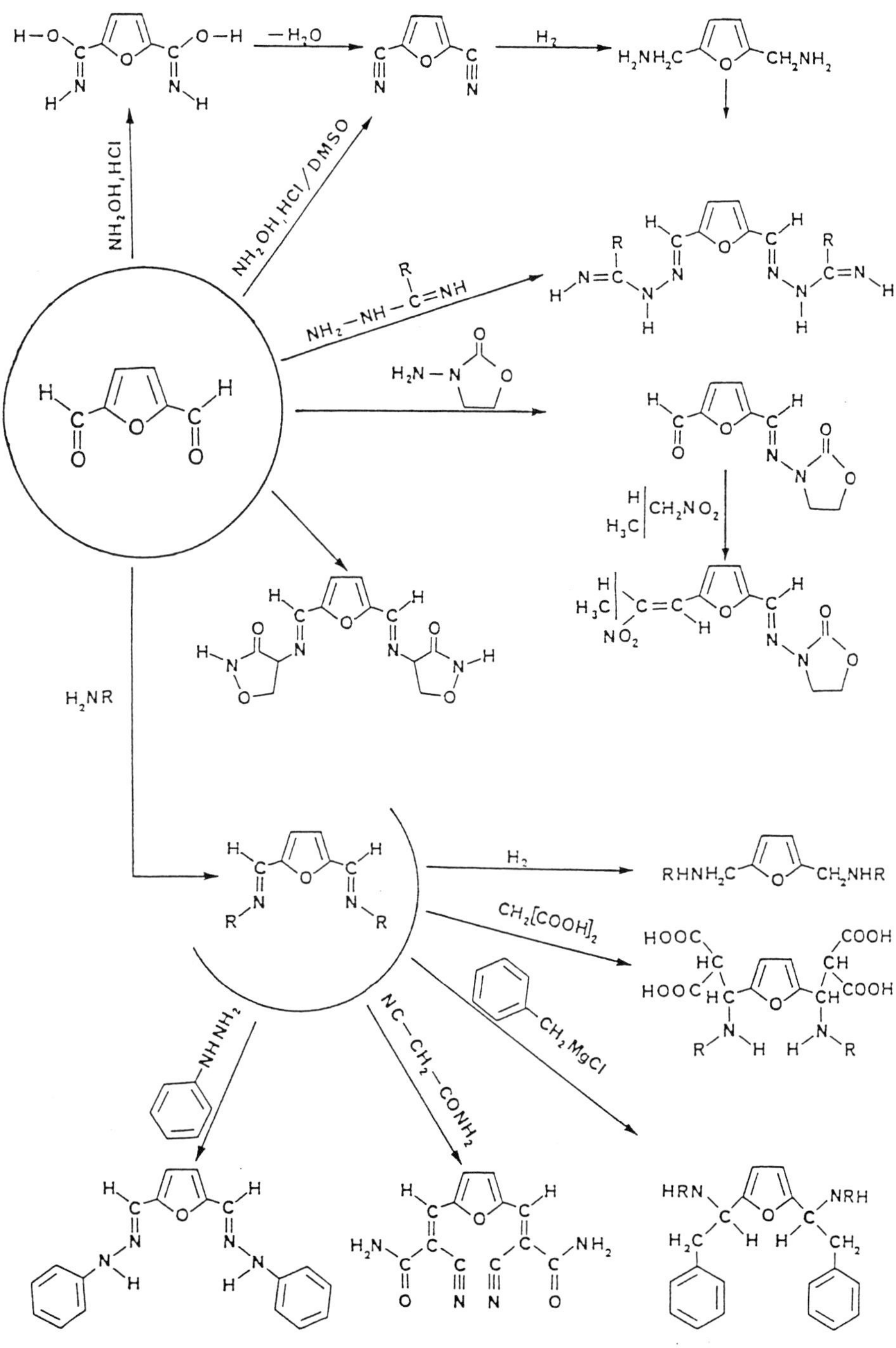

Figure 134. Derivatives of Furan Dialdehyde by Linkages via Nitrogen.

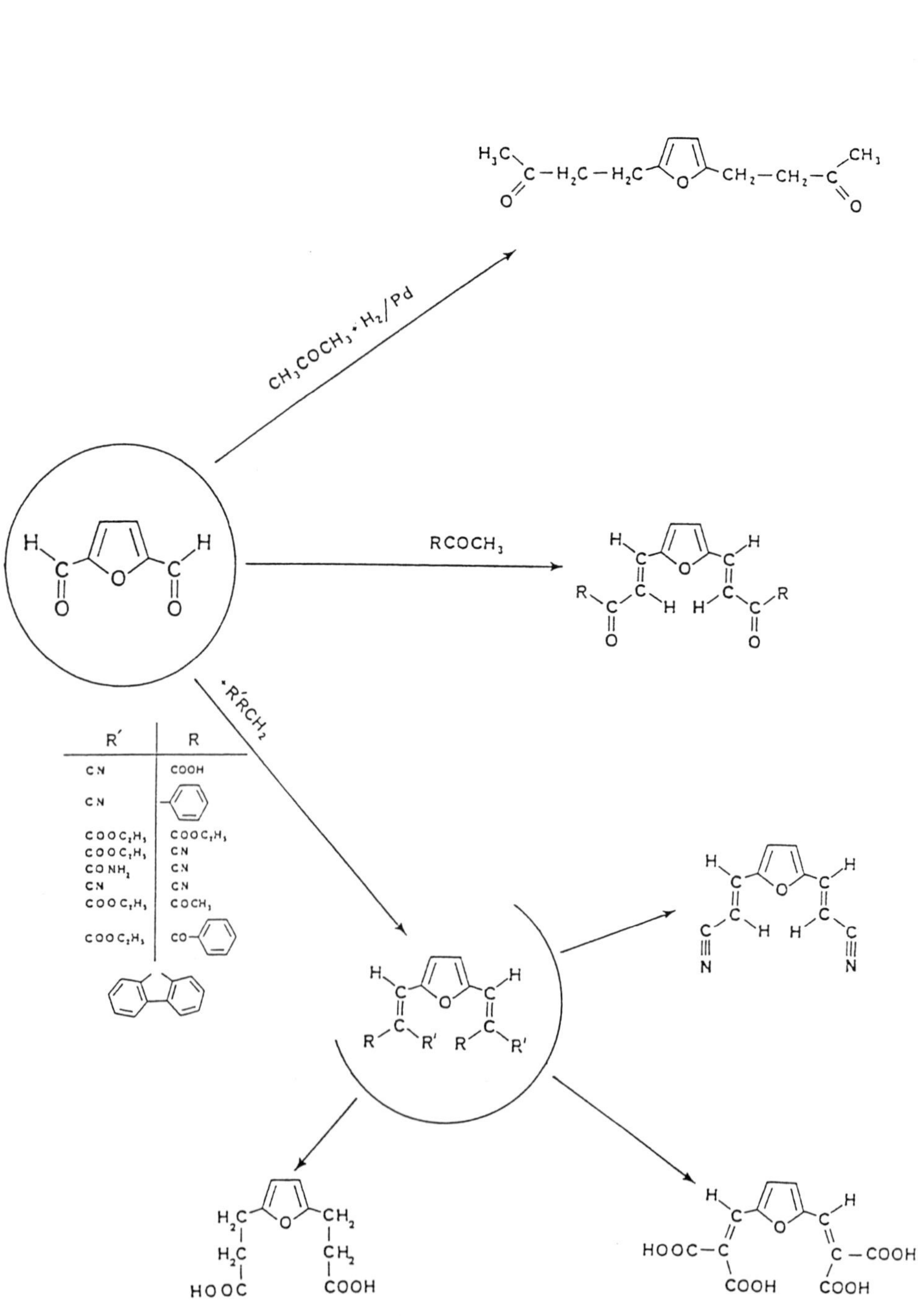

Figure 135. Derivatives of Furan Dialdehyde
by CLAISEN and KNOEVENAGEL Condensations.

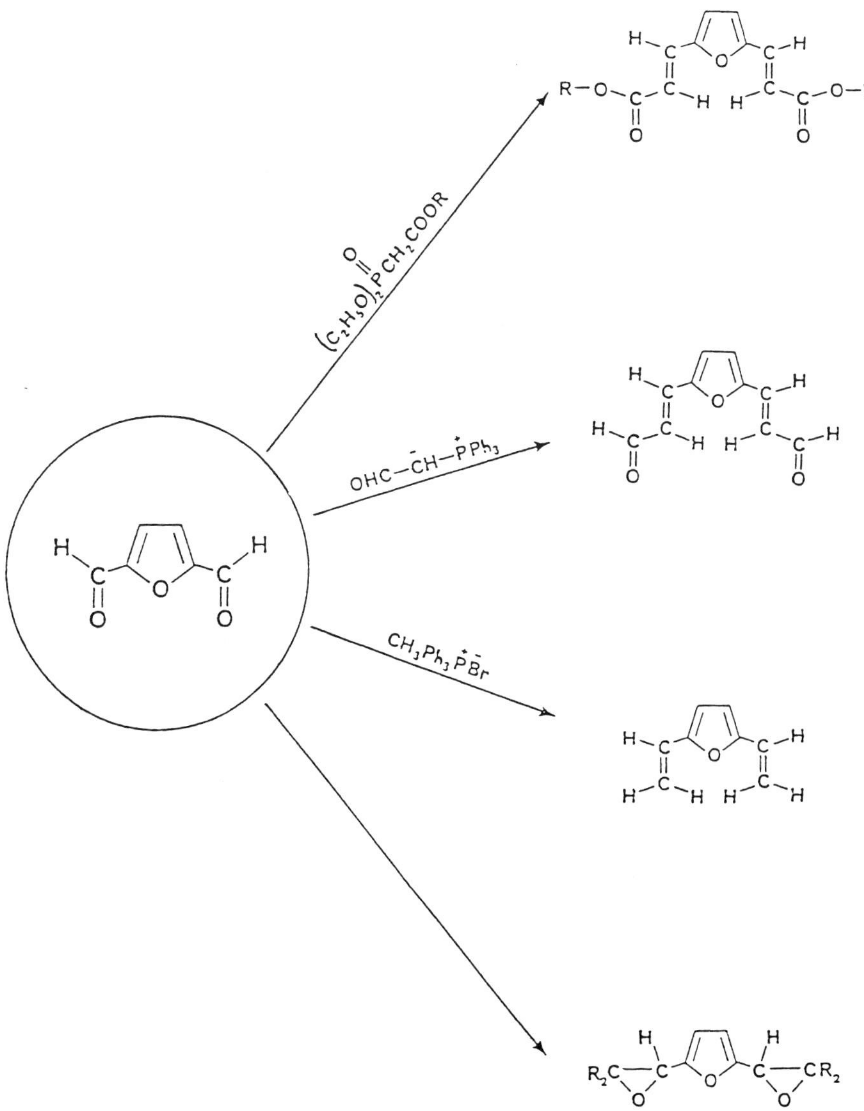

Figure 136. Derivatives of Furan Dialdehyde
by WITTIG Condensations.

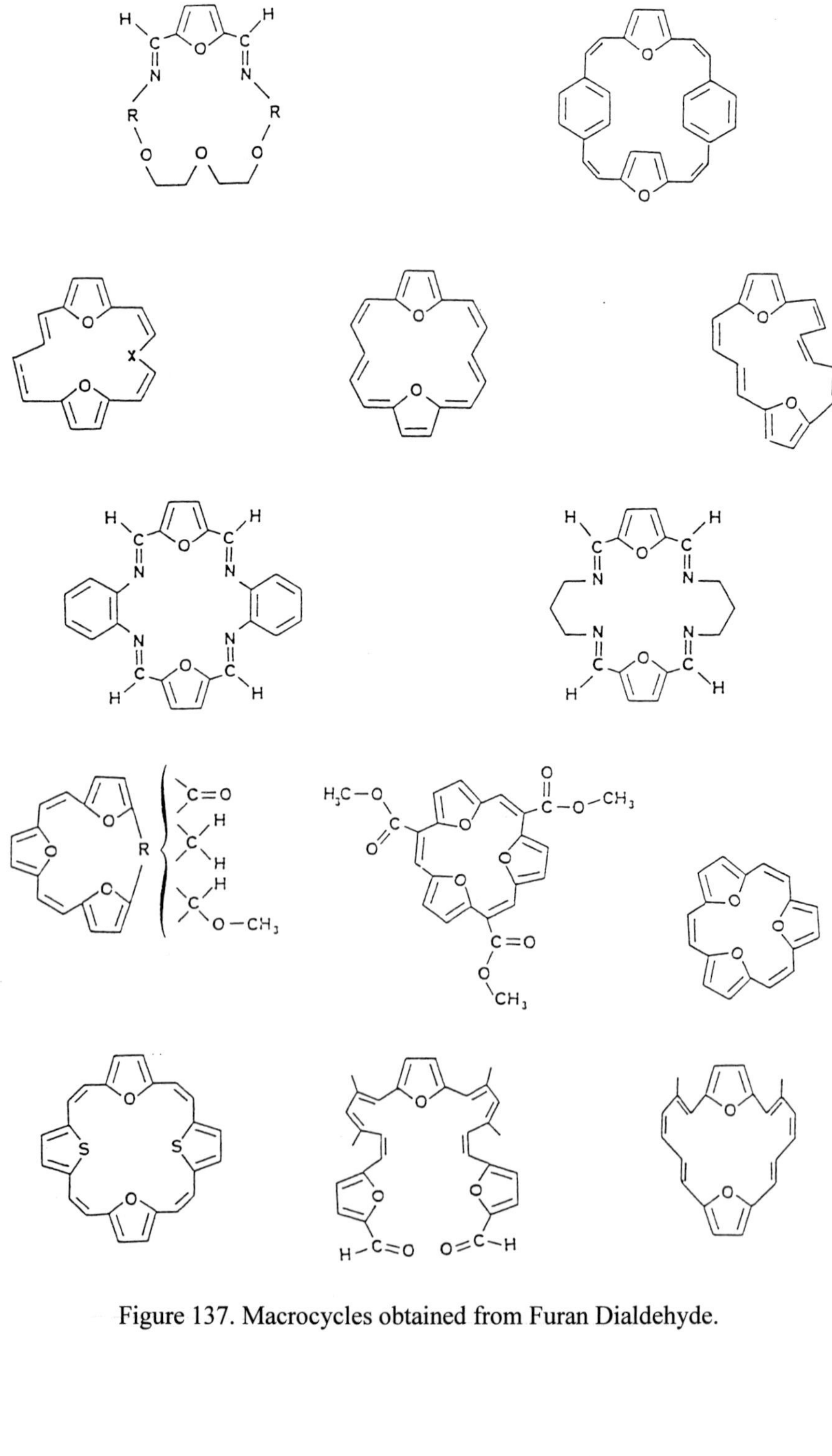

Figure 137. Macrocycles obtained from Furan Dialdehyde.

augur interesting applications as chelating or complexing agents for alkali metals, alkaline-earth metals, and transition metals, as ion transfer agents for membranes, and as bactericidal agents.

References

[127] H. F. Winterkorn, Ind. Eng. Chem. 30 (1938) 1362-1368.

[128] M. Marval, Procédé de Synthèse du Furannedicarboxaldéhyde-2,5 à partir du D-Fructose, Ph.D. Thesis, Institut National Polytechnique, Toulouse, 1985.

Epilogue

So far, there has never been a book on furfural, the reason being that the furfural industry has been traditionally secretive to the point of appearing shrouded in clouds of mystery. As discussed in chapter 16, a vivid proof for this clandestineness is the totally erroneous information, in a renowned encyclopedia, on the formation of diacetin (glyceryl diacetate) in the ROSENLEW process while in reality it is diacetyl (2,3-butanedione) which is formed.

This surreptitious atmosphere has not been conducive to progress as evidenced by the fact that the very first industrial furfural process, launched 78 years ago with old equipment of a defunct cereal plant, is still used today although its yield is poor, without necessarily being so.

With this book, the author hopes to rouse the interest and the inventiveness of a new generation, to the end of moving the field of furfural and its by-products into the limelight it deserves.

Subject Index

The figures after the entries are page numbers.